Every Third Woman in America

How Legal Abortion Transformed Our Nation

Also by David Grimes

Teenage Sexual Health, with A.M. Withington AM, and R. A. Hatcher: New York, Irvington Publishers, 1983

Arztliche Aspekte des Legalen Schwangerschaftsabbruchs, with H. H. Brautigan: Stuttgart, Ferdinand Enke Verlag, 1984

Modern Methods of Inducing Abortion, Edited, with D. T. Baird and P. F. A.Van Look: Oxford, Blackwell Science, 1995

U.S. Preventive Services Task Force. Guide to Clinical Preventive Services, 2nd ed. :Baltimore, Williams & Wilkins, 1995

Modern Contraception. Updates from the Contraception Report, with M. Wallach, E. J. Chaney, E. B. Connell, S. J. Emans, J. W. Goldzieher, P. J. A. Hillard, L. Mastroianni, Jr. : Totowa, NJ, Emron, 1997

A Clinician's Guide to Medical and Surgical Abortion, Edited, with M. Paul, E. S. Lichtenberg, L. Borgatta, P. G. Stubblefield: New York: Churchill Livingstone, 1999

Summary of contraindications to oral contraceptives, with S. C. M. Knijff, E. M. Goorissen, E. J. M Velthuis-te Wierik, T. Korver: New York, Parthenon Publishing, 2000

Modern Oral Contraception. Updates from the Contraception Report, with M. Wallach, E. J. Chaney, E. B. Connell, M. D. Creinin, S. J. Emans, J. W. Goldzieher, P. J. A. Hillard, L. Mastroianni, Jr.: Totowa, NJ, Emron, 2000

Lancet handbook of essential concepts in clinical research, with K. F. Schulz: London, Elsevier, 2006

Management of unintended and abnormal pregnancy: comprehensive abortion care, Edited, with M. Paul, S. Lichtenberg, L. Borgatta, P. Stubblefield, M. Creinin: New York, Wiley-Blackwell, 2009

Every Third Woman in America

How Legal Abortion Transformed Our Nation

David A. Grimes, M.D.

with Linda G. Brandon

Daymark Publishing
Carolina Beach, North Carolina

First Printing: 2014

ISBN: 978-0-9908-3360-4 (sc)
ISBN: 978-0-9908-3361-1 (hc)
ISBN: 978-1-4834-1421-8 (e)

Library of Congress Control Number: 2014914932

Lulu Publishing Services rev. date: 12/09/2014

To women everywhere
faced with difficult decisions,
and to those who help them.

Contents

Preface

One in three women in the U.S. will have an induced abortion during her lifetime. Despite this common experience in women's lives, abortion remains one of the most corrosive social issues in America. In most industrialized nations the legalization of abortion blended smoothly into contemporary society. Not here. Conflict over abortion - strident, violent and sometimes murderous - is uniquely American. The epidemic of bombing, arson and murder of health-care providers is unparalleled. Like stem-cell research, artificial prolongation of life, and assisted reproductive technologies, abortion involves deeply held beliefs. These ethical considerations are inherently subjective and personal; they will never be resolved. However, scientific and medical evidence is clear and incontrovertible. Rather than dealing with theories and beliefs, this book examines the extensive medical and public health evidence amassed over four decades. It chronicles an extraordinary chapter in the annals of public health.

The book has five parts. Section I describes the three eras of abortion availability in the U.S. Section II explores the medical aspects of common pregnancy outcomes: miscarriage, child birth, and abortion. Section III summarizes the impact of legal abortion on women and their families. In Section IV, some of the "hot-button" issues of our times are considered. Section V uses the past to predict the future of abortion in America.

Four decades after *Roe v. Wade*, many have forgotten why the restrictive laws were changed: women. Our mothers, wives, sisters, and daughters were suffering and dying in large numbers - needlessly. After legalization, the carnage stopped. A generation has grown up unaware of the horrors of the bad old days (https://archive.org/details/when_abortion_was_illegal).

Ironically, legal abortion has become a victim of its own success. Because of legal abortion, America has lost its collective memory of the "bad old days" of illegal abortion. That experience is largely forgotten or ignored four decades later: the political - and sometimes judicial - abortion debate rages in a vacuum. Abortion, however, does not occur in a vacuum. It always occurs in context: what alternatives exist for the pregnant woman? This book explores some of those choices and their safety.

Although I have aimed for accuracy in content and citations, in a book of this length some errors are likely. Please let me know of any (everythirdwoman@earthlink.net), and I will quickly correct them in the electronic version of the book. Without the support and editorial help of my family, this book would not have been possible.

A grandfather, I lived through the bad old days in America. We must not revisit them. Women deserve better from us.

David A. Grimes, M.D., FACOG, FACPM, FRCOG (Hon.)

Section I
Three eras of abortion

Gerri

Born in 1935, Gerri Santoro was one of 14 children who grew up on a farm in rural Connecticut. Married at age 18 years, she and her husband had two children. A victim of domestic violence, Gerri left her husband in 1963. She had an affair with a married man and became pregnant again. When she learned that her husband was coming from California to visit their children, she feared for her life.

More than six months pregnant, she and her lover checked into a Norwich motel on June 8, 1964. They attempted to perform an abortion with surgical instruments, guided by a medical textbook. Hemorrhage ensued, and her lover abandoned her in the motel. She was found dead on the bathroom floor the next morning by a maid. A police photograph of Gerri on the floor, with bloody towels between her legs, was published in *Ms.* Magazine in April of 1973 and became an enduring symbol of the "bad old days" before *Roe v. Wade.*[1]

1. Anonymous. Gerri Santoro. http://en.wikipedia.org/wiki/Gerri_santoro, accessed December 26, 2013.

Chapter 1
The bad old days

Abortion has been with us as long as has pregnancy. The question for society has always been the price women will be forced to pay for their abortions in terms of dollars, disease, degradation, and death.[2] Until the 1970s, the price in the United States was frightfully high. Similarly, the price of caring for the complications of unsafe, clandestine abortion was vastly higher than is the cost of safe, legal abortion today.[3-5] Those who wish to restrict or eliminate access to legal abortion[6] must be willing to accept responsibility for these costs, both financial and human.

> *"Someone gave me the phone number of a person who did abortions and I made the arrangements. I borrowed about $300 from my roommate and went alone to a dirty, run-down bungalow in a dangerous neighborhood in East Los Angeles. A greasy looking man came to the door and asked for the money as soon as I walked in. He told me to take off all my clothes except my blouse; there was a towel to wrap around myself. I got up on a cold metal kitchen table. He performed a procedure, using something sharp. He didn't give me anything for the pain – he just did it. He said that he had packed me with some gauze, that I should expect some cramping, and that I would be fine. I left."*
>
> *Actress Polly Bergen, on the illegal abortion in the 1940s that left her infertile.*[1]

Three eras of abortion

The United States has progressed through three eras of abortion in recent decades.[7] Until the middle of 1970, legal abortions were largely unavailable. From the middle of 1970 until January of 1973, abortions were available regionally. After January 22, 1973, legal abortions were theoretically available nationwide, although large disparities in access persist four decades later.

The bad old days: illegal abortion in America

In the early days of the United States, abortions were widely available, although of limited success and safety. In the mid-1800s organized medicine became concerned by the carnage resulting from inept attempts at induced abortion performed by women themselves or by unqualified persons. To reduce the suffering and death related to unskilled abortion, physicians and others urged State legislatures to outlaw abortion altogether.

These laws remained in place until the 1960s, when, again, physicians, public health officials, religious leaders, and women's groups argued for the repeal of the laws. Ironically, they called for the restrictive State abortion laws to be repealed for the same reason they were originally enacted: to protect the health of women. Initially crafted in the 1870s to protect women from quacks,[8,9] these laws by the mid-1900s were paradoxically denying women access to physicians with the training and equipment to provide safe abortions.[10] In that era, contraceptive methods were limited to barrier methods of modest effectiveness or fertility-awareness methods (for example, the rhythm method). Even today, the latter has limited popularity and even more limited effectiveness.[11]

Faced with decades of fertility and inadequate means of avoiding pregnancy, millions of desperate women terminated unplanned pregnancies to preserve the health and well-being of their families. Thousands of women suffered and died in the process.[8,12-15] Physicians born in the latter half of the Twentieth Century did not professionally encounter the "bad old days" of abortion in America; indeed, the medical profession is at risk of losing its collective memory of the era. More than a generation of American women and their families has now grown up unaware of the circumstances their mothers and grandmothers faced. Women today expect safe, legal abortion to be part of the full range of health services.

Burden of suffering

Because of its clandestine nature, estimates of the scope of illegal abortion are necessarily imprecise. Nevertheless, based on survey data, the best estimates in the 1950s were that somewhere between 200,000 and 1.2 million illegal—and generally unsafe—abortions took place annually[16,17] in this country alone.

As recently as the decade when I was born (the 1940s), more than 1000 women were known to have died each year from complications of illegal abortion.[18] The true number was considerably higher. Every large municipal or county hospital had a "septic abortion ward," and infected induced abortion was the most common reason for admission to gynecology services nationwide during those years. Reports from large

> *"My husband reluctantly agreed to take me to the local back-alley abortionist – an alcoholic who had buried more than one of his mistakes...After I had swallowed my two-aspirin 'anesthetic,' I was told to climb up on what resembled a dirty kitchen table and hoist up my skirt....Then the pain. Eyeball popping pain. Lots and lots of it. Far more, I'm sure than was necessary....Another trip to the hospital, another ten-day stay, a little bout with peritonitis, a half-dozen [blood] transfusions...and the old girl was as good as new..."*
>
> *Sherry Matulis, on her illegal abortion in 1954 after having been stabbed and raped*[38]

public hospitals chronicled the suffering. Among 1,248 women admitted to Bellevue Hospital, New York City, with infected incomplete abortions from 1934 to 1937, 108 admitted to taking a drug, and 126 acknowledged introduction of an instrument into the uterus to cause abortion; 117 reported that trauma, such as a fall down the stairs, caused the "miscarriage."[19] On the same gynecology service from 1940 to 1954, more than 7000 cases of incomplete abortion were treated, and more than a third of these were complicated by infection. This high complication rate strongly indicated attempts at induced abortion, since spontaneous miscarriages rarely get infected.[20] Twenty-two women died of infection.

Details were available for several cases: two had a catheter inserted through the cervix, three reported a "fall," and one attributed the loss to a child jumping on her abdomen.[21] A later report from the same New York hospital indicated that 60% of all incomplete abortions were illegally induced.[22] At Los Angeles County Medical Center, the septic abortion ward had about 20 beds in a horseshoe-shaped pattern, with two private rooms. The latter were provided so that women could be alone with their families when they died. The beds stayed full.

These were desperate women in dangerous times. Abortions were available in hospital only for life-threatening conditions. Many women suffered from serious psychiatric disease or social deprivation. A consecutive series of 199 New York City patients thoroughly evaluated by a psychiatrist from 1968 to 1970 portrayed a bleak picture.[23] Fifty-seven percent of women requesting abortion at Bellevue Hospital had concrete evidence of psychiatric disturbance (Figure 1-1). More than one-third had attempted suicide previously or during the current pregnancy. Five percent of the pregnancies reportedly stemmed from rape, and 79% of the women lacked emotional support from their male partner. Seventeen women were victims of domestic violence, and ten had children with psychiatric disorders or intellectual handicaps. While these 199 women were not representative of all women seeking abortions, they reflected the difficult circumstances of many women seeking abortions in that era.

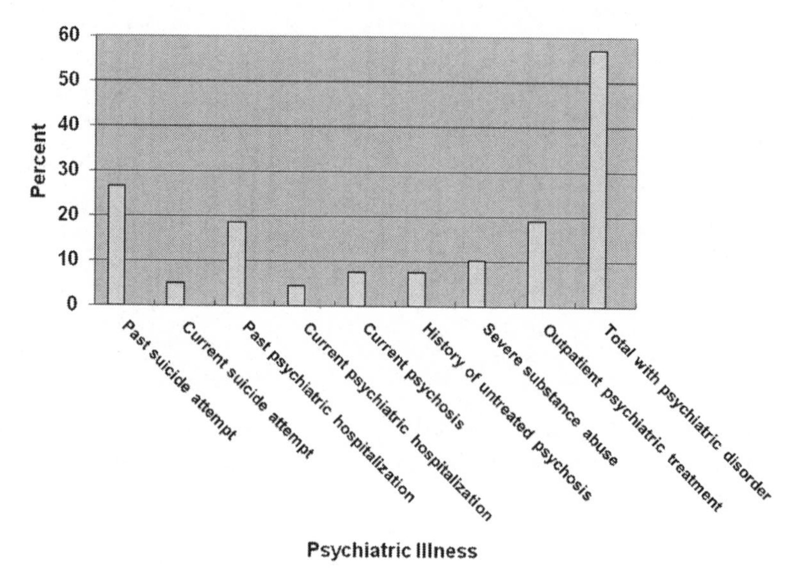

Figure 1-1
Psychiatric history of poor women seeking abortions, Bellevue Hospital, New York City, 1969-1979
Source: Belsky[23]

Tools of the trade

The primitive tools used for abortion reflected the grim determination of the women. Surveys conducted in New York City by the National Opinion Research Center in 1965 and 1967 documented the methods in common use.[24] Of 899 women interviewed, 74 reported having attempted to abort one or more pregnancies; 338 noted that one of their friends, relatives, or acquaintances had done so. Of those reported abortion attempts, 80% tried to abort themselves. As shown in Figure 1-2 and Table 1-1, the methods ranged from oral preparations to instrumentation of the cervix and uterus. Nearly 40% of women used a combination of approaches. In general, the more invasive the technique, the more dangerous it was to the woman and the more likely it was to disrupt the pregnancy. As shown in Figure 1-3, invasive methods, such as insertion of tubes or liquids into the uterus, were more successful than other approaches. Coat hangers, knitting needles, and slippery elm bark were common insertion methods; the bark would expand when moistened, causing the cervix to open. An old method was to place a flexible rubber catheter (a hollow tube) into the uterus to stimulate labor.

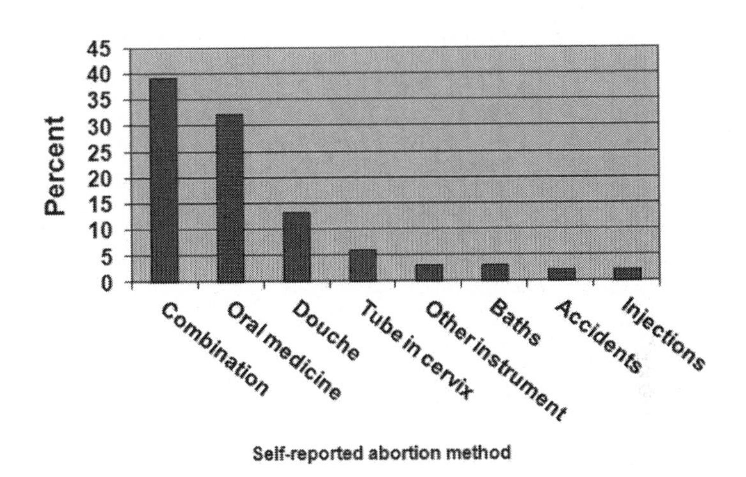

Figure 1-2
Percent distribution of self-reported methods used for abortion, 122 respondents, New York City, 1965 and 1967
Source: Polgar[24]

Table 1-1
Partial inventory of unsafe abortion methods by route of administration

Treatments taken by mouth

- Toxic solutions
  - Turpentine
  - Laundry bleach
  - Detergent solutions
  - Acid
  - Laundry bluing
  - Cottonseed oil
  - Arak (a strong liquor)

Teas and herbal remedies
Strong tea
Tea made of livestock manure
Boiled and ground avocado or basil leaves
Wine boiled with raisins and cinnamon
Black beer boiled with soap, oregano, and parsley
Boiled apio (celery plant) water with aspirin
Tea with apio, avocado bark, ginger, etc.
"Bitter concoction"
Assorted herbal medications

Drugs

Uterine stimulants, such as misoprostol or oxytocin (used in obstetrics)
Quinine and chloroquine (used for treating malaria)
Oral contraceptive pills (ineffective in causing abortion)
Treatments placed in the vagina or cervix
Potassium permanganate tablets
Herbal preparations
Misoprostol
Intramuscular injections
Two cholera immunizations

Foreign bodies placed into the uterus through the cervix

Stick, sometimes dipped in oil
Lump of sugar
Hard green bean
Root or leaf of plant
Wire
Knitting needle
Rubber catheter
Bougie (large rubber catheter)
Intrauterine contraceptive device
Coat hanger
Ball-point pen
Chicken bone
Bicycle spoke
Air blown in by a syringe or turkey baster
Enemas
Soap
Shih tea (wormwood)

Trauma

Abdominal or back massage
Lifting heavy weights
Jumping from top of stairs or roof

Sources: Grimes,[25] Lane,[2] Salter,[26] Sambhi,[27] Liskin,[28] Goyaux,[29] Thapa,[30] Ankomah,[31] Okonofua[32]

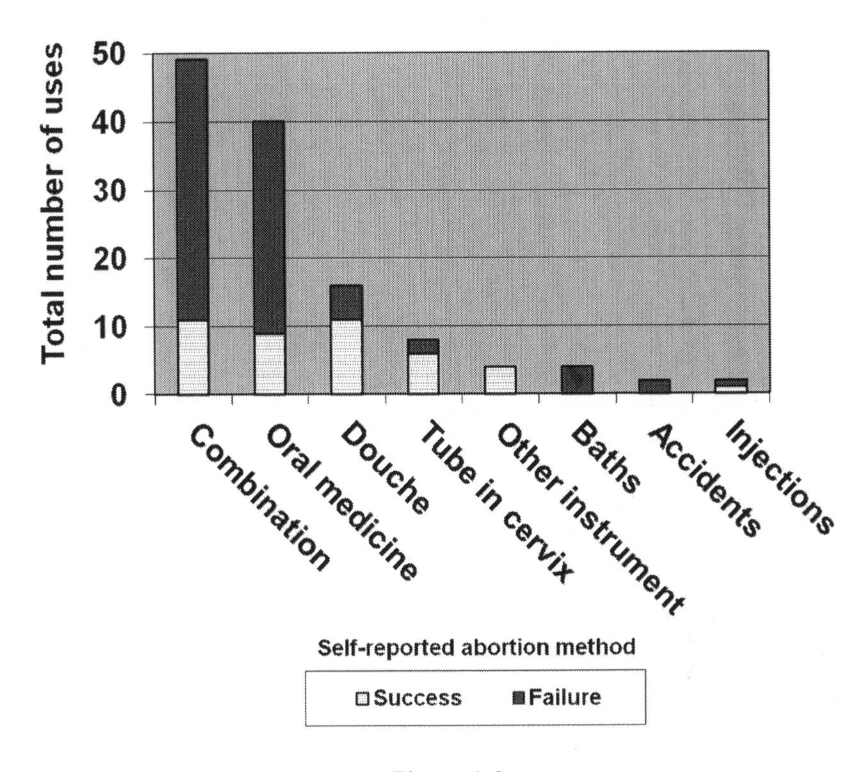

Figure 1-3
Self-reported abortion methods by success, 126 respondents, New York City, 1965 and 1967
Source: Polgar[24]

I cared for women suffering from complications of illegal abortion as a young physician in training. One afternoon, my hospital's emergency room paged me to see a gynecology patient with a temperature of 106° Fahrenheit. I presumed the reported fever was a mistake. Regrettably, it was not. The flushed woman with a racing pulse was indeed that hot. During the pelvic examination, I found a red rubber catheter protruding from her cervix, the opening to her uterus. She reported with embarrassment that a dietitian in her hometown had inserted the catheter to cause an abortion. I quickly emptied her uterus by vacuum aspiration and gave intravenous antibiotics; she recovered without incident and left the hospital a few days later.

Surveys suggested that miscellaneous methods and oral medications, such as laundry bleach, turpentine, and massive doses of quinine (a drug used to treat malaria), were the most commonly used[24] approaches.

> *"...The girl was 16 weeks pregnant. She suffered complications consisting of perforation of the vaginal wall through the utero-vesical space into the abdominal cavity with gangrenous loops of small intestine herniating through it [her vagina]..."*
>
> *Oye-Adeniran*[46]

Injecting toxic solutions into the uterus using douche bags (as was done by the protagonist in the movie *Vera Drake*) or turkey basters was common. Absorption of soap solutions, turpentine, antiseptics, and other toxins into the woman's blood stream could poison the kidneys, lead to kidney failure, and ultimately kill the woman.[33] Potassium permanganate tablets placed in the vagina were popular as well; these did not induce abortion but could cause severe chemical burns to the vagina, sometimes eroding through to the bowel.[34-37]

And the poor get buried . . .

As might be expected, affluent women fared better than did the poor. Women with money and connections were often able to find a willing physician or were able to travel to countries like Cuba and Sweden, which offered easier access to abortion. The poor were left to their own devices. This disparity in abortion access continues today in countries where abortion is illegal.[2] Data from the 1950s document that access to safe abortions was directly related to socioeconomic status. From 1951 to 1953, the ratio of "therapeutic" abortions performed in New York City hospitals ranged from 1.2 abortions per 1,000 live births in municipal hospitals caring for the poor to 6.3 per 1,000 in private hospitals. Affluent patients were better able to find physicians who could document grounds for therapeutic abortions than were poor women. Clearly, money could buy safety.

In the bad old days, abortion provision was racist as well. By 1960-1962, the ratio of "therapeutic" abortions in New York City had fallen to 1.8 per 1,000 live births. However, large ethnic disparities persisted: ratios ranged from 0.1 per 1,000 live births among Puerto Rican women to 0.5 among African-American women to 2.6 among white women. Again, these large differences reflected money and access to care, rather than the prevalence of medical and psychiatric illness necessitating abortion.[39]

Unsafe abortion: the Third World's silent scourge

Consider the international media attention to the faulty lithium-ion batteries on Boeing's new 787 Dreamliner airplane.[40] Fortunately, no one has been hurt or killed as a result of this problem. Now imagine that a jumbo jetliner loaded with 400 passengers plummeted from the sky over Long Island, New York today, with all on board killed. Within hours, news media and safety investigators would be poring over the burning wreckage. News clips from the site and interviews with grieving relatives at airports would be carried worldwide by our twenty-four hour news networks.

Imagine the international response if yet another jumbo jet, fully loaded, crashed in the United Kingdom. Again, all the passengers killed were younger than 45 years. And a few days later another airplane crashed in Australia, another in France, and another in Sweden. International outrage would force governments to ground these deadly airplanes (as was done preemptively with Boeing's Dreamliner) until the cause could be determined and remedied.

Instead, assume that each of these jumbo jets was filled with women of reproductive age, and the crashes occurred in Nigeria, Pakistan, and Brazil. Imagine further that over the course of a year, 118 giant airliners (made by the same hypothetical company) met the same fate. This is the carnage today in the Third World from unsafe abortion.[25,41,42] How much media or governmental attention has this generated? Very little.

According to estimates from the World Health Organization, of the estimated 20 million desperate women who risk their lives through unsafe abortion each year, about 47,000 die. Stated alternatively, about 1 woman in 425 dies trying to control her fertility, her body, and her destiny this way.

That this silent epidemic remains ignored reflects the nature of the victims: all women, mostly young, mostly of color. They live in places like Khartoum, not Kansas City; Ouagadougou, not Oshkosh.[42]

"First, I had two injections of Methergin [a uterine stimulant]. Afterwards, for three days, I drank before breakfast red wine boiled with borage and rue, to which I added nine aspirins. My body was full of pimples but I did not abort. A few days later I drank cement water. It did not work either. Then I went to a lady who inserted a rubber catheter into me….."[47]

Each death is a tragedy, and each leaves behind a mourning family, often including motherless children who may then die from neglect.

In 1981, I saw my first case of tetanus. The young woman's colorful sari (gown) was a stark contrast to the rubber sheet beneath her, in a rusty hospital bed in a grim Bangladesh hospital. Paralyzed, she lay in a puddle of urine. Tetanus ("lockjaw") is a constant threat in developing countries, where immunization against the disease is not universal. She had developed tetanus after having a stick inserted into her uterus by her village "dai" (midwife) to induce abortion. When I asked how long she had been lying paralyzed, "six weeks" was her doctor's response.

Conditions for women today in Third-World countries, where abortion is either usually illegal—or legal but not generally available (e.g., Zambia and Burundi)[43,44]—recall the plight of American women before the 1960s. Although modern methods of contraception are increasingly used in developing countries, millions of women still have limited or no access to effective methods. In the United States, 69% of couples use modern, effective methods of contraception. In Chad the corresponding figure is 2%, in Guinea-Bissau 4%, and Afghanistan 9%.[45] These women, unable to prevent unwanted pregnancy, frequently turn to abortion as a last resort to control their fertility.

Third-World abortion techniques

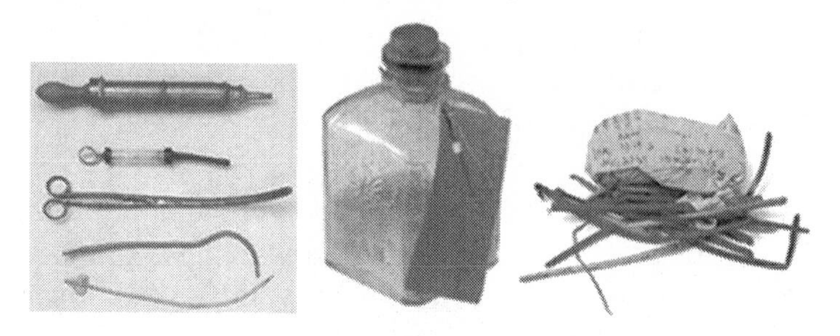

Figure 1-4
Items used to induce abortion
Source: Museum of Contraception and Abortion, www.muvs.org

Third-World abortion methods today are similar to those in the U.S. before the 1970s (Table 1-1).[25,28-32,47-50] These can be divided into broad classes: oral medications and injectable medicines, vaginal preparations, intrauterine foreign bodies, and trauma to the abdomen. In addition to the detergents, solvents, and bleach used in the U.S., women in the

Third World still rely on teas and brews made from local plant or animal products, including dung. Misoprostol, a drug that causes the uterus to contract and that can be an effective abortifacient, has been widely available on the black market in Brazil. Its use has reduced the mortality risk from illegal abortion.[51]

Foreign bodies inserted into the uterus to disrupt the pregnancy often damage the uterus and internal organs, including the bowel (Figure 1-4). Hemorrhage and infection related to these crude methods may require hysterectomy as a life-saving measure, leaving the women sterile - and often castrated - early in life. When pelvic infection is severe after unsafe abortion, hysterectomy plus removal of the ovaries (thus, castration) and fallopian tubes may be needed to save the woman's life.[52] In settings as diverse as the South Pacific and equatorial Africa, abortion by abdominal massage is still used by untrained practitioners.[27] The vigorous pummeling of the woman's lower abdomen is designed to disrupt the pregnancy but sometimes bursts the uterus and kills the pregnant woman instead.[53]

In the 1960s, physicians and clergy led the quest for safe abortion. They argued that the suffering from unsafe abortion could no longer be tacitly sanctioned. By the late 1960s, most Americans were eager to abandon the bad old days of abortion as well. The public health consequences of illegal - or legal but inaccessible - abortion were clear and well-documented. [25,28-32,47-50] Women clearly deserved better.

Take home messages

- *Before Roe v. Wade, an estimated 200,000 to 1.2 million illegal abortions occurred each year in the U.S.*
- *More than 1,000 U.S. women died each year from unsafe abortion as recently as the 1940's*
- *Globally, unsafe abortion leads to the preventable deaths of about 47,000 women each year*
- *Before Roe v. Wade, poor women of minority races disproportionately suffered and died*

Chapter 2
Mass medical tourism – and resulting health benefits

During the second phase of abortion in America, availability was described as a "western sandwich:" service on both coasts, but none in the wide open spaces in between. Indeed, during the "sandwich" years, three-quarters of all abortions in America took place in just two states — New York and California. Women took to the road in great numbers — a massive medical tourism unprecedented in American history. The resulting improvements in the health of women and infants were quickly and consistently documented in the nation's largest hospitals.

The legal background

In 1957, the American Law Institute proposed a Model Penal Code, which expanded the indications for legal abortion.[1,2] Beginning in 1967, 13 states adopted fairly restrictive abortion legislation that incorporated most of the recommendations of the Institute. In November, 1967, California enacted a Therapeutic Abortion Act extending indications for abortion to include the mental or physical health of the woman.[3] An important motivation for this liberalization was that in the prior year, nine San Francisco gynecologists were charged with unprofessional conduct by the State Board of Medical Examiners for having performed hospital committee-approved abortions on women with first-trimester exposure to rubella (German measles), which can cause birth defects.[4] The turning point came in 1970, when four States (Alaska, Hawaii, New York, and Washington) enacted non-restrictive legislation that ushered in abortion on request.[2] The most important such change took place when New York's liberal law became effective on July 1, 1970; the effect reverberated across America, and New York City, in particular, was quickly inundated with women seeking service.

Surveillance starts

Recognizing that abortion was emerging as an important public health issue and that no national data were available, researchers at the Centers for Disease Control and Prevention (CDC) began to track abortions nationwide. The CDC's annual Abortion Surveillance Reports chronicled the early days of legal abortion in America.

For 1970, 14 states and the District of Columbia voluntarily provided data to the CDC on numbers and characteristics of women having abortions.[5] Laws were rapidly changing, and State surveillance systems were incomplete. Nevertheless, 176,000 abortions were reported for the year. The following year, 24 States and the District reported a total of 480,000 abortions.[6] For 1972, the totals increased to 27 States plus the District and 587,000 procedures.[7] Thus, well before the pivotal *Roe v. Wade* decision, abortion services were expanding dramatically in response to grassroots support for legal abortion at the State level.

Medical tourism to find care

New York and California immediately dominated abortion provision in America; in the year before *Roe v. Wade,* these two States accounted for three-fourths of all abortions nationwide. In 1970, these two States accounted for 86% of all abortions nationwide (Figure 2-1).[5] The following year, as more States began providing services, the proportion in New York and California declined to 79%,[6] and in 1972, to 75%.[7]

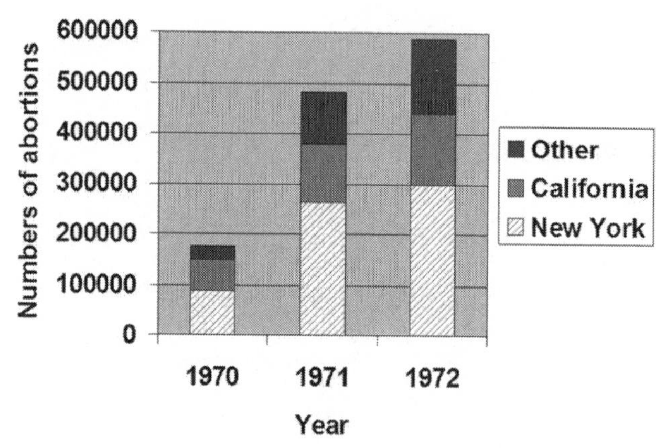

Figure 2-1
Numbers of abortions, United States, 1970-1972, by area
Source: Center for Disease Control [5-7]

New York City quickly became an abortion center, and development of outpatient clinics not affiliated with hospitals drew national attention.[8] Twenty-four freestanding abortion clinics had opened by 1972, and these accounted for the largest number of procedures. Women from across the country flocked to the city in the early days of legal abortion. Indeed, most women (60%) having abortions in New York in 1971 and 1972 did not live in the State (Figure 2-2). A smaller but growing proportion of women who received care in California were from out-of-State.

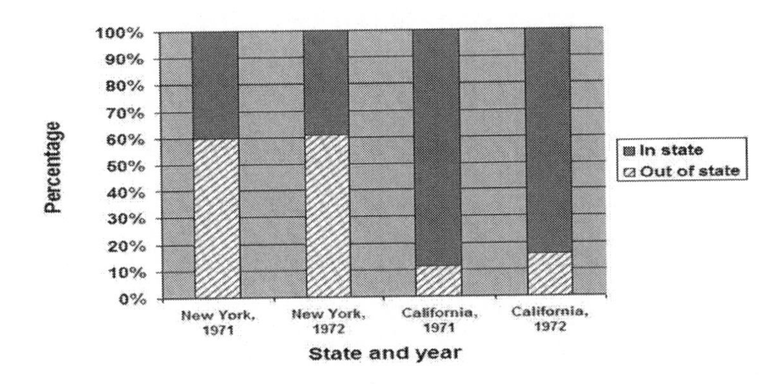

Figure 2-2
Proportion of abortions for in- and out-of-State residents, New York and California, 1971 and 1972
Source: Center for Disease Control[5-7]

Women seeking care in New York City came from nearly every State and from some other countries, especially Canada.[9] As might be anticipated, New Jersey residents comprised the largest number of out-of-State patients. However, distant States in the top ten out-of-State areas included, in descending order, Illinois, Michigan, Ohio, and Florida. In buses, cars, planes, and trains, more than 71,000 women traveled from these four States to New York City for abortions between 1970 and 1972.

In the year prior to *Roe v. Wade*, nearly half of all abortions in the U.S. were performed on out-of-State residents. In 1972, CDC reported that 44% of all women having abortions nationwide received care in a State where they did not live. The highest proportions of out-of-State patients were seen in the District of Columbia (75%), New Mexico (69%), Kansas (63%), and New York (61%).[7] Charged with evaluating the early public health impact of the legalization of abortion, the Institute of Medicine concluded that laws mattered: women traveled in large numbers to States with less restrictive abortion laws.[2] As a result, States with more liberal laws had disproportionately high abortion rates.

This massive medical tourism to seek health services was unprecedented in American history. Never before had tens of thousands of patients traveled such great distances for medical services. The cost of transportation, food, and lodging while in transit delayed services and added to the cost. As a result, follow-up care in case of complications became more complex.[9] Women endured substantial financial and personal hardships during these journeys in order to obtain safe, legal abortions that were not available locally. Investigations by the Centers for Disease Control and Prevention linked travel to obtain abortion services with preventable complications[10] and death.[11]

Travel patterns provide some insight into the sacrifices involved (Table 2-1).[9] Women from Illinois logged the most miles, an estimated 33 million. For the 160,153 women from these ten areas, the total travel distance was about 120 million miles. This estimate is conservative, since it uses straight-line, rather than actual road, distances.

Table 2-1

Estimated cumulative travel distances for non-resident abortion patients, New York City, by top ten homes of record, 1970-1972

Home of record	Round-trip distance* to New York City (mi.)	Number of patients making trip	Total travel distance (million miles)
New Jersey	104	31,424	3
Illinois	1650	20,179	33
Michigan	1122	19,324	22
Ohio	952	18,713	18
Pennsylvania	308	18,191	6
Florida	1826	13,152	24
Massachusetts	380	11,578	4
New York State	268	10,787	3
Connecticut	200	9,073	2
Canada, other countries	672	7,732	5
Total			120

*As the crow flies, from the State capital or Ottawa, Ontario, for Canada and other countries

Source: Pakter[9]

One hundred twenty million miles is a long way to travel....about 251 round trips between the earth and moon or 4,800 circumnavigations of the world at the equator....just to get medical care. The collective trip was expensive as well. At $0.56 per mile (2014 mileage reimbursement rate allowed by the Internal Revenue Service), a trip of this distance would cost $67 million in today's dollars—a figure that excludes both food and lodging. The financial sacrifices involved often meant hardships for women and their children.

International medical tourism for abortion

After the restrictive Polish abortion law was enacted in 1993, an estimated 16,000 women traveled to neighboring countries for care. The cost of the procedure alone generally was greater than a month's salary.[12] From 2001 to 2008, more than 45,000 women from the Republic of Ireland obtained care in Britain.[13] In 2005, the total costs to the woman, including travel, lodging, and the procedure, were estimated to be from €965 to €1,750 (about $1,300 to $2,400). Another 2,000 women from Northern Ireland made this trip annually.[14] Mexican women continue to seek care in U.S. abortion facilities,[15] where the costs of the procedure are approximately two months' salary.[16] Women from Calgary, Alberta, seeking abortions in Seattle, Washington in the 1970s traveled about 1200 km, taking 12 hr by car or 24 hr by bus.[17] Today, women still travel long distances to get care even within countries.[18-20] Young women and those with limited incomes are disproportionately put on the road for care.[18,19]

The health of women and babies improves

Key hospitals in New York foretold what lay ahead with *Roe v. Wade*. New York City[21-24] and New York State[25] became the largest providers nationwide. Physicians in large municipal hospitals provided early reports on the prompt, dramatic health benefits of legalizing abortion.

Kings County-State University Hospitals, five other affiliated Brooklyn hospitals, and the Harlem Hospital Center all chronicled unprecedented changes in obstetric events. Kings County, the largest of the city's municipal hospitals, and its affiliates reported more than a 5800% increase in the number of elective abortions performed in the year after the law changed. In that year the number of "spontaneous" abortions dropped by 20%. This led the authors to conclude that many of the allegedly spontaneous abortions in the past had been illegally induced.

Even more striking was the 36% decline in immature births (infants weighing 500 to 999g [1.1 to 2.2 lb]). This abrupt change suggested a rapid decrease in illegal termination procedures that yielded tiny newborns, often with early death or the devastating long-term consequences of severe immaturity, such as cerebral palsy. The proportion of unwanted newborn infants abandoned by their mothers plummeted from 15 per 1,000 deliveries to 7.[21] A follow-up report from Brooklyn documented a continuing decline in the numbers of "spontaneous" abortions through 1972 (net reduction of 46% at Kings County-State University Hospitals) and in immature births.[22] The researchers concluded that the dramatic rise in elective abortions (and subsequent improvements in obstetrical outcomes) reflected the replacement of illegal abortions with safer, hospital abortions.[22]

Similar salutary changes took place at Harlem Hospital Center, an 1100-bed municipal hospital on the Upper East Side of Manhattan. Perinatal mortality (stillbirths and infant deaths in the first week of life) dropped from 57 deaths per 1,000 live births in 1969 to 40 in 1971. This was largely attributable to the marked decline in the number of immature births (weighing from 500 to 750g), decreasing from 22 to 8 per 1,000 births. Nearly all these infants died soon after birth. The authors commented that the high proportion of immature births stemmed from community lore that insertion of rigid or soft catheters (favorite abortion methods in Harlem) into the uterus was more effective after four months of pregnancy than before.[23,24] In the remainder of New York State, excluding New York City, similar benefits were seen, as deaths from illegal abortion fell dramatically after 1970.[25]

Subsiding sepsis

Similar obstetrical changes were underway on both coasts. Georgia had permitted abortions since 1968, with the law patterned after the American Law Institute Model Penal Code.[1] In 1970, the procedural requirements were relaxed.[26] At Grady Memorial Hospital, the municipal hospital serving Atlanta, liberalization of abortion was temporally related to a decline in complications of illegal abortion, although the effect was not immediate. The researchers speculated that not until more legal abortions took place (a threshold effect) did the decline occur.

California municipal hospitals had similar favorable experiences. At Los Angeles County-University of Southern California (the nation's largest municipal hospital), its Women's Hospital unit found an inverse relationship between numbers of legal abortions performed and numbers of women admitted because of septic (infected) abortions.[27] In the late 1960s, more than 600 women were admitted annually for this complication, which often included infection spread beyond the uterus to include the abdomen and bloodstream. By 1970, the number had dropped to 305. In contrast, the number of legal abortions grew from 2 in calendar year 1966 to 3,469 in 1970.

To the north, San Francisco General Hospital documented the inverse pattern, although published data were more limited.[3,28] Ninety-five legal abortions were provided in 1968, and 256 in 1969. As the legal abortion service began, the number of admissions to the hospital for septic abortion dropped from 69 in 1967 to 22 in 1969.

A consistent inverse relationship was seen in these three different cities between the provision of legal abortion to local women and hospitalizations for infected abortions. Most of the latter were presumed to be illegally induced. This temporal relationship, in the absence of alternative explanations for the marked declines in septic abortion, provides strong indirect evidence of an early, important health benefit.[29] As abortion moved from the back alley to municipal hospitals, life-threatening complications of abortion diminished.

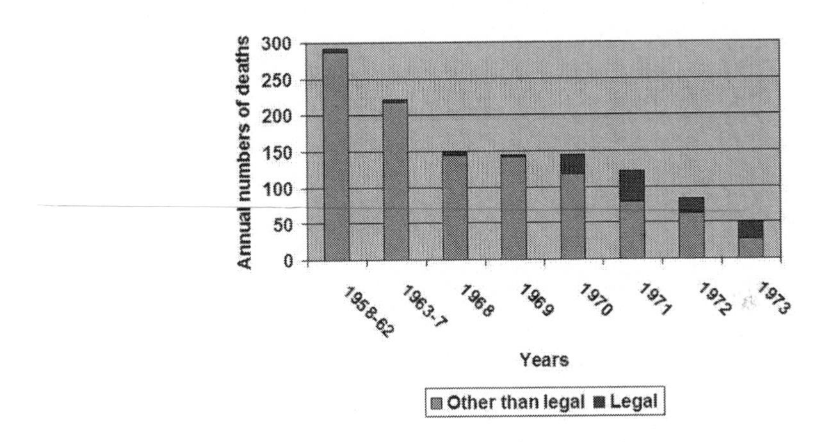

Figure 2-3
Annual numbers of abortion deaths, United States, 1958-1973, by type of abortion
Source: Tietze[31]

Deaths from abortion nationwide corroborated this trend (Figure 2-3). Legal abortions began to replace illegal procedures,[30] and the numbers of deaths reported through vital statistics nationwide dropped rapidly, after a plateau in the late 1960s.[31] Epidemiologists and demographers estimated that during the "sandwich" years, 70% of legal abortions replaced clandestine, unsafe abortions.[32]

Declines in unintended fertility

States that were early to liberalize abortion laws saw a prompt though minor decline in fertility. Births in States with early legalization of abortion fell by 4% relative to States without liberalized laws between 1971 and 1973. The differences diminished in the next two years and disappeared by 1976.[1] This figure substantially underestimates the true effect on fertility, however, since it does not account for interstate travel, which was extensive. In New York, California, and the few other States, early legalization reduced the fertility rate by about 11%.

Moreover, the demographic impact of the legalization of abortion was blunted by the substitution of legal for illegal abortion. Each year, large numbers of pregnancies were already being terminated illegally. Hence, legalization of abortion improved women's health proportionately more than it reduced fertility.[32] As 1972 came to a close, the medical community had a glimpse of the public health benefits of legal abortion.[2] A societal sea change lay just weeks ahead.

Take-home messages

- *In the early 1970s, most abortions took place in just two States*
- *Tens of thousands of women were forced to travel long distances to get care*
- *Safe, legal abortions replaced large numbers of dangerous, illegal abortions*
- *The health of women and children improved as a result of this transition to safe care*
- *Septic abortion wards in large hospitals emptied, then closed*

Chapter 3
Legalization and medicalization

Roe v. Wade ushered in the third phase of abortion in America. The changes wrought by legalization were as profound as they were prompt. Although the health benefits of liberalized abortion laws were already manifest by January 22, 1973, the *Roe v. Wade* decision added momentum to this sea change in health. *Roe* struck down restrictive abortion statutes, like that in Texas, because they violated the Due Process Clause of the 14th Amendment to the Constitution.

> *"Acknowledging his strong 'pro-life' bias, Koop testified that any public-health problem associated with abortion is 'minimal.'"*
>
> *Surgeon General C. Everett Koop*[1]

Technology transitions

Abortion became not only legalized but also medicalized; it came out of the shadows and into the bright light of scientific scrutiny. After decades of stagnation, abortion technology advanced at an unprecedented pace. Before legalization, the most common abortion procedure used by physicians had been sharp curettage (also known as dilation and curettage, or D&C). In this operation, the surgeon dilated the cervix by introducing a series of progressively larger tapered metal rods into the uterus. Then, the surgeon emptied the uterine contents with a metal curette, a crude scraping tool invented by Recamier in 1843. Initially designed to scrape off uterine "fungosities," this inelegant instrument was embroiled in controversy from its inception.[2]

In the early days of legal abortion, suction curettage quickly supplanted sharp curettage as the dominant method. With this approach, suction, rather than scraping, evacuates the uterus. (An analogy would be a vacuum cleaner vs. a garden hoe to clean a delicate carpet.) Numerous early studies documented that suction curettage (also called vacuum aspiration) was safer, faster, and more comfortable.[3] In 1970, 46% of abortions were performed by sharp curettage; by 1995, this operation had nearly vanished as an abortion method (2%).[4]

Freestanding abortion clinics quickly emerged as a new model of outpatient surgery. Because most U.S. nonfederal hospitals failed to start providing abortions after *Roe v. Wade*, a huge need for abortion services remained unmet. To fill the void, clinics rapidly opened, becoming the prototype for those clinics that today provide a range of outpatient surgical services previously unavailable outside of hospitals. Today, one can have operations ranging from ophthalmology to podiatry in freestanding clinics. This evolution in health care would have occurred eventually for economic and other reasons, but it occurred sooner because of the early, excellent track record of abortion clinics. Contrary to conventional wisdom, abortions performed in freestanding clinics proved safer than those provided in hospitals.[5,6] In fact, despite a few notable exceptions,[7,8] freestanding clinics have set the standard in safe abortion practice.[9]

Paradoxically, innovation in abortion practice shifted from academic medical centers to the community. Traditionally, research in medical centers leads to improved care, and these innovations then spread centrifugally into wider clinical use. In contrast, important breakthroughs in abortion technology, such as dilation and evacuation (D&E) for midtrimester abortion, came from individual practitioners[10-12] and spread centripetally to academe. Ironically, academic research with prostaglandin $F_{2\alpha}$, first hailed as a major advance in labor-induction abortion, ultimately found this "Cinderella drug" inferior to common table salt as an abortifacient.[13,14]

Conventional wisdom held that abortions could not be done safely in the "grey zone" of 13 to 15 weeks' gestation, so women who requested procedures at this stage of pregnancy were customarily delayed for several weeks. D&E eradicated this invalid concept of a gestational-age threshold for surgical evacuation. Conventional wisdom held that curettage abortions could not be done safely after 12 weeks' gestation. Instead, curettage operations in the midtrimester proved not only feasible but safer than the accepted standard of labor-induction abortion.[15-17]

As abortions moved from hospitals to freestanding clinics, anesthesia practice shifted as well. Dilation and curettage had traditionally been done under general (unconscious) anesthesia in a hospital operating room; the new norm became suction curettage under local anesthesia in a clinic procedure room. This change reaped important public health benefits as well. Local anesthesia proved safer, simpler, and less expensive than general anesthesia.[18] The same held true with anesthesia for later D&E procedures.[19] Although local anesthesia carries risk when administered improperly,[20] it still is safer than general anesthesia. Indeed, complications of general anesthesia - not the procedure itself—have emerged as a leading cause of death with first-trimester abortion.[21]

Abortion as a social equalizer

Demographic changes in who had abortions were as dramatic as were changes in how abortions were done. During the years before *Roe v. Wade*, affluent women could often find safe abortion care. This was not the case for women disadvantaged by poverty and other hardships. After *Roe v. Wade*, the proportion of African-American women having abortions quickly increased. In 1972, the proportion of African-American abortion patients was 23%; by 1995, the figure had nearly doubled. African-American women continue to choose to have abortions at a disproportionately high rate.[22] The claim from anti-abortion groups (e.g., www.BlackGenocide.org and www.KlannedParenthood.com) that this reflects "genocide" is demeaning to these women.[23] Similarly, the proportion of unmarried abortion patients grew from 70% to 81% from 1972 to 1995.

Roe v. Wade also facilitated care closer to home. In 1972, 56% of all abortions were performed outside of women's home States. In contrast, by the 1980s and 1990s, 90% of women received abortion care in their State of residence.[24] Additionally, *Roe v. Wade* accelerated a shift in the timing of abortion to earlier, safer stages of pregnancy. In 1970, nearly a quarter of all abortions nationwide took place at 13 weeks' gestation or later. Within a decade, that figure had dropped to 10%. *Roe* led to abortions performed earlier and closer to home, thus improving the safety and lowering the costs for women.

Deaths plummet

In the span of a few years, illegal abortion nearly disappeared from the American scene (Figure 3-1). Because of their clandestine nature, illegal abortions could not be enumerated…only estimated. One method of estimating annual numbers of illegal abortions involved multiplying the number of illegal abortion deaths by the presumed mortality rate. Based on this approach, the projected numbers of illegal abortions quickly fell to low levels.

As of 1970, government statistics found the risk of death from abortion and from childbirth comparable (Figure 3-2). As legalization of abortion spread nationwide in the early 1970s, death rates from these two pregnancy outcomes diverged. This wide disparity has persisted for more than three decades now. The likelihood of dying from childbirth (top line) continued to decline gradually, as it had done for decades. In contrast, the risk of dying from abortion (bottom line) dropped precipitously. Given that the same health care providers, hospitals, antibiotics, and blood banks were available to pregnant women for childbirth or abortion, the inescapable conclusion was that the legalization of abortion was the critical difference. Because it replaced the dangerous, clandestine procedures of the past, legal abortion's health benefits were both immediate and profound.

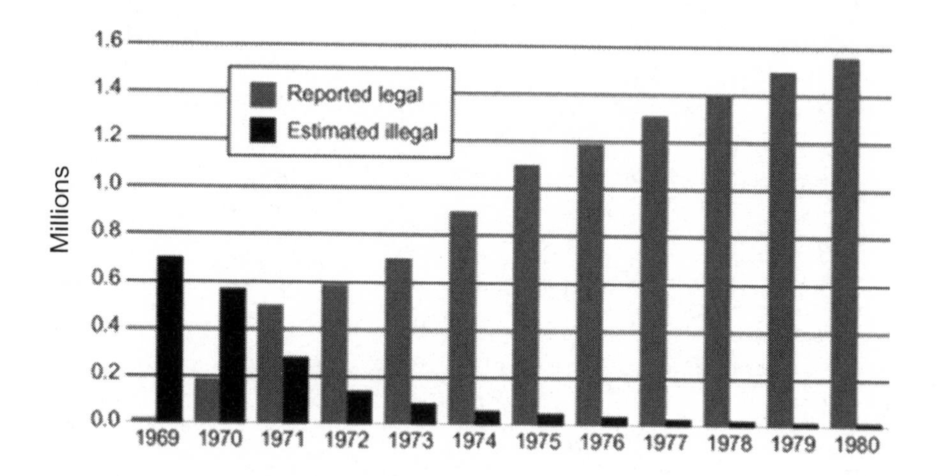

Figure 3-1
Estimated numbers of legal and illegal abortions, United States, 1969-1980
Source: Cates[24] Reproduced with permission of the Guttmacher Institute.

Several concurrent trends were responsible for this dramatic achievement. First, safer procedures became standard practice. As noted previously, suction curettage largely replaced sharp curettage, and D&E replaced labor-induction abortions and hysterotomy (a mini-cesarean) in the second trimester.[26] Pain relief shifted from general to local anesthesia. Women obtained their abortions at earlier, safer times in pregnancy.[3] Clinicians became more skillful in providing abortions and in managing their complications, as abortion training in obstetrics and gynecology residency proliferated.[27]

Experience in the U.S. echoed that in Europe. For example, the improvement of abortion safety in Sweden had three principal sources. The introduction of antibiotics in the 1940s reduced the risk of septic abortions from illegal procedures. Abortion acts in 1938, 1946, and 1975, which progressively liberalized abortion, both eliminated illegal abortions and shifted legal abortions to earlier, safer gestational ages. Finally, suction curettage became the standard of care.[28]

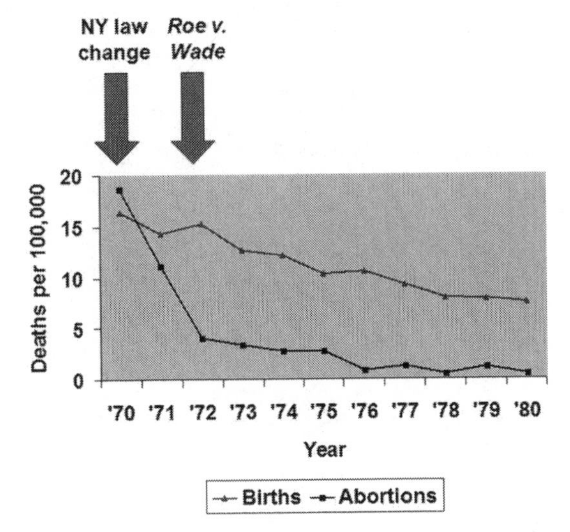

Figure 3-2
Abortion- and birth-related mortality, United States, 1970-1980
Source: Tietze[25]

Compared to what?

Within a decade after *Roe v. Wade*, the risk of death from legal abortion had fallen below one death per 100,000 abortions. To put that remote risk in perspective, the risk of death from an allergic reaction (anaphylaxis) from an injection of penicillin has been estimated to be around 2 deaths per 100,000 injections.[29,30] Thus, by the late 1970s, this surgical procedure had become safer than an injection of penicillin.

Other risks in daily life provide additional useful benchmarks. In 1980, Dinman pointed out that voluntary activities of daily life carry varying risks (Figure 3-3),[31] and Trussell updated these estimates more recently.[32] Their results show that death risks range from as high as 100 deaths per 100,000 for those who ride motorcycles to as low as 1 per 100,000 for those who paddle canoes. The 1980 risk of death from legal abortion was similar to that from canoeing.[25]

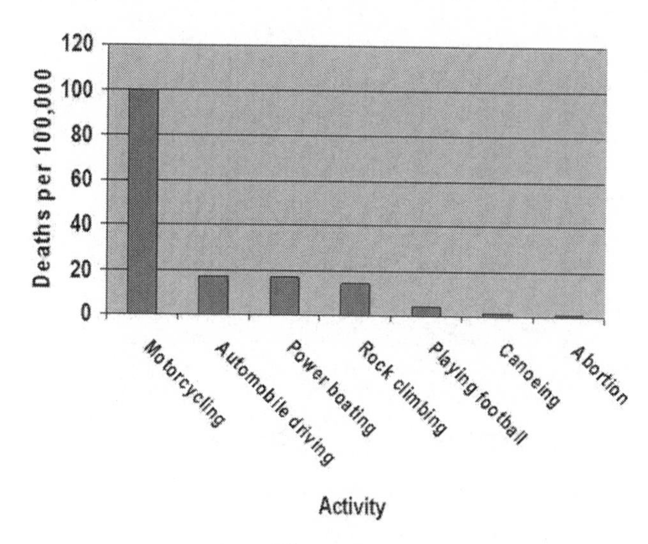

Figure 3-3
Estimated risks of death per year for men and women of all ages who participate in selected activities, 1980
Sources: Trussell,[32] Dinman[31]

By 1975, the Institute of Medicine, "...an independent, nonprofit organization that works outside of government to provide unbiased and authoritative advice to decision makers and the public,"[33] had documented that legal abortion improved the health of Americans.[34] Several other confirmations followed soon thereafter. Vital statistics[35] confirm that abortion became a rare cause of death among women in the U.S. (Table 3-1). According to the National Center for Health Statistics, in 1980, 318 women died of appendicitis (usually treated by an appendectomy), and 62 died of syphilis, an infection easily cured by penicillin. More than 5,000 women were murdered, and nearly 7,000 took their own lives. Unusual variants of motor-vehicle accidents killed still other women. Riding on the rear seat of a motorcycle proved fatal for nearly 300 women, and another 194 lost their lives when their cars collided with a train. How many died from legal abortion in 1980? Nine – the same number who died in snowmobile accidents.[36] Because these numbers of deaths lack denominators (numbers of women who engaged in these various activities), they do not allow estimates of death rates. They do provide, however, a crude assessment of the national loss of life. I am unaware of any State legislatures that have passed "Woman's Right to Know" bills[37] describing risks and benefits of snowmobile use by women.

Table 3-1

Some Uncommon Causes of Death Among Women of All Ages in the United States, 1980

Cause	Numbers of deaths
Infections	
Kidney infection	1772
Influenza	1709
Tuberculosis	647
Meningitis	619
Appendicitis	318
Syphilis	62
Suicide and homicide	
Suicide	6964
Homicide	5190
Vehicular accidents	
Passenger on motorcycle	271
Collision of car with train	194
Snowmobile accident	9
Legal abortion	9

Sources: National Center for Health Statistics,[35] *Centers for Disease Control and Prevention*[36]

According to the Centers for Disease Control and Prevention, by the mid-1970s, abortion had become far safer than childbirth. This has remained true for three decades. Abortion has lower risks of complications and the resultant need for an abdominal operation to treat complications. For example, the risk of requiring a cesarean delivery for childbirth is several hundred times higher than the risk of requiring an exploratory laparotomy (opening the abdomen surgically) to treat a complication of suction curettage.[24]

The Koop (Non)Report

Other independent assessments of the comparative safety of abortion confirmed the vital statistics. In 1987, Surgeon General C. Everett Koop was directed by President Ronald Reagan to examine the medical and psychological consequences of legal abortion.[38] The implicit objective was to document alleged harm that could then be used to justify overturning *Roe v. Wade*. Koop was an outspoken opponent of abortion and had written a book opposing elective abortion. A scientist of integrity, he conducted a diligent search through the literature and met with members of twenty-seven groups on both sides of the abortion issue. Part of a Planned Parenthood Federation of America contingent, I met with him during the preparation of his report.

That report was squelched. In January, 1989, Koop announced he could not release the report because the data were flawed and inconclusive (despite the fact that more was known about the epidemiology of abortion than about any other operation).[24] Subpoenaed by a House of Representatives subcommittee chaired by Rep. Ted Weiss (D-NY), Koop's report concluded that "abortion imposes a relatively low physical risk." The subcommittee found that Koop had "expressed doubts about the existence of Postabortion Stress Syndrome" as well.[38] Koop reported that emotional problems resulting from abortion are "miniscule from a public health perspective." Reagan's directive failed to uncover any significant medical or psychological consequences of abortion.

The American Medical Association weighs in

A third independent assessment of the public health impact of legal abortion was done by the American Medical Association (AMA). At its 1991 annual meeting, the House of Delegates passed Resolution 17, which called for the AMA to "...perform an objective study of available data on the mortality and morbidity associated with illegally induced abortions prior to the US Supreme Court's *Roe v. Wade* decision and compare it with the mortality and morbidity incurred by abortions performed today."[39]

The motivation for this study was more political than scientific. The House of Delegates specified that "...the results of this study be published in a manner accessible to *legislators* [emphasis mine] and the public by the 1992 Annual Meeting."[39] As the AMA noted, this review came at a time of "continued heated national debate on abortion," and it cited numerous examples of States enacting restrictive abortion laws. A secondary goal of the review was to predict the health effect of such restrictive laws.

The AMA report confirmed the greater safety of legal abortion compared with either illegal abortion or childbirth. Concerning the latter, the report concluded that, "This figure [abortion mortality rate] is more than 10 times lower than the maternal mortality ratio. The discrepancy is even larger when adjustments are made for age, race, and preexisting conditions."[39]

"Mandatory waiting periods, parental or spousal consent and notification statutes, a reduction in the number and geographic availability of abortion providers, and a reduction in the number of physicians who are trained and willing to perform first- and second-trimester abortions increase the gestational age at which the induced pregnancy termination occurs, thereby also increasing the risk associated with the procedure."

American Medical Association[39]

As requested by the Board of Delegates, the AMA's Council on Scientific Affairs commented on the public health impact of anti-abortion legislation at the State and federal level.

The report also predicted that women would be forced back to interstate travel for care: "...if some States maintain nonrestrictive abortion laws and offer the procedures to nonresidents, many women will travel to States with more moderate abortion laws."

Should access to abortion become more limited, the AMA predicted that "...more women are likely to bear unwanted children, continue a potentially health-threatening pregnancy to term, or undergo abortion procedures that would endanger their health. As access to safer, earlier legal abortion becomes increasingly restricted, there is likely to be a small but measurable increase in mortality and morbidity among women in the United States."[39] The AMA report both confirmed the public health benefits of legal abortion and foresaw the public health danger of legislative attempts to turn back the clock on American women.

"Thus, the actual risk of death from pregnancy and childbirth has been underestimated over time and when compared with the risks from abortion."

American Medical Association[39]

Take-home messages

After Roe v. Wade

- *Vacuum aspiration quickly replaced sharp curettage (D&C) for first-trimester abortion*
- *Outpatient clinics became the dominant abortion provider*
- *Surgical abortion supplanted labor-induction abortion for second-trimester procedures*
- *Abortion was consistently found to be safer than continuation of pregnancy*

Section II
All pregnancies terminate

Tamesha

On December 1, 2010, Tamesha Means had spontaneous rupture of her membranes at 18 weeks of pregnancy. Having labor pains, she called a friend for a ride to the only hospital in Muskegon County, Michigan: Mercy Health Partners, a Catholic institution. Her condition at that time should have been diagnosed and managed as an "inevitable abortion:" the miscarriage was already underway and could not be stopped. The appropriate medical care in this situation is prompt emptying of the uterus, either surgically or medically.

Tamesha claims that she was not advised of the diagnosis and proper management; instead, she was sent home with pain medicine and told to keep her scheduled doctor's appointment on December 9.[1]

Tamesha returned to the hospital early on December 2, in pain and with a low-grade fever. She was sent home and returned that night in labor. While waiting to be discharged yet again, she miscarried in the early hours of December 3. Her immature infant died in a few hours. She was then told that she was responsible for arranging burial or cremation of her dead infant.

Tamesha survived the miscarriage. She is now suing the United States Conference of Catholic Bishops for dictating suboptimal care in Catholic hospitals for women experiencing miscarriages, viz., valuing the life of the fetus over that of the pregnant woman.[1] Given that the fetus at 18 weeks had no chance for survival, failing to end the pregnancy after rupture of membranes was dangerous and medically irresponsible. Commenting on the case, former President of the American College of Obstetricians and Gynecologists Dr. Douglas Laube described her care as "basic neglect." "It could have turned into a disaster, with both baby and mother dying."[2]

1. American Civil Liberties Union. Tamesha Means vs. United States Conference of Catholic Bishops. https://www.aclu.org/reproductive-freedom-womens-rights/tamesha-means-v-united-states-conference-catholic-bishops, accessed January 5, 2014.

2. Eckholm E. Bishops sued over anti-abortion policies at Catholic hospitals. http://www.nytimes.com/2013/12/03/us/lawsuit-challenges-anti-abortion-policies-at-catholic-hospitals.html, accessed December 3, 2013.

Chapter 4
Miscarriage: the healthy winnowing of pregnancy

"The ultimate outcome of all human conception is death. Less well known, but no less true, the majority of those deaths occur before birth."

Boklage[1]

Human reproduction is a surprisingly clumsy enterprise. If adults tripped during every third walk or vomited after every other meal, our species likely would have gone the way of the dinosaurs. Yet that gross inefficiency characterizes how we humans reproduce.[2]

Pregnancy losses

Human reproduction features profligate wasted potential. Women in the U.S. have an average of two births over their lifetimes (actually, 2.007 births as of 2009).[3] What precedes or follows these two births? Before being born, women have a maximum of six to seven million egg precursors, called oogonia.[4] Attrition then proceeds at a brisk pace, so that by the time of birth, around two million primordial follicles (potential human eggs) remain. After the woman is born, the number of eggs continues to dwindle: when she reaches puberty, only 300,000 remain, and still the losses continue. By the time of menopause, from zero to a few hundred follicles persist.

The loss of male reproductive potential is even greater. Beginning at puberty, the testicles produce extravagant numbers of sperm. Sperm production declines as the man ages but continues until death. Assuming a conservative daily sperm production of 45 to 207 million,[5,6] that estimate translates into 16 to 76 billion sperm per year over an adult lifetime. Of more than one trillion sperm produced in a man's lifetime, just two are ultimately needed to fertilize the typical two eggs to create the average two pregnancies per woman's lifetime. At first glance, such waste seems senseless. However, given the normal winnowing

"...once a pregnancy has been established, all will proceed normally until parturition [childbirth]; in point of fact, this is far from the truth, because in many species early embryonic mortality takes a major toll."

Short[10]

of pregnancy, this redundancy in reproductive potential ensures that our species not only endures, but thrives.

Losses after fertilization

Up to one-third of all recognizable pregnancies ends through miscarriage (also called spontaneous abortion). However, extrapolation back to the earliest days after fertilization suggests that most conceptions do not even survive until six weeks of age. Of those conceptions that actually implant in the uterus, a large proportion is lost before the woman even recognizes that she is pregnant (called occult, or chemical, pregnancies). While this huge loss, which carries adverse medical[7] and psychological consequences,[8,9] seems cruel and senseless, it is not. Miscarriage plays a crucial biological role in promoting the well-being of our children - and thus our species.

> *"Whatever the reasons, natural reproduction occurs in such a way that more than one half (some estimates range as high as 78%) of fertilizations do not result in live births."*
>
> *American College of Obstetricians and Gynecologists*[11]

Awkward steps toward pregnancy

The tentative steps toward establishment of a pregnancy are complex and shrouded in mystery. Indeed, much of the early biology of pregnancy is unknowable with present technology. Just as early astronomers could only guess about the appearance of the dark side of the moon, biologists today have no way of studying the early days after human fertilization in the fallopian tube. The preembryo cannot be seen or detected for two weeks, until its presence can be sensed by its unique biochemical footprint (Figure 4-1).

Fertilization is a necessary but insufficient step on the perilous path to pregnancy. Two important consequences of fertilization are mixing of the genetic material and prompting division of the fertilized egg. Union of egg and sperm usually takes place in the fallopian tube, near its trumpet-like end. The process takes about 24 hours to complete. After merger of the genetic material from the parents, the haploid chromosome sets (23 chromosomes from each parent) become diploid, with a full complement of 46 chromosomes. The second consequence of fertilization is activation of cell division. The single fertilized egg (now called a zygote) begins to divide, truly a humble start for the billions of cells ultimately needed.

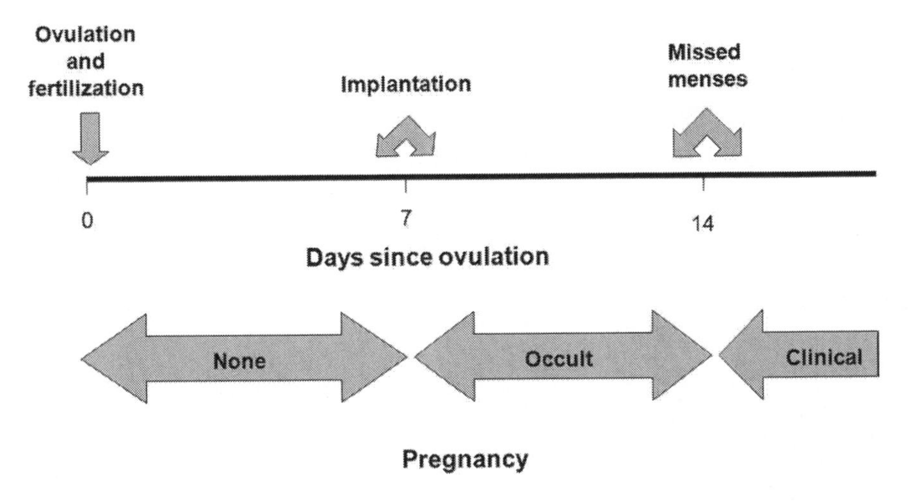

Figure 4-1
A timeline of the steps to pregnancy

By three days after fertilization, the preembryo (as it is now called) has reached an eight-cell structure. However, the cells are not fused together into a single organism. Instead, they are loosely bundled together by outside material. The free-floating clump of cells makes a leisurely passage through the fallopian tube toward the uterus: several days are required to travel several inches.

Around four to five days after merger of egg and sperm, the preembryo enters the uterine cavity. About five to seven days after fertilization, the preembryo begins burrowing into the wall of the uterus. By eight to ten days after fertilization, the process of implantation is complete. This is an important milestone, for now a pregnancy has begun, and the preembryo graduates to become an embryo.[11] As the early placenta (afterbirth) grows to nourish the embryo, it produces a hormone (human chorionic gonadotropin, abbreviated hCG) that can be detected by commercial pregnancy tests. This is the basis for tests that can be bought in pharmacies as well as the sensitive blood and urine tests that health care providers use.

Many women are surprised to learn that pregnancy does not occur in the uterine cavity. Instead, the entire process takes place in the uterine *wall*. After burrowing into the wall, the growing pregnancy progressively bulges outward, eventually filling the cavity, until the pregnancy sac reaches the opposite wall. *Nevertheless, the pregnancy remains in the wall for the duration of the pregnancy.* (Because of this, an IUD can be removed from the uterine cavity during pregnancy without disrupting the sac, which is within the wall.)

The existence of a preembryo cannot be detected (Figure 4-1). No hCG is made. After implantation, the embryo produces the hCG that enables the early pregnancy to be detected chemically. Since the woman has not yet missed a menstrual period, she does not suspect pregnancy. At the time of the missed menses, the woman has her first overt indication of pregnancy.

Biological foundations for miscarriage

Miscarriage plays several important roles in human biology. In species like ours, in which a single fetus is the normal pregnancy, miscarriage may help to space pregnancies. This benefits the family. For example, pregnancies that occur in rapid succession carry increased risks of illness and death to both the newborn and the mother.[10,12] Hence, miscarriage can be considered an innate, early form of family planning in our species. Like the child-spacing stemming from prolonged breastfeeding, miscarriage may have helped humans prevail in primitive times.[13,14]

While this evolutionary benefit of child-spacing is speculative, miscarriage effectively insures against the birth of defective infants with genetic abnormalities. Human fertilization tolerates an extraordinary array of bizarre genetic complements, not just the normal 46 XX (female) or 46 XY (male) patterns. Soon after fertilization, a mysterious screening process culls the large majority of these abnormal entities, such as monosomy, trisomy, polyploidy, mosaicism, and chromosome re-arrangements (Please see Glossary).[15] In a recent series of miscarriages with abnormal chromosomes, the frequencies were trisomy, 62%; triploidy, 12%; monosomy X, 11%; tetraploidy, 9%; and structural anomalies, 5%.[16] This quality-assurance mechanism ensures that nearly all live births are normal and healthy.

This screening process is amazingly efficient. At least *half* of all pregnancies that miscarry have either visible abnormalities or chromosome defects.[17] Moreover, even minor defects, such as cleft lip that would not seem to impair the survival ability of the newborn, increase the risk of miscarriage.

Some genetic defects are more deadly than others. For example, embryos with trisomy 16 (an extra chromosome #16) and tetraploidy (four haploid sets of chromosomes) all die during pregnancy; no infants have ever been born with these anomalies, although they are commonly seen in miscarriage. Turner syndrome (45X) causes high losses during pregnancy. Embryos with Down syndrome (an extra chromosome #21), the most common cause of severe mental retardation, are selectively miscarried in early pregnancy.

While occult pregnancy losses are presumably high, losses after clinical recognition of pregnancy are high as well. Prenatal diagnosis studies in women 35 years and older indicate that 32% of Down syndrome pregnancies are lost between 10 and 16 weeks' gestation, and 54% are lost by the time of delivery.[18] Stated alternatively, infants born with Down syndrome are the minority of their cohort with this genetic abnormality who were missed by the screening process.

Spontaneous abortion is the norm for Down syndrome and most other chromosome anomalies. The proportions of recognized pregnancies with chromosome defects that end through miscarriage are trisomy 13, 83%; trisomy 16, 100%; trisomy 17-18, 84%, trisomy 21, 70%; 45X, 99.5%, and triploidy, 99.8%.[17] Neural tube defects also suffer high losses. An estimated 92% of pregnancies with anencephaly (absent brain) are lost.[17]

The most dangerous voyage

Obstetricians used to call birth the most dangerous voyage humans ever take. However, the trip the preembryo makes down the fallopian tube before implantation is even more hazardous. The risk of loss is inversely related to the time since fertilization (Figure 4-2). In the earliest days of the preembryo, during its lazy sojourn down the fallopian tube and into the uterus, losses predicted by mathematical models are very high. That likelihood declines exponentially with time. Once an embryo can be seen in the uterus and exhibits fetal heart activity, the risk of miscarriage is low for most women. In contrast, women with infertility problems, late in their childbearing years, have a significantly worse prognosis. In one study with confirmation of fetal heart activity by vaginal ultrasound, the miscarriage rate was eight-fold higher for women 36 years and older compared with younger women.[19]

Several recent studies using frequent, sensitive pregnancy tests to look for early pregnancies document a high miscarriage rate (Figure 4-3). In a widely cited study of 221 healthy women in North Carolina who were trying to get pregnant, investigators found that 22% of the pregnancies were lost before the woman even suspected she was pregnant (occult pregnancies).[20] Another 9% were lost as "clinical" pregnancies, for a total of 31%. A similar study in New York among 217 women trying to conceive had similar findings. Depending on the definitions used for occult pregnancy, the rates of occult pregnancy loss were 11% to 27%. Clinical pregnancy losses were 14%.[21] Among 200 couples in Washington, D.C., the occult pregnancy miscarriage rate was 13%, and the clinical pregnancy loss rate was 18%, for a total of 31%.[22] A larger study of 518 women in China who were trying to conceive reported an occult pregnancy loss of 25% and a clinical pregnancy loss of 8%.[23] Reviews of the literature suggest an early pregnancy loss of about 50%, ranging from 15% to 62%.[24]

A study performed among women in Vermont and New York reported findings separately for women with a history of infertility (subfertile) and those without such a history (fertile). Miscarriage patterns differed greatly. Among subfertile women, the occult pregnancy loss rate was 79%, with an additional 9% lost as clinical pregnancies. Among fertile women, the corresponding figures were 21% and 17%.[25] This study suggests that the large majority of pregnancies in women with infertility problems is lost through miscarriage, most before the pregnancy would have been suspected.

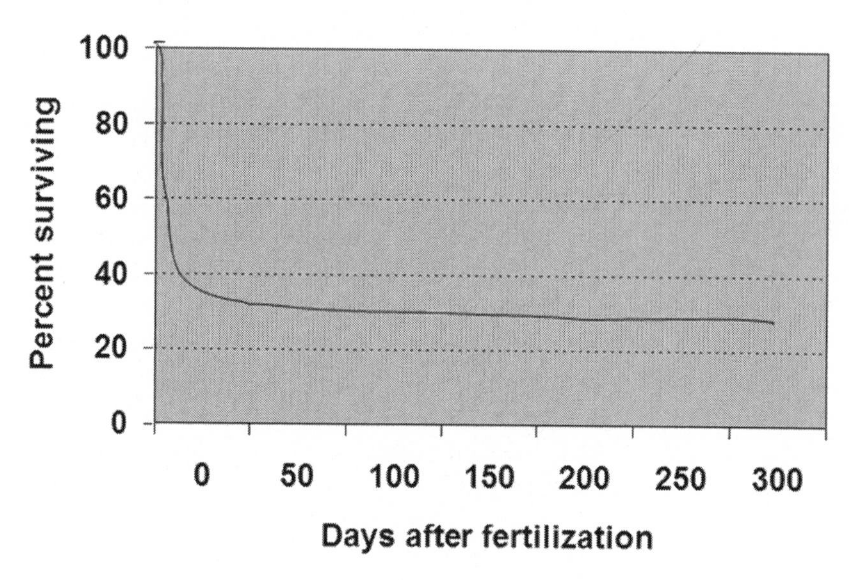

Figure 4-2
Probability of survival by time since fertilization of the egg
Source: Boklage[1]

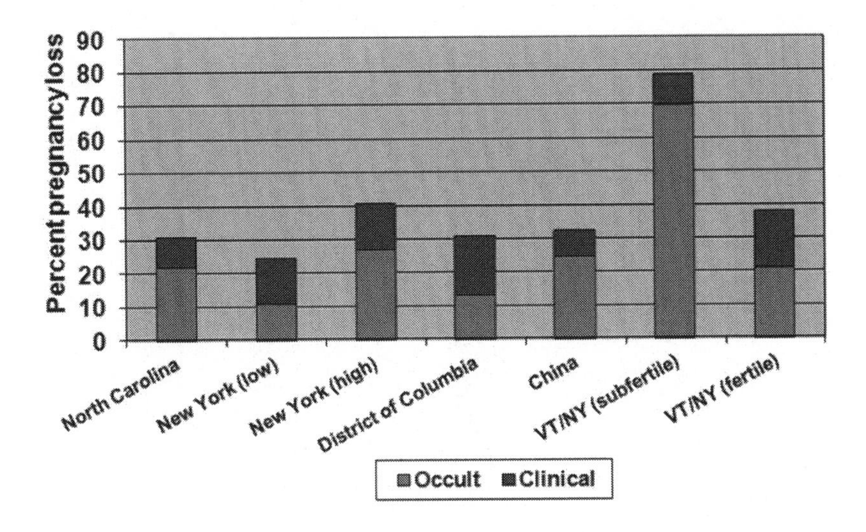

Figure 4-3
The distribution of miscarriages by type in representative studies. Occult pregnancies are those detectable only by hormone tests; clinical pregnancies are those suspected by a missed menstrual period.
Sources: Wilcox,[20] Ellish,[21] Zinaman,[22] Wang,[23] Hakim[25]

Disappearing womb-mates: the vanishing twin[26]

Multiple births, such as twins, are an anomaly in our species. Human reproduction is designed to produce one newborn at a time; the more fetuses, the greater the danger to mother and fetus.

As with chromosomal anomalies, twin gestations are commonly eliminated before birth. Most twins and higher-order multiples (e.g., triplets, quadruplets) disappear during pregnancy. One analysis of 325 twins identified by ultrasound examination in early pregnancy revealed the following outcomes: twin births in 61 (19%), singleton births in 125 (39%), and loss of both twins in 139 cases (43%).[1] Stated alternatively, the overall individual survival was 38%. The likelihood of death was inversely related to the duration of the pregnancy.

The figures are dramatic: four-fifths of all twin pairs spontaneously reduce to singletons, or both twins perish during the perilous path to birth. The prevalence of twins among miscarriages is three times that of infants at birth.[1] For every live-born twin pair, at least six singleton births occur that represent the surviving twin, whose partner died during pregnancy. Seen in this context, "multifetal pregnancy reduction" (the intentional destruction of surplus fetuses resulting from infertility treatment) simulates the normal winnowing of pregnancy.[26] By reducing the number of fetuses from four or five to only two, the prognosis for the woman and her offspring improves...but does not equal the more favorable prospects of spontaneously occurring twins.[27]

Stale eggs and tardy transit

Factors other than abnormal chromosomes are associated with miscarriage as well. Aging of the egg after release from the ovary has been shown to damage reproduction in rodents, suggesting the possibility of a similar effect in humans. In a careful human study with daily chemical surveillance for pregnancy and daily diaries of sexual activity, the "freshest" eggs at fertilization had an early pregnancy loss of 24%, in contrast to 67% with the highest likelihood of aged eggs.[28] In contrast, no such association was evident between reproductive loss and aging sperm.

> *"The ability of the maternal organism to detect and reject abnormal embryos is so efficient, so remarkable, and so vital a process for the well-being of the species that it deserves detailed investigation."*
>
> *Short*[10]

The timing of implantation is critical as well. Most preembryos implant in the uterine wall on days eight through ten after ovulation. The risk of early pregnancy loss increases if implantation does not occur within this narrow window. The likelihood of loss increased progressively with greater delay, from 13% loss by the ninth day, to 26% with implantation on day 10, to 52% on day 11, and to 82% on days thereafter.[29]

When life begins

Some political and religious figures hold that a new, unique human life begins with the junction of sperm and egg. Accordingly, some propose that such new "humans" be afforded full Constitutional protections.

The theory breaks down

Is union of egg and sperm the defining moment, the *sine qua non*, of a new human life? Biology suggests otherwise. First, while the fertilized egg, now called a preembryo or zygote,[11] is clearly new and alive, it did not spring from an inanimate source. The egg had always been alive in the ovary of the woman, since her own days as a fetus, and so forth back through generations. Thus, "new life" is a misnomer. Biologists and clinicians can define when life ends, but they cannot pinpoint its beginnings. The living material in our bodies – including our germ cells – has been passed down, alive, from our ancestors. All life is a continuation from the remote past. We are indeed one with our predecessors: they live on through us, their offspring, today and into the indefinite future. The stuff of life, like a torch handed from runner to runner, is passed on from parent to child.

Several quirks of biology, generally misunderstood or ignored, indicate that the zygote is not a unique new individual. First, fertilization is not a momentary event but, rather, a two-day process. Thus, claims that "life begins at conception," and accompanying demands for Constitutional protection from the "moment of fertilization," become simply irrelevant, since no such discrete moment exists. With the exception of *in vitro* fertilization, we cannot know the exact timing. Second, the zygote cannot express its own genetic information; instead, instructions from the egg control initial growth and development. While the zygote has a unique genetic complement, this information is inaccessible. Third, until it reaches the eight-cell stage of development, the preembryo is divisible: one or more cells can be removed, and the remaining cells will still grow into a whole fetus. Similarly, a cell can be removed from the mass, and that single cell can grow into a complete fetus. Interestingly, preembryo cells from different parents can combine and grow into a chimera (an individual with cells from two or more zygotes) – in this case, an entity containing genetic material from four parents!

Spontaneous chimeras (Figure 4-4) occur rarely in our species.[30] Recent examples include a woman who resulted from the merger of two zygotes or the early fusion of two genetically distinct embryos.[31] Another fertile woman was found to have a 46,XY (male) chromosome pattern; she had apparently received male blood cells from her twin brother during their time in utero.[32] A girl with Down syndrome lacked internal female reproductive organs due to a fetus-to-fetus spontaneous transfusion from her normal brother.[33] Sometimes two eggs are fertilized by two sperm; these zygotes then fuse, creating a tetragametic (four-germ-cell) chimera.[34] If the zygotes fusing are of opposite sexes, this oddity can cause true hermaphroditism (an individual having both male and female sexual organs). However, some chimeras with both female and male chromosomes can reproduce normally.[30] Each of these chimeric individuals resulted from biological events that occurred after fertilization.

Figure 4-4
The mythical chimera
The chimera (also spelled chimaera) was a monster with three heads: a lion, a goat, and a snake. The body was that of a lion, the legs that of a goat, and the tail was the serpent.
Source: http://en.wikipedia.org/wiki/Chimera_(mythology)

Identical twins further refute the notion that the junction of egg and sperm defines the beginning of a new human life (perhaps imbued with a soul at this point). Identical twins are an uncommon but fascinating biological oddity. After fertilization of the egg, it begins to grow. Sometime thereafter, the mass of the developing embryo splits. The resulting two masses of primordial cells have all the genetic information needed, and twins with identical chromosomes are created days after fertilization. If the union of egg and sperm is the defining moment, the *sine qua non* of new human life, is only one of the twins—formed after actual fertilization occurs—actually alive? Since both are clearly alive, the exact moment when the egg and sperm join *cannot* be the definition of new human life. For at least two weeks after fertilization, a unique human entity destined to become an individual does not yet exist.[11]

Hydatidiform mole (medical term meaning grapelike meal - as in oatmeal) is another obstetrical oddity that belies the notion that fertilization defines new human life. This condition arises from a sperm fertilizing an egg with a missing or inactivated nucleus. (The cell's nucleus is where the chromosomes reside.) Hence, in classic hydatidiform mole, all the genetic material comes from the father. Grapelike mush fills the uterus. A small proportion of these moles becomes a highly deadly cancer (choriocarcinoma). Another type of mole stems from the fertilization of a normal egg by two, rather than just one, sperm (called dispermy). In this case (a partial mole), a fetus develops adjacent to the grapelike material, with grim prospects for the fetus.[35.]

Fifteen years ago, a woman was admitted to a hospital with a partial mole and a live fetus at 22 weeks' gestation. She had developed severe, life-threatening preeclampsia (common with partial moles) with HELLP (hemolysis, elevated liver enzymes, and low platelets) syndrome. This severe variant of preeclampsia includes destruction of the mother's red blood cells, liver failure, and loss of blood-clotting ability. She was admitted to the gynecological cancer service, which usually cared for women with molar disease. However, her attending physician was opposed to abortion. Rather than emptying her uterus (the only way to save her life), he chose to do nothing. Only when she was near death did a courageous colleague intervene and perform the life-saving abortion.[36]

When pregnancy begins

While the beginnings of human life will forever remain indeterminate, the beginnings of pregnancy can be identified within a narrow window. However, considerable confusion prevails in lay circles concerning this event. Major medical organizations[37,38] and even the U.S. Federal government[39] agree that pregnancy begins when the preembryo is implanted in the woman's body. Implantation usually occurs in the uterus, but it can happen in other locations as well, such as the fallopian tubes (an ectopic pregnancy).

Fertilization is a necessary but insufficient step on the path to a pregnancy. A fertilized egg that does not implant will simply *never* become a viable fetus. A common event shows why implantation defines the start of human pregnancy. Imagine that an infertile couple in

Oakland, California, is undergoing *in vitro* fertilization in San Francisco. To their great delight, the physicians in San Francisco get an egg and sperm to unite in the laboratory dish. Genetic material unites, and cell division begins.

Having heard the good news, can the woman in Oakland announce to her neighbors that she is now pregnant? Should she call her obstetrician to begin prenatal care? (Her preembryo still resides across the Bay.) Few neighbors would sponsor a baby shower at this point; she simply is not pregnant. Similarly, her obstetrician would not offer a new obstetrical visit while her preembryo resides in another zip code.[40] Not until the preembryo implants (and becomes an embryo) can she rightfully claim to be pregnant.

The role of miscarriage

Miscarriage compensates for the remarkable inefficiency of human reproduction. Many, if not most, conceptions never become a pregnancy, and many pregnancies are spontaneously lost. *Thus, spontaneous abortion is a normal, healthy part of reproduction.* It ensures that a high proportion of babies born are healthy, with good prospects for long-term survival. Body wisdom, evolutionary pressure, or Divine intervention dictates that a large percentage of conceptions should not develop, and that many pregnancies should die. Spontaneous abortion is part of being human.

Take-home messages

- *Spontaneous abortion plays an important quality-control role in pregnancy*
- *Early pregnancy losses are common*
- *Implantation, not fertilization, defines the beginning of a pregnancy*
- *The claim that a new, unique life begins at the moment of fertilization is refuted by identical twins, hydatidiform moles, and chimeras*

Chapter 5
Induced abortion: where and how

In addition to spontaneous miscarriage, another common outcome of pregnancy is induced abortion. This entails "removal of a fetus or embryo from the uterus before the state of viability, further defined as '20 weeks' gestation or fetal weight <500g.'"[1] This standard definition is inadequate and misleading, since fetal viability at 20 weeks has not been reported, and weight is a poor predictor of viability.[2]

This chapter describes where abortions are performed, by whom, and how. It relies on national surveys of abortion providers and their patients, and on health statistics from the federal government and other independent sources.

Who provides abortions?

Older, male obstetricians/gynecologists have traditionally performed most abortions. Surveys of National Abortion Federation members described providers. In 2001, 77% of male providers of first-trimester surgical abortions were 50 years and older, in contrast to 43% of female providers.[3] This reflects the changing demographics of obstetricians/gynecologists in recent decades: most residents training in this discipline today are women. While most male providers were obstetricians/gynecologists (77%), 44% of female providers in 2001 were family physicians, and only 38% were obstetricians/gynecologists.

Where are services provided?

Most abortions take place in freestanding clinics. The most recent national survey of the Guttmacher Institute found that in 2011, abortion clinics comprised 19% of all providers, but they performed 63% of all abortions in that year.[4] Other clinics accounted for 30% of providers and performed 31% of abortions nationwide. Hospitals were common providers (more than one-third) but accounted for few procedures (4%). Physicians' offices (17% of providers) performed few abortions (1%).

Most abortion services are located in metropolitan areas. In 2011, 89% of counties lacked a provider, and 38% of reproductive-age women lived in these counties.[4] Given prevailing abortion rates (15 to 20 per 1,000 women of reproductive age, or about 1.5 to 2% per year),[4,5] this translates into travel burdens for tens of thousands of women.

How are abortions done?

Most first-trimester abortions are done by suction curettage, also known as vacuum aspiration.[5] Suction curettage opens the cervical canal to a small diameter, as it would open during the early stages of childbirth, and then suctions the uterine contents by inserting a small hollow tube, called a cannula. The cannula is attached to a suction source, and the uterus is emptied in a matter of several minutes. Since the introduction of suction technology into North America in the 1960s,[6,7] it has dominated abortion practice (Figure 5-1). In 2010, 72% of abortions were performed by suction curettage at ≤ 13 weeks' gestation, and 8% were performed by curettage (D&E) at later gestational ages.[5]

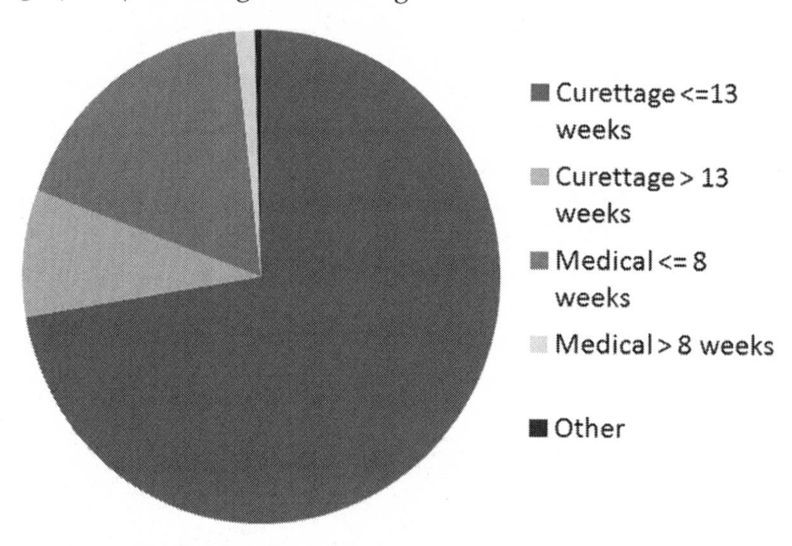

Figure 5-1
Distribution of abortion methods, United States, 2010
Source: Pazol[5]

Medical abortion is accounting for a growing proportion of early abortions in the United States.[5,8] The most common regimen is mifepristone (RU 486) followed by misoprostol. Mifepristone is an antiprogestin, which blocks the action of progesterone, a hormone that sustains early pregnancy. This induces a medical miscarriage. Misoprostol is a prostaglandin drug, which causes the uterus to contract, expediting the process. The regimen approved by the FDA is mifepristone 600 mg (three tablets) followed in two days by misoprostol 400 mcg (two tablets) for women up to 49 days from their last menses. This produces complete abortion in 92% to 99% of women. The FDA-approved dose of mifepristone has been found to be excessive, so many clinicians use an equally effective smaller dose (200 mg). In addition to being swallowed, misoprostol can also be placed in the vagina, under the tongue, or against the cheek; different doses are commonly used as well.[9] When mifepristone is not available, some clinicians use a combination of methotrexate (a cancer drug) followed by misoprostol; others use repeated doses of misoprostol alone, although this is less effective in producing abortion.[10]

In 2011, about 23% of all non-hospital abortions were done medically, up from 17% in 2008.[4] The CDC reported that in 2010, 18% of abortions nationwide were early medical abortions.[5] In 2007 and 2008, estimates of 14%[8] and 15%[11] were given. A medical degree is certainly not necessary to perform a medical abortion safely;[12] the availability of this method should theoretically have increased access to abortion. However, although the proportion of medical abortions continues to increase each year, the potential for a wider variety of providers offering this service has not been met. Family physicians continue to provide relatively small numbers of medical abortions, and obtaining liability coverage may be an important impediment to expanding access.[13]

Nearly two-thirds (66%) of abortions in the U.S. occur within eight weeks from the last menstrual period.[5] At this stage of pregnancy, the embryo is so tiny that it cannot be seen with the naked eye. As shown in Figure 5-2, the number of abortions performed nationwide is inversely related to gestational age. Most take place at 8 weeks or earlier, and the proportions fall off steeply thereafter. About 92% of abortions take place within 13 weeks, and 99% within 20 weeks of the last menses. Abortions at 21 weeks and later are rare.

Abortion providers are increasingly offering very early abortions. As of 2001, 65% of providers of surgical abortions offered them before six weeks' gestation.[3] This represents a large increase over 1997, when only 42% of providers offered procedures this early in pregnancy.

Ultrasound (sonar) examination plays an important role in evaluating patients and, to a lesser extent, in guiding operations. Many providers (91%) use this tool to confirm pregnancy or to confirm the duration of the pregnancy. Other commonly cited reasons for using ultrasound were to evaluate pregnancies earlier than 8 weeks or more than 13 weeks, to assist when pelvic examination was difficult (e.g., due to obesity), and to evaluate women with a history of ectopic (tubal) pregnancy.

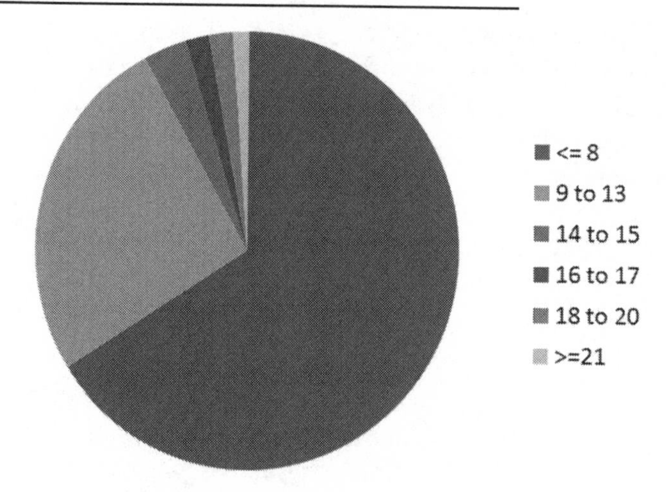

Figure 5-2
Distribution of abortions by weeks of pregnancy,
United States, 2010
Source: Pazol[5]

Misoprostol, the prostaglandin drug, is increasingly being used to prepare the cervix for the operation (cervical "ripening"). In addition, providers use the drug to help the uterus contract afterwards in an effort to minimize bleeding. While the ripening effect of misoprostol has been known for years,[14] a recent randomized controlled trial by the World Health Organization has demonstrated that this translates into a lower complication rate for women.[15]

Local anesthesia, usually an injection of lidocaine (a numbing medicine) around the cervix, was the most common method of pain relief. About a third of providers used local anesthesia supplemented by a sedative given intravenously. About 21% used deep sedation or general anesthesia.[3] Among the oral drugs offered to patients, common medications included nonsteroidal anti-inflammatory drugs (NSAIDs, like ibuprofen), tranquilizers (like diazepam), acetaminophen (an over-the-counter pain remedy), or codeine (a narcotic). Ancillary techniques included focused breathing (similar to Lamaze techniques) and visualization.

Most clinicians dilate the cervical canal with metal or plastic rods, which come in progressively larger sizes. For example, to perform an abortion at 8 weeks' gestation, the surgeon might dilate the cervix to an opening of 7 mm (less than a third of an inch; a fully dilated cervix reaches approximately 10 cm [about 4 inches] during childbirth). Most then use a rigid plastic cannula inserted into the uterus to remove the pregnancy;[16] thereafter, many (47%) check the completeness of the evacuation with a sharp curette.[3] Nearly all (93%) then inspect the aspirated tissue immediately on site;[3] this precaution helps prevent the rare ectopic pregnancy escaping diagnosis.[17] Most clinics do not routinely give medicines to help the uterus contract (to reduce bleeding) after the procedure. In contrast, 88% provide preventive antibiotics to women (usually a tetracycline called doxycycline);[3] this reduces the small risk of infection even further.[18]

> *"From the patient's perspective one aspect of abortion that is not an issue of debate is quality of care: women who have abortions say they are very satisfied with the care they receive."*
>
> *Tina Hoff*
> *Director of Public Health Information and Communication*
> *Henry J. Kaiser Family Foundation*[19]

Accolades for care

An independent survey of former patients has found U.S. women pleased with their abortion care. Sponsored by the Henry J. Kaiser Family Foundation, the survey of 2215 women aged 18 years and older was conducted by The Picker Institute of Boston, Massachusetts, in 1998; the women were representative of those having abortions nationwide.[19] This was the first comprehensive, national assessment of perceived quality of care, and the findings were reassuring.

Women gave high marks to providers for the completeness of their counseling. When queried three to four weeks after the abortion, 98% said the procedure had been adequately described to them. Nearly all (99%) said the information was clear and understandable. This contrasted sharply with surveys after other types of medical services studied by The Picker Institute, for which patients commonly want more information provided. Attention to privacy was rated "excellent," "very good," or "good" by 94% of patients. Most (88%) felt they had been treated respectfully, and 78% had "a lot of confidence and trust" in the clinic staff.[19]

The survey also identified areas needing improvement. These included more attention to privacy, especially in the recovery room (most recovery rooms are shared by several patients at a time). Confidentiality concerns in the waiting room were another issue raised by 19%. Of the minority of women who called the clinic after the procedure with questions or problems, one in four did not get all the help they wanted.

Complications of abortion

Because of its public health importance, abortion has been studied extensively by the federal government and other health organizations, such as the World Health Organization. The CDC conducted one of the world's largest studies of abortion complications during the 1970s, the Joint Program for the Study of Abortion. It gathered detailed, individual case report forms on more than 160,000 abortions performed in hospitals and clinics.[20] Indeed, no other operation has ever been subjected to such intense medical scrutiny. While these data are now old, they established early on that legal abortion was safe.

As shown in Table 5-1, suction curettage had the lowest risk of major complications, followed by dilation and evacuation (D&E), saline instillation, and prostaglandin instillation. Indeed, data from this pivotal study popularized safer methods of second-trimester abortion[21] and helped eliminate those less safe.[22] The study also showed the safety of abortion among teenagers, which had been a source of concern for some.[23] This study ended in 1978, since the federal government could not justify continuing expensive studies of a procedure that posed no public health threat. Instead, the Abortion Surveillance Branch at the CDC became the Pregnancy Epidemiology Branch and turned its attention to the greater hazards of ectopic pregnancy,[24-26] hydatidiform mole,[27,28] and childbirth.[29-31]

Table 5-1

Major complications* per 100 abortions, by method and weeks of pregnancy, 1971-1975

	Suction curettage			Dilation and evacuation		Saline instillation		Prostaglandin instillation	
	≤8	9-10	11-12	13-16	≥17	13-16	≥17	13-16	≥17
Rate	0.23	0.36	0.44	0.69	0.69	1.76	1.83	3.00	2.80

*One of 15 specified complications, such as fever for 3 or more days, hemorrhage requiring blood transfusion, or unintended major operation

Source: Cates[32]

As good as it gets?

Doctors, like other people, tend to do well what they do often. Surgery is no exception. If one has highly experienced physicians operating on uncomplicated, low-risk patients, one can approach the lowest possible complication rates. A large series of 170,000 first-trimester abortions performed at Planned Parenthood of New York City from 1971 to 1987 documented exceptional safety.[33] No death occurred; 121 women required hospitalization for treatment of complications (0.7 per 1,000 cases). A total of 1438 "minor" complications occurred, for a rate of 8 per 1,000 cases. If the complications requiring hospitalization are combined with the less serious issues, the total complication rate was 9 per 1,000 cases. Stated alternatively, less than 1% of women reported a complication of any kind. A partial inventory of complications follows.

Specific complications

Incomplete abortion

Retained pregnancy tissue can occur after abortion as it does after childbirth. With suction curettage, the most common abortion method, the risk of requiring a repeat aspiration to remove residual tissue in U.S. studies has been less than 1 per 100 cases.[34] In contrast, with labor-induction abortion, failure of the placenta to be expelled shortly after the fetus is common. This can be managed with medications to make the uterus contract or with instrumental removal of the placenta.

Acute hematometra

Occasionally, for unknown reasons, blood backs up in the uterus immediately after a suction curettage abortion. For example, a blood clot might obstruct the outflow through the cervix, akin to a ball valve. The uterus can become tense and enlarged with backed up liquid and clotted blood (sometimes a cup or two); repeat aspiration of the uterus quickly relieves the problem, which tends not to recur. The frequency of this condition is about 2 per 1,000 cases.[34]

Hemorrhage

Bleeding can complicate any abortion, though the risks increase with gestational age and with labor-induction methods. Physicians' estimates of vaginal blood loss are unreliable; hence, estimates of hemorrhage are difficult to interpret. A better determinant for medically important bleeding is the requirement of a blood transfusion. In such circumstances, the medical risk posed by the bleeding exceeds the cost and risk involved with transfusing blood products. Blood transfusion associated with contemporary abortion methods is rare, around 1 per 1,000 cases.[34]

Atony

The principal means by which the uterus stops bleeding after abortion or birth is contraction. This squeezes off the blood vessels that course through the muscle wall. Should the uterus fail to contract adequately, hemorrhage can occur. This condition, called atony, can be managed by massage of the uterus and by administration of drugs to induce contractions.

Coagulation defects

Rarely, defects in the blood's ability to form clots can develop during an abortion. This can be caused by severe hemorrhage (in which the clotting factors are lost with the blood) or by some of the amniotic fluid surrounding the fetus gaining entry to the woman's blood circulation (known as amniotic fluid embolism). Other possible causes are severe infection and rare placental problems. The risk of coagulation defects in first-trimester abortion is less than 1 per 1,000 cases, increasing to 5 to 7 per 1,000 cases in the second trimester. Coagulation defects are managed by replacing the missing coagulation factors with blood products.[34]

Infection

In operations through the cervix, "sterile technique" is not feasible. Because bacteria in large numbers live in the cervical canal (which cannot be sterilized by antiseptics), all intrauterine operations are contaminated by microbes. Nevertheless, the body's defense mechanisms are so robust that infection is uncommon after both abortion and childbirth. In U.S. reports, the risk of infection after first-trimester abortion has been less than 1 per 100 cases; administration of antibiotics to prevent this postoperative complication is common practice.[34]

Trauma to the uterus

Perforation of the wall of the uterus can occur during suction curettage or D&E. The true rate of perforation is unknown but is higher than reported rates, since most perforations are innocuous, unsuspected, and undetected. In a study in which laparoscopy for tubal sterilization was performed immediately after a suction curettage abortion, the rate of perforation was found to be about 2 per 100 cases.[35] However, some perforations can result in injury to uterine blood vessels, bowel, or other abdominal organs. Hence, this rare complication is potentially one of the most serious.

Cervical injury

The most common cause of cervical injury is the tenaculum (a holding instrument to stabilize the cervix) pulling off during suction curettage. In the early years of legal abortion, the rate of cervical injury during suction curettage ranged from 0.2 to 1 per 100 cases.[36] Since then, use of osmotic dilators, such as laminaria,[37] or drugs, such as misoprostol,[38] to prepare the cervix for dilation have likely reduced the risk of injury. Osmotic dilators, such as naturally occurring laminaria (stalks of seaweed) or synthetic sponges containing Epsom salts, absorb water and open the cervix slowly and safely. Similarly, pre-operative use of laminaria before dilation and evacuation has become common practice over the past three decades.

Potential late complications

Having been so extensively studied, the long-term safety of abortion is well-documented. Stated alternatively, many alleged adverse effects[39] of abortion have been refuted by published evidence. Those who claim that abortion causes long-term damage of any kind are simply incorrect. The literature is clear: induced abortion does not impair a woman's ability to have children in the future. Prematurity, infertility, ectopic pregnancy, miscarriage, and pregnancy complications are *not* more frequent after abortion, once women's personal characteristics are taken into account. Women choosing abortion are often less healthy physically and psychologically than are women who choose to continue their pregnancies, and these personal differences bias comparisons. The question of placenta previa (attachment of the afterbirth low in the uterus, covering the cervical opening) in later pregnancies remains unsettled. Published studies are divided on this point.[40,41] Cesarean delivery, especially when the operation is repeated over several pregnancies, has an adverse effect on placental attachment.[42]

Summing up

Because of the controversial nature of abortion and because it is often done on young women whose childbearing years lie ahead, abortion has been intensively and extensively studied. More is known today about the safety of legal abortion than most other operations in surgical practice. The complication rate with current methods is low, and concern about potential late complications has largely vanished.[43]

Take-home messages

- *Most abortions are done early in pregnancy by vacuum aspiration in a freestanding clinic*
- *An increasing proportion of early abortions are performed medically with mifepristone followed by misoprostol*
- *The complication rate for both surgical and medical abortion is low*
- *Women are generally pleased with the quality of care they receive*

Chapter 6
Giving birth: still risky business

One of the most famous - and beautiful - buildings in the world is a monument to maternal mortality (Figure 6-1). When Mumtaz Mahal died in 1631 after giving birth to her 13th or 14th child (accounts differ), her heartbroken husband, Emperor Shah Jahan, set out to construct a suitable mausoleum for his beloved wife. Twenty-two years later the Taj Mahal was completed on the banks of the Yamuna River in Agra, India. Some accounts claim that Mumtaz Mahal died of post-partum hemorrhage; others say the cause of death is unknown. What is clear is that pregnancy then and now carries serious health risks for women.

Figure 6-1
World's most famous maternal mortality monument
Source: Amal Mongia, from Wikimedia Commons

A stroll through old graveyards in Boston or Charleston reveals that many tombstones carry the same terse epitaph "Died in childbirth." Death and birth have been common travelers. Three centuries after these tombstones were erected in the Colonies, giving birth in the U.S. remains a risky business. Despite important improvements in safety, pregnancy and childbirth remain hazardous. The risk of maternal death is low, but it has been steadily increasing in recent years. This may be an artifact of more complete reporting of such deaths, or, alternatively, pregnancy may be getting more dangerous.

Moreover, the likelihood of suffering complications of pregnancy and childbirth remains surprisingly high. The general perception is that pregnancy and childbirth are "natural" events, implying "healthy" (as in "natural foods"). "Natural" though it may be, once a woman becomes pregnant, her risk of illness and death increases and remains elevated until some weeks after the pregnancy ends. In some countries in Africa where women have limited access to obstetrical care, the risk of death associated with each pregnancy exceeds 1%.[1]

The burden of suffering

About four million births occur each year in the U.S. Preliminary data for 2012, the most recent year available as of this writing, indicate 3,953,000 births.[2] The general fertility rate was 63 births per 1,000 women of reproductive age; stated alternatively, about 6% of women of reproductive age gave birth that year. In 2012, about one in three births was by cesarean delivery… an operation with a substantial complication rate.

Each year in the U.S., more than 500 women die from pregnancy and its complications. More than 2,000,000 women suffer complications of pregnancy and childbirth, of which about 20,000 are life-threatening. The costs, both direct and indirect, are staggering. In the 1990s, the total expenditure for premature labor—only one of a host of possible complications—was estimated to be more than $800 million.[3]

Deaths from pregnancy and childbirth

According to the Centers for Disease Control and Prevention (CDC), about one woman in 10,000 dies from pregnancy, childbirth, or related complications. The CDC defines a pregnancy-related death as the death of a woman during pregnancy or within one year of its end, from any cause related to or aggravated by the pregnancy or its management. Deaths from accidents and incidental causes are excluded.[4] From 1998 to 2005, 4,693 women were reported to have died in the U.S. because of pregnancy complications; the aggregate pregnancy-related mortality ratio was 15 deaths per 100,000 live births, the highest ratio in twenty years.[5] If corrected for under-reporting of maternal deaths,[6-8] the true figure may be twice as high. Sadly, about 40% of these deaths could have been prevented, principally through better obstetrical care.[9]

Age and race influence the risk of maternal death in important ways (Figure 6-2). The risk of death increases with age; starting at age 30 years, however, the risk rises dramatically. Women 40 years and older have more than five times the risk of a pregnancy-related death than do teenagers. A large racial disparity in risk persists as well. African-American women have nearly a four-fold higher risk of death than do white women. Both age and race synergize to increase risks. For example, African-American women aged 40 years and older have a pregnancy-related mortality rate of 167 deaths per 100,000 live births. This is four times that of a white woman of the same age and ten times that of an African-American teenager.[5]

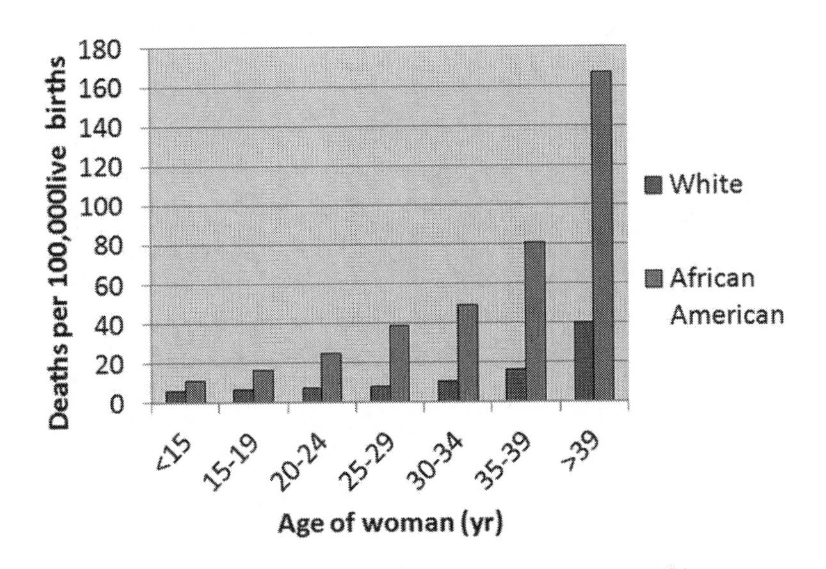

Figure 6-2

Pregnancy-related mortality rates by age and race, United States, 1998-2005

Source: Berg[5]

The most common pregnancy outcome resulting in maternal death is childbirth. From 1998-2005, 71% of pregnancy-related deaths followed delivery of a live infant.[5] Fifteen percent of deaths occurred before delivery, 6% were associated with stillbirth, 4% after an ectopic (usually tubal) pregnancy, and 3% after a spontaneous or induced abortion. The outcome of the pregnancy could not be confirmed for 15% of women who died.

Deaths associated with live births were most commonly caused by high blood pressure disorders, cardiomyopathy (disease of the heart muscle), other medical conditions not related to the heart, and heart conditions.[5] Pregnancy often worsens pre-existing medical conditions, such as diabetes, high blood pressure, or heart disease, resulting in death. For example, women with sickle cell disease have high risks of complications during pregnancy and a risk of death six times higher than healthy women.[10] Similarly, women who are depressed have worse pregnancy outcomes for themselves and their babies than do healthy women.[11]

Comparing risks of death

How a pregnancy ends influences a woman's risk of death. Comparing the risk of death associated with different pregnancy outcomes is important, but difficult. In addition to the problem of under-reporting of such deaths, determining the number of pregnant women at risk is challenging. For example, ectopic (tubal) pregnancy used to require hospitalization, which enabled identification of such events through hospital records. Today, laparoscopy (an operating telescope placed into the abdomen) and medical therapy of ectopic pregnancy (with a chemotherapy drug, methotrexate) have eliminated the need for many hospitalizations.[12] Similarly, estimating the number of women who have a miscarriage is difficult, since many early losses are not recognized as pregnancies, and many women do not seek care for uncomplicated miscarriages.

To address this problem, the CDC developed national estimates of the numbers of pregnancy outcomes each year in the United States. With numbers of deaths reported by CDC and numbers of women at risk of death from CDC estimates,[13] one can make crude comparisons of the risk of death by type of pregnancy outcome (Figure 6-3).[14]

As Figure 6-3 reveals, different pregnancy outcomes have widely varying safety. Both legal abortion and miscarriage have very low risks of death: around 1 death per 100,000 abortions. Live birth carries an intermediate risk, about 7 deaths per 100,000 live births. Ectopic pregnancies are far riskier, with a rate around 32 deaths per 100,000 ectopic pregnancies. The most dangerous pregnancy outcome by far is fetal death, with more than 90 maternal deaths per 100,000 fetal deaths. Fetal deaths are often associated with severe maternal illness, such as dangerously high blood pressure related to pregnancy or premature detachment of the placenta from the uterus. Such events threaten both the woman and her fetus. At the most extreme, the risk of a pregnant woman's death associated with fetal death is 170 times higher than that with a legal abortion.

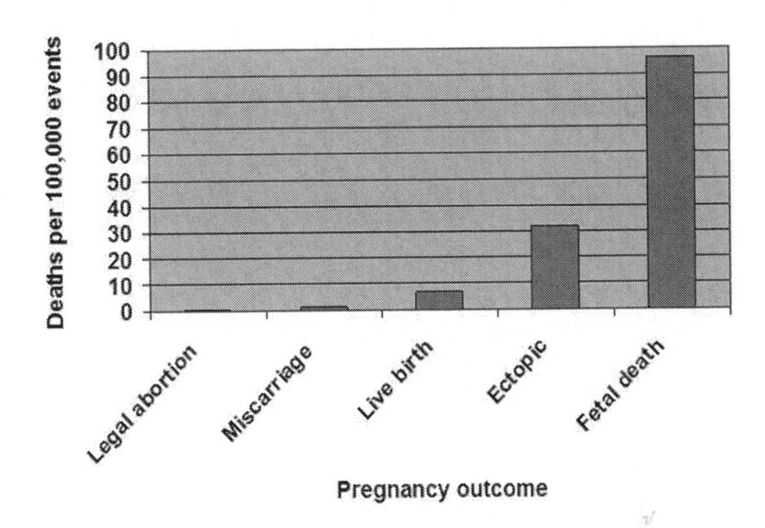

Figure 6-3
Estimated risk of pregnancy-related death by pregnancy outcome, United States, 1991-1999
Source: Grimes[14]

These estimates have several important limitations.[14] These rates are useful only for comparison purposes. A quarter of all deaths involved molar pregnancy, undelivered pregnancy, or unknown pregnancy outcome; these were not included in the calculations. Except for the numbers of live births, all the denominators were estimates using CDC data.[13] Nonetheless, the results are consistent with earlier reports from the CDC using different methods of calculations.[15,16]

The most recent update of the comparative risks of death from childbirth and induced abortion confirmed a widening disparity.[17] From 1998-2005, CDC data indicated a risk of death related to birth of a live infant of 9 per 100,000 live births. The corresponding figure for induced abortion was 0.6 per 100,000 abortions. During this interval, the risk of death from childbirth was 14 times that of legal abortion. While some underreporting of deaths is likely in both groups, selective underreporting of abortion deaths cannot account for a difference this large. For example, assume that only one in five abortion-related deaths was correctly identified. If that were the case, then the pregnancy-related mortality risk for live birth would be three times higher than abortion, not 14 times. Even assuming gross underreporting of abortion deaths, it remains far safer than childbirth.

Complications of pregnancy and childbirth

Several recent studies from health maintenance organizations have documented that pregnancy in the U.S. remains risky business. CDC investigators used the Kaiser Permanente Northwest database to examine pregnancy-related complications from 1998-2001. For women who had live births during that interval, the overall complication rate was 60%.[18] Using a similar Kaiser Permanente database from Georgia in a more recent interval (2000-2006), CDC researchers found that 61% of women having a live birth had one or more complications of pregnancy and childbirth (Figure 6-4).[19] Fifteen percent of pregnant women experienced a urinary tract infection, 13% anemia, 7% mental health problems, and 9% pelvic and perineal (vulvar) problems, such as a tear during delivery. If these complication rates are extrapolated to the nation, then more than 2,400,000 women out of the 4,000,000 or so total pregnancies suffer complications of pregnancy and childbirth each year.

Complications during the delivery hospitalization

Other studies have focused on complications during the birth hospitalization. Using National Hospital Discharge Survey data from California, researchers examined complications associated with delivery.[20] From 2005-2007, about one in four women (24%) had a complication associated with delivery. The most common were episiotomy (an incision to enlarge the vagina for delivery), pelvic trauma, infection, hemorrhage, and severe laceration (tear).

CDC investigators performed a similar analysis of the National Hospital Discharge Survey data in 2001-2005 for the entire country.[21] Twenty-nine percent of women had one or more complications associated with delivery. Regrettably, this percentage has remained unchanged since 1993-1997.[22]

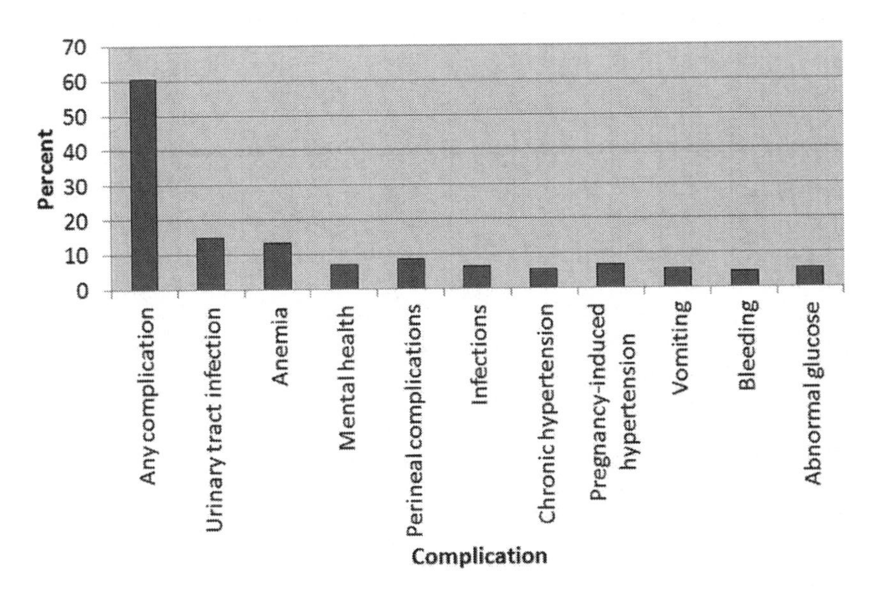

Figure 6-4
Prevalence of ten most common pregnancy-related complications associated with live births, 2000-2006
Source: Bruce[19]

Severe complications and "near misses"

Deaths are the tip of the iceberg; far more women sustain life-threatening complications but survive. Their survival is often a tribute to their youthfulness, health, and competent obstetrical care. Several groups have assessed the frequency of severe maternal complications. CDC investigators, using a national database from 1998-2005, reported that the frequency of severe complications increased from 0.6% to 0.8% from 1998-1999 to 2004-2005. Severe or "near-miss" complications included a list developed by the World Health Organization.[23] Significant increases were observed for kidney failure, pulmonary embolism (a clot in the lungs), adult respiratory distress syndrome, shock (low blood pressure), blood transfusion, and need for a ventilator to support breathing.

Risk factors for severe complications are predictable. A case-control study from Washington State found that increasing age, minority race, either no or many prior births, multiple fetuses, and prior cesarean delivery all raised the risk of a severe complication.[24] Importantly, women with pre-existing medical conditions were twice as likely to have a brush with death as were healthy women.

Others have confirmed that pregnancy is riskier for women with pre-existing medical conditions. Using a national database, researchers examined "near-misses" in the U.S. from 2003-2006. About 1 in 1,000 hospitalizations for delivery had such an event. Most of these life-threatening complications occurred in women with pre-existing medical problems or identified obstetrical problems. The greatest risk occurred among women with pulmonary hypertension (high blood pressure in the lung arteries), cancer, and systemic lupus erythematosus (an autoimmune disease).[25]

In another study, CDC researchers found that life-threatening complications occurred in 5 per 1,000 deliveries.[26] The higher figure than above[25] reflects different definitions of "severe" and use of a different database for the years 1991-2003. The most common severe complications included hemorrhage requiring blood transfusion, a complication requiring hysterectomy, and eclampsia (toxemia with seizures). Severe complications were more common at the extremes of reproductive age (very young or very old women) and more common for African-American women than for white women. These "near-misses" were fifty times more frequent than maternal deaths; about 20,000 women have obstetrical "near misses" each year in the U.S.

Hospitalization for complications

About half a million pregnant women are hospitalized during pregnancy each year for a problem unrelated to delivery. In 1999-2000, the proportion being hospitalized was about 13%.[27] The likelihood of being hospitalized for complications was greatest for women who were young, African-American, and without private insurance. Premature labor, nausea and vomiting, and urinary tract infections accounted for about half of these admissions.

A similar CDC analysis of a large managed care database for 1997 found that 9% of women were hospitalized during their pregnancy.[28] The most common reasons for admission were premature labor, severe vomiting, high blood pressure, kidney infection, and premature rupture of the membranes.

Prolonged or recurrent hospitalization after delivery

Prolonged hospitalization or re-admission to a hospital reflects serious complications. Epidemiologists have provided estimates using 1991 Medicaid data from Tennessee.[29] About 3% of all women who delivered vaginally had a prolonged hospitalization or early re-admission. In contrast, for women who delivered by cesarean, the percentage was three times higher, about 9%. Given the rising proportion of women in the U.S. who are having cesarean deliveries, this three-fold differential has important medical and economic implications.

Infection was the most common serious complication leading to a prolonged stay or re-admission. Overall, infection was four times more common among women delivering by cesarean operation than by vaginal birth.[29] The other two major complications were high blood pressure and hemorrhage. As with hospitalization during pregnancy, prolonged or repeat hospitalization after delivery demonstrated a "U"-shaped distribution by age. Complication rates were higher in women at the extremes of the reproductive years (16 years and younger or 33 and older) than in women of other ages. In addition, African-American women and those giving birth for the first time had higher risks.

Injuries to bowel and bladder

Vaginal delivery can injure the urinary tract and rectum, with long-lasting disability. Obstetrical tears during delivery (or with episiotomy) may extend into the anal sphincter (the muscle that controls bowel movements). When recognized, these injuries are promptly repaired. Despite repair, from 4 to 50% of women will later note inability to control bowel movements or gas.[30]

Unsuspected injury to the anal sphincter may be more common than previously thought. New ultrasound technology enables physicians to evaluate the integrity of the anal sphincter by inserting a cylindrical ultrasound probe into the rectum. These studies indicate a surprisingly high frequency of hidden tears to this critical muscle. A review of five reports found that 27% of women giving birth vaginally to their first child had an anal sphincter defect, and new defects were noted in 9% of women giving birth for the second or greater time.[31] Fortunately, only a minority (30%) of these injuries caused symptoms. A second cause of difficulty controlling bowel movements and flatus may be obstetrical injury to the nerves supplying the muscle.[30]

Accidental loss of urine (incontinence) occurs in up to 30% of pregnant women. This complaint usually resolves afterwards without intervention. In contrast, among women who lose urine for the first time *after* delivery, the problem may persist. In one report, 24% of such women were still having difficulty with urine control a year afterwards.[30] Injury to the support of the urethra and bladder during vaginal delivery is associated with urinary incontinence, and injury to the nerve supply has been implicated as well.

Hemorrhage

The rising proportion of births by cesarean nationwide has been accompanied by an increasing frequency of hemorrhage after childbirth.[32] Using a national database, CDC epidemiologists found that in 2006 about 3% of deliveries were complicated by heavy bleeding. Repeat cesarean deliveries are associated with abnormal attachment of the placenta to the uterine wall, which can result in life-threatening bleeding sometimes requiring hysterectomy as treatment.[33]

Parenthood is one of the greatest joys—and privileges—of the human experience. Hence, women who want children are willing—indeed eager—to assume pregnancy's risks and discomforts. The same is not true for other women. For them, mandatory motherhood looms as cruel punishment for the "crime" of forgotten birth control pills[34] or condom failure.[35]

Take-home messages

Each year in the U.S:

- *More than 500 women die from pregnancy and childbirth*
- *About 20,000 women suffer a "near-miss," a close brush with death*
- *One in three births occurs by cesarean delivery*
- *More than 2,000,000 women suffer pregnancy complications*
- *About 500,000 women are hospitalized during pregnancy*

Section III
Abortion in grassroots America

Hanh

Hanh and her husband, immigrants from Vietnam, discovered during a routine ultrasound examination at 18 weeks that their fetus had a complex heart defect. It was oriented backwards in the chest, with abnormal connections to it. Their first child, now five years old, has severe autism and is in a special education school. All their efforts are directed to providing the resources for their son to achieve his potential. He speaks very little. They are saddened that his cousins, who are several years younger, have better verbal skills already.

After several consultations with pediatric cardiologists and maternal-fetal medicine specialists over the ensuing week, they came to realize that if this fetus were born, it would require multiple heart operations. Even after that, the baby's survival would likely be brief, and his short life would be characterized by multiple operations, pain, and suffering. During this time, the parents would be unable to provide adequate care for their handicapped child. After careful consideration and many tears, they reluctantly decided that abortion was the best option to protect their family and offer their son the emotional and financial resources he needs. She declined contraception afterwards, since they planned to have another pregnancy soon.

Chapter 7: Every third woman in America

"She isn't a lawyer or a journalist or a politician or a consultant or a PR agent. But she is a first-class media hound, blessed with a savage wit, good looks, and – as one enemy put it – 'the tact of a bulldozer.'"

Plotz[1]

Susan Carpenter-McMillan reached the zenith of her media fame in 1997 as self-appointed spokesperson for Paula Jones, who had accused President Bill Clinton of sexual impropriety.[1] I had crossed paths (and swords) with Ms. Carpenter-McMillan a few years before Paula Jones did. In the early days of the Montel Williams Show, when it was filming in Los Angeles, the topic of the show was the French abortion-inducing medicine, mifepristone (RU 486). She had labeled it "human insecticide."[1] At that time, I was one of the few U.S. physicians who had clinical experience with the novel pill; Ms. Carpenter-McMillan was the media representative for Right-to-Life League of Southern California.[2]

During the show, she attacked both abortion and the physicians who provide abortions. She did *not*, however, disclose to the nationwide audience that she had reportedly undergone two induced abortions herself. The first was in 1970 when she was a 21-year-old, unmarried undergraduate at the University of Southern California[1,3] and the second in 1983 when she was married. The latter procedure, reportedly for medical reasons, took place three years after she became an anti-abortion activist.[3-5]

She was "outed" by Paul Dean, a reporter for the Los Angeles Times.[2,3] She had kept this secret for decades. Her departure from the Right-to-Life League came soon after this disclosure by the Los Angeles Times.[7] She moved on to represent Paula Jones.

Some commentators noted the hypocrisy: the spokesperson for an anti-abortion group had been an abortion patient herself....twice.[4,5] Abortion was "murder" for others but, when queried, she reported that, "It was my own private life...."[4] By 1997, she reported that she had left the movement because it was being led by "misogynists who don't care about women" and "crazies who murder doctors."[7]

"I lied at first....."

Susan Carpenter-McMillan[6]

Many physicians in busy clinics have had similar experiences, performing an abortion on an anti-abortion picketer or her daughter. These abortions typically take place mid-week, when no other picketers are present. Next Saturday, the anti-abortion patient is seen picketing in front of the clinic again, calling those inside "murderers." Confidentiality prevents us from disclosing the inconsistency.

Carpenter-McMillan's discrepancy between words and actions strikes some as profoundly hypocritical; however, it exemplifies the ambivalence that many have about abortion. Substantial numbers of women requesting—and having—abortions report that they are "opposed to abortion." In a recent study from Vancouver, 61% of women requesting medication abortion claimed to be "anti-abortion." Nevertheless, they chose to have one. These women had higher levels of anxiety afterwards than did women who were "pro-choice," likely reflecting unresolved ambivalence.[8] Abortion patients opposed to abortion also tend to be critical of other women's reasons for abortion.[9] Most women seeking abortion are unhappy about the situation that led them to it. I have yet to meet a woman who was "pro-abortion;" the same is true for "pro-hysterectomy," or "pro-mastectomy." Unfortunately, these procedures are sometimes what a woman needs.

Every third woman

Induced abortion is an everyday occurrence in the lives of grassroots American women. If current rates prevail, about one of every three women in the U.S. will have had one or more abortions by the time she reaches menopause.[10] Had earlier abortion rates continued instead of declining, 43% of all U.S. women would have had one or more abortions during their reproductive lives.[11] From 1973 (the *Roe v. Wade* decision) to 2008, the number of abortions performed in America totals about 50 million[12]...greater than the population of Spain today.

Several indices measure the frequency of abortion. The simplest is the total count. Two organizations, one federal and one private, have been tracking these trends for many years. The Centers for Disease Control and Prevention (CDC) relies on reports provided voluntarily from health agencies, typically State health departments. However, since no uniform reporting requirement exists nationwide, the data provided are inconsistent, and, in recent years, increasingly incomplete. For 2010, 49 areas (excluding California, Maryland, and New Hampshire) provided abortion statistics.[13] California is a large provider not included in the CDC statistics. The Guttmacher Institute in New York City has been conducting periodic surveys of known abortion providers; its total abortion count is consistently higher than that of the CDC and is deemed more accurate.

According to the CDC, the 49 areas reported about 766,000 abortions in 2010.[13] In contrast, the more complete enumeration from the Guttmacher Institute estimated 1.1 million abortions in 2011.[10] By contrast, only about half a million women have a hysterectomy each year. Thus, abortion has touched the life of most American women and their families—either directly or indirectly.

Measuring abortion frequency

In addition to total numbers, two other measures of abortion frequency provide important benchmarks: the abortion ratio and rate. The abortion ratio, as defined by the CDC, is the number of abortions per 1,000 live births.[13] According to the CDC, the national abortion ratio in 2010 was 228 abortions per 1,000 live births.[13] Stated alternatively, about one in six pregnancies ends by induced abortion (one abortion for every five live births).The ratio measures the "unwantedness" of pregnancies. A high ratio of abortions to births indicates that many pregnancies are unwanted, and a low ratio implies the opposite.

The abortion rate, as defined by the CDC, is the number of abortions per 1,000 women aged 15-44 years.[13] According to the CDC, the national abortion rate in 2010 was 14.6 abortions per 1,000 women of reproductive age. According to Guttmacher Institute figures, the national abortion rate in 2011 was 16.9 per 1,000 women of reproductive age.[10] Thus, about one in 50 women aged 15 to 44 years has an abortion each year.

The abortion rate reflects the underlying fertility of the women. Thus, abortion ratios and rates measure different features. For example, teenagers younger than 15 years have the highest abortion ratios (851) but the lowest abortion rates (1).[13] Stated alternatively, few very young teenagers become pregnant, but if they do, nearly as many choose abortions as choose to continue the pregnancy. This reflects the undesirability of middle-school motherhood. A tearful 12-year-old rape victim, assaulted by her mother's boyfriend, recently explained to me that she did not want to have a baby while in the sixth grade.

Who has abortions?

Unintended pregnancy is epidemic in the U.S., with nearly half of all pregnancies unintended and, generally, unwanted. Disadvantaged women have the greatest risk of such a pregnancy: these are young women with low educational attainment, of minority race, poor, and cohabiting.[14] While women from all walks of life have abortions for a myriad of reasons, the prototypical patient is a mother in her twenties, unmarried, Christian, with limited financial means,[15] and in early pregnancy.[13] The same risk factors exist for second-trimester abortions: non-white race, less education, and three or more disruptive events in the past year.[16]

Abortion ratios vary widely by age, reflecting the relative desirability of childbearing at different times in women's lives (Figure 7-1). Young teens have the highest ratios; ratios then decline to their lowest at age 30-34 years, after which they climb again. Pregnancies among the very young are often "timing failures," that is, pregnancies that are ultimately wanted, but not this early. In contrast, pregnancies among women nearing the end of their reproductive years are often deemed "number failures," indicating that older women's family sizes have been achieved, and no further pregnancies are desired.

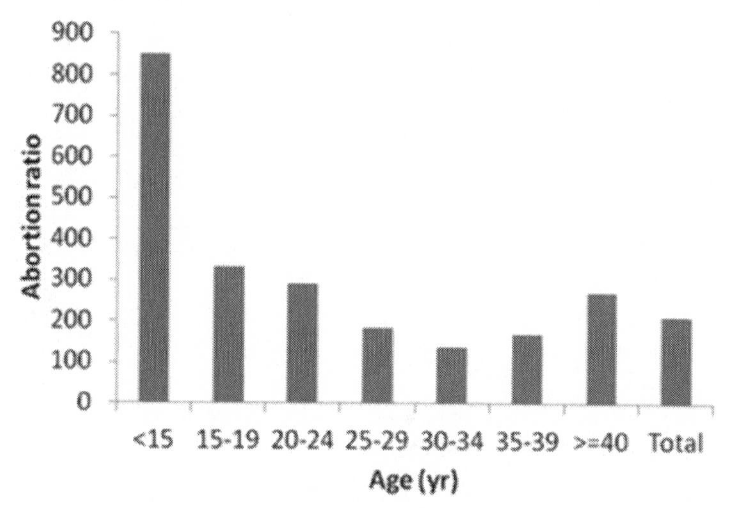

Figure 7-1
Abortion ratios* by age group, United States, 2010
* Number of abortions among women of given age group per 1,000 live births
Source: Pazol[13]

The pattern of abortion rates by age differs from that of ratios, resembling an inverted "U" (Figure 7-2). Here, the highest number of abortions per 1,000 women (27) occurs among women aged 20-24 years, a time of peak fertility. Rates for younger and older women are lower.[13]

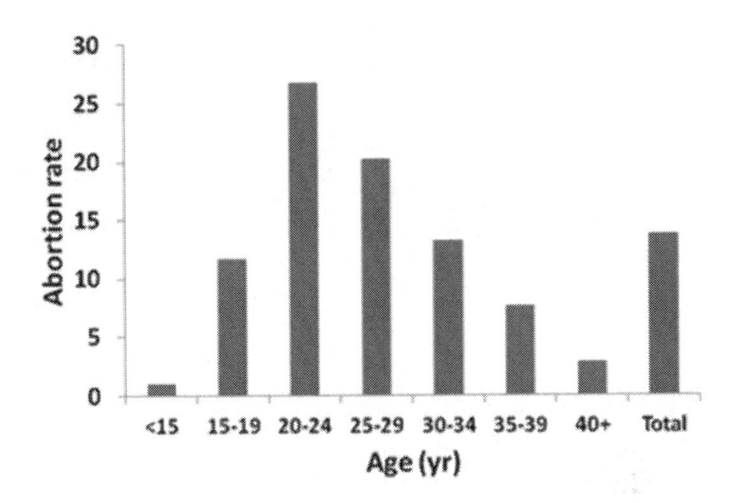

Figure 7-2
Abortion rates* by age group, United States, 2010
*Number of abortions among women of given age group per 1,000 women in the same age group
Source: Pazol[13]

Women of minority race rely on abortion to control their fertility far more than do white women.[13] African-American women have abortion ratios three times that of white women (Figure 7-3), while those of other minority races are twice those of white women. Abortion rates by race reveal the same pattern: African-American women have the highest rates (32 abortions per 1,000 women aged 15-44 years), Hispanic women intermediate (19), and white women the lowest (9).[13] African-American women have both higher fertility than do women of other races and more unwanted fertility. Inadequate access to contraceptive services explains much of these discrepancies.

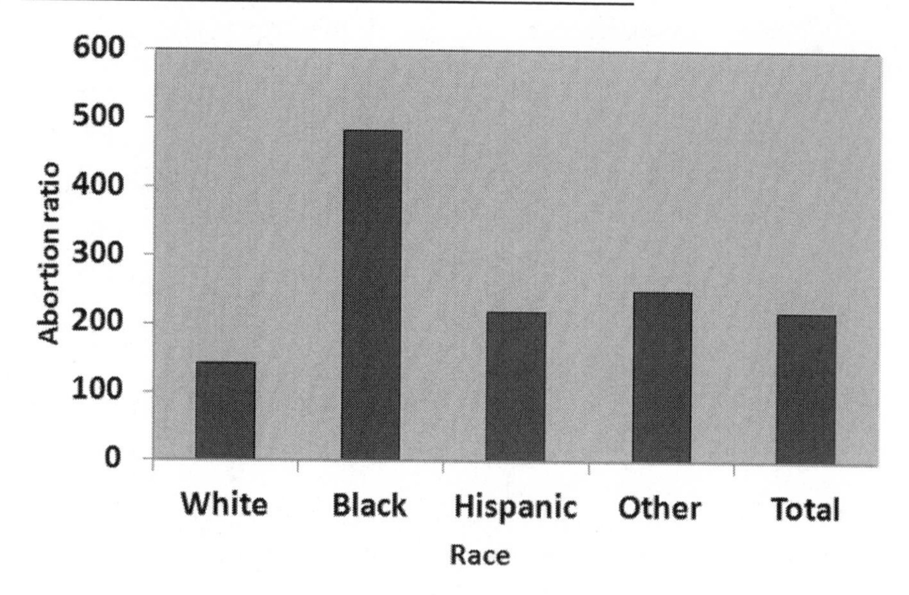

Figure 7-3
Abortion ratios* by race, United States, 2010
*Number of abortions among women of a given race per 1,000 live births to women of the same race
Source: Pazol[13]

Marital status, as might be expected, has a powerful effect on abortion use. Unmarried women have abortion ratios eight times higher than do married women.[13] Data from 38 areas for 2010 reveal that pregnancies among unmarried women accounted for 85% of all abortions.[13] This reflects the relative challenges—whether societal or financial— of childbearing when unmarried, compared with marital fertility.

A 2008 nationwide survey of more than 10,000 abortion patients by the Guttmacher Institute complemented the CDC data: unmarried women of minority races and those with limited income simply rely on abortion more than do other women.[14]

Most are mothers

Perhaps surprising to some, most women having abortions are already mothers.[13] In 2008, 61% of abortion patients had already given birth to one or more children.[15] CDC data indicate that most patients (55%) had not had an abortion before.[13]

Many are poor

Poverty profoundly influences the decision to abort a pregnancy: rates are inversely related to income (Figure 7-4). In 2008, women with incomes less than 100% of federal poverty level had abortion rates six times that of women with incomes ≥ 200% of the poverty level. In 2008, poor women (with family incomes at less than 100% of the federal poverty level) accounted for 42% of all abortions.[15] (For comparison, the 2008 Federal poverty threshold for a woman with one dependent child was $14,000; in 2014, that figure rose to $15,730).[17,18] Medicaid eligibility is another useful marker: women poor enough to receive Medicaid for general health care have abortion rates three times that of women not eligible for Medicaid benefits. Again, these differences are, in large part, due to the higher fertility of poor women.[19]

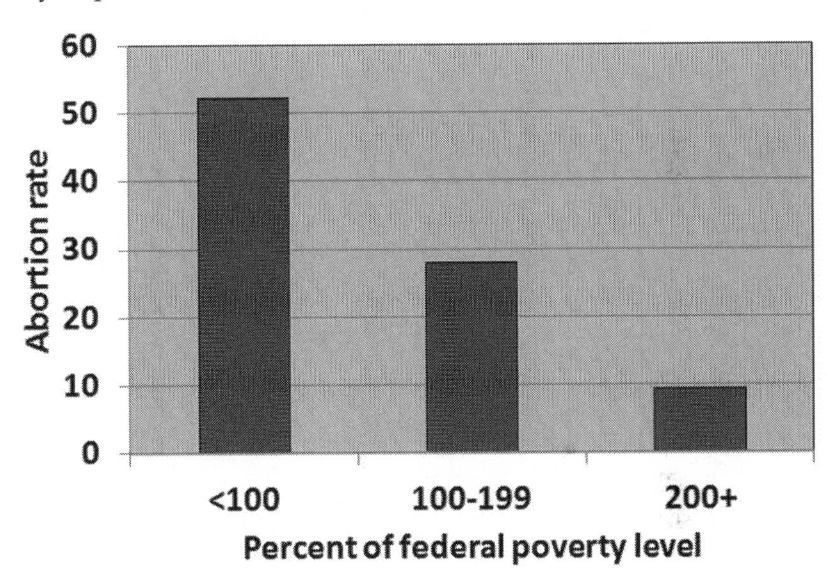

Figure 7-4
Abortion rates by poverty levels, United States, 2008
Source: Jones[15]

Most are religious

Most abortion patients report a religious affiliation. In 2008, the highest proportion (37%) identified themselves as Protestant. Despite the staunch opposition of the Roman Catholic Church to abortion, 28% of abortion patients reported being Catholic. Fifteen percent reported being "born-again, Evangelical, or Fundamentalist."[15] Comparing abortion rates by religious affiliation reveals an irony: among women reporting a religious affiliation, Roman Catholics had the highest abortion rates in the U.S.: 22 per 1,000 women of reproductive age. This represented an 11% increase since 2000, despite the opposition of the Church to abortion except to save the life of the mother.[15]

Most were using contraception

Most abortion patients (54%) reported that they were actively using contraception in the cycle in which they became pregnant.[20] Among those women who were not using contraception, the two most common explanations were an assumed low risk of pregnancy and concerns about contraception. Among those who were using contraception at the time of the unintended pregnancy, the condom and oral contraceptive were the most common methods. For both, inconsistent use was the dominant cause of failure.

Women seeking abortion often face several hurdles. Many women travel long distances to receive care.[12,21] Non-hospital providers surveyed by the Guttmacher Institute in 2001-2002 estimated that 16% of patients traveled 50 to 100 miles, while an additional 8% traveled further.[21] Cost is another obstacle. In 2009, the median charge for a surgical abortion at 10 weeks' gestation with local anesthesia was $470, while that for a medication abortion was $490.[12] Unlike other medical care, most women pay out-of-pocket. Non-hospital providers estimated that 74% of patients paid for services with their own money or that from their partner, spouse, or family.[21]

Why do women have abortions?

The reasons women opt for abortion are as varied as are the women themselves. Several common themes are apparent, however. Studies of the indications for abortion largely stopped after *Roe v. Wade* made such determinations moot; the decision became a private matter between the woman and her physician.

In 2003-2004, the Guttmacher Institute surveyed 1209 abortion patients in 11 large facilities.[22] The most common reasons for choosing abortion were that having a baby would interfere with a woman's education, work, or ability to care for her dependents (74%); inability to support a child (73%); and relationship problems or desire not to become a single mother (48%) (Figure 7-5).

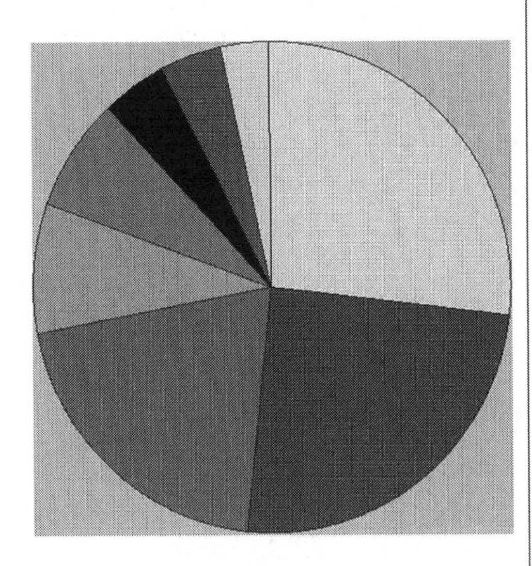

Figure 7-5
Most important reasons for abortion cited by women, United States, 2004
Source: Finer[22]

About one-third of women were not ready to have a child, and about 40% reported having completed childbearing. As in the earlier survey, most women described multiple, interrelated reasons for abortion, with leitmotifs being responsibility to others (such as children), limited resources, and lack of partner support.

> *"I understand that it's better to have an abortion than to have a baby — bring life to this earth that you cannot handle. You cannot buy him food, or you cannot buy him anything he wants. But just to bring another life to struggle with you when you are already struggling yourself."*
>
> Ms. B.[23]

Rape

Rape is a compelling cause for abortion, one of the few for which Medicaid will pay for care. Claims that "the

> *"I am on my own, and financially and mentally, I can't stand it now. That is one whole reason...It's a sin to bring the child here and not be able to provide for it...This is just in the best interest for me and the children – no, my children and this child."*
>
> *19-year-old with three children, living in poverty*[22]

female body has ways to try to shut that whole thing down"[24] notwithstanding, the risk of pregnancy from a single act of rape is estimated to be 5%. Because rape is so common in contemporary society, an estimated 32,000 women became pregnant this way in 1996.[25] Due to the decline in reported rape nationwide over the past decade, an estimated 25,000 pregnancies resulted from rape in 2000.[26] In a series of rape victims who became pregnant and were treated at the Medical University of South Carolina in Charleston,[25] only 12% had sought care immediately, and nearly half received no medical care related to the rape. A third did not discover they were pregnant until the second trimester. Half of those women had abortions, 32% continued the pregnancy and kept the child, and only 6% placed the infant for adoption. Pregnancies resulting from rape have higher risks of poor outcomes for both the mother and baby.[27,28]

Domestic violence

Strongly related to coercive sex, domestic violence is common in the U.S. and around the world. As many as 4 million U.S. women may have been assaulted by their intimate partners.[29] Abusive, violent relationships are especially common among women seeking abortions.[30] Several studies have attempted to estimate the frequency.

A small, early study from a medical school abortion clinic found that a third of patients had a history of domestic abuse.[29] Within the prior year, 22% had been assaulted, and 8% reported having been attacked during the current pregnancy. A second, larger study from another medical school found an even higher lifetime history of abuse: 40%.[31] Women who had been abused were significantly more likely to cite relationship issues as the reason for abortion than were women without this history.

Many young adolescents are sexually victimized by older men. For example, in Alaska, from 39% to 66% of all births to unmarried teenagers younger than 16 years in 1991-1994 resulted from second-degree statutory rape.[32] Among 818 abortion patients at a Houston, Texas, clinic, 17% of women did not disclose the abortion to their partner or spouse; 14% reported abuse within the past year.[33] Physical or sexual abuse was twice as common among women who did not disclose the abortion as among those who shared this information. Some women reported that fear of further violence led them to conceal the abortion from their intimate partners. A recent survey indicated that as many as 20% of abortion patients have experienced sexual or physical abuse.[30]

> *"If he [father] comes to know, he will cut me up. He will kill me."*
>
> *16-year-old unmarried woman*[38]

Domestic abuse is a global problem as well. In a study done in one clinic in Vancouver, British Columbia, 15% of 254 abortion patients reported abuse within the past year.[34] Of these, 8% had been physically abused, 7% sexually abused, and 8% reported being afraid of their partners. In Uganda, where abortion is illegal but widespread, 311 women were admitted to a teaching hospital because of complications.[35] Over half (57%) had experienced domestic violence during their first pregnancy; this included physical, sexual, and psychological abuse.

Childhood sexual abuse was common among these women as well. Among the 70 women who admitted to an illegal abortion, 39% cited domestic violence as the main reason they chose this dangerous option. In China, where abortion has been legal for many years, 1215 women seeking abortion were queried about domestic violence.[36] A history of violence was reported by 23%, including sexual abuse, physical abuse, and emotional violence. As might be anticipated, abused women reported having more abortions than did women without a history of violence. Living in fear as well as in emotional and financial deprivation, these women have little control over sexual activity and contraception.[36,37]

While women of all types seek abortions, disadvantaged women use it disproportionately. Those who are very young, of minority race, poor, and abused turn to abortion more often than do others. Traumatic childhood experiences are associated with repeat abortions.[39] Personal upheavals are common as well: loss of employment, breakup of a relationship, failing to keep up with rent or mortgage payments, and moving multiple times.[40] Lack of access to contraception, coupled with limited ability to control one's sexual life, results in both higher fertility and higher *unwanted* fertility among such women.

In the ideal world, every child born would be healthy and wanted. However, as suggested above, the daily lives of millions of women are characterized by poverty, poor health, and violence. The prospect of bringing children into settings reminiscent of a tale from Dickens is unacceptable to many.

Illness

Serious medical or psychiatric illness sometimes leads to abortion. Physicians may counsel such women that abortion may be the safer pregnancy option for them. Recent patients referred to me provide some insight into the challenges these women face. The pediatrician of a 15-year-old advised her patient to have a mid-trimester abortion, because she had become pregnant while her phenylketonuria was wildly out of control. This is an inherited metabolic disease that requires life-long dietary control; because of her noncompliance with her diet, the fetus would be profoundly retarded. A 22-year-old enlisted woman in the military was diagnosed with acute leukemia at 8 weeks' gestation, and her oncologist did not want to expose her embryo to the high-dose chemotherapy about to begin. A 25-year-old woman with newly diagnosed Marfan syndrome was found to have, as a result, a huge aneurysm (widening) of her aorta (the artery coming out of her heart). Because her aorta was on the brink of rupturing, we performed her abortion in a cardiothoracic operating room with a surgical team standing by.

> *"The Supreme Court gave voice to what the mothers in Romania and Scandinavia – and elsewhere – had long known: when a woman does not want to have a child, she usually has good reason. She may be unmarried or in a bad marriage. She may consider herself too poor to raise a child. She may think her life is too unstable or unhappy, or she may think that her drinking or drug use will damage the baby's health. She may believe that she is too young or hasn't yet received enough education. She may want a child badly but in a few years, not now. For any of a hundred reasons, she may feel that she cannot provide a home environment that is conducive to raising a healthy and productive child."*
>
> *Levitt*[44]

Childbirth is dangerous for very obese women; a recent abortion patient was 5 feet, 2 inches tall and weighed 485 pounds. Another woman in congestive heart failure was too sick to leave the intensive-care unit; we did the suction curettage in her bed.

An adult woman with severe intellectual impairment as a result of head trauma sustained in an automobile accident was impregnated at her group home. Poorly-controlled schizophrenia sometimes leads to requests for abortion; psychiatrists sometimes advise that delivery and relinquishment of a child will aggravate the mental illness.

The reluctant abortion patient

Large discrepancies exist between what women say about abortion and what they do. Susan Carpenter-McMillan, twice an abortion patient, became an outspoken critic of the procedure from which she benefited. That she later regretted her decision to abort[6] does not suggest that other women should be denied the chance to decide for themselves. African-American women are consistently more likely to report being philosophically opposed to abortion,[9,41] despite the fact that they disproportionately rely on it.[42] Many Americans ultimately regret their choice of spouse (only half of first marriages remain intact after twenty years).[43] Later regret about marriage does not mean that the opportunity to choose a marital partner should be outlawed. Situations change, as do attitudes. The only unchanging feature may be the inevitable ambivalence surrounding this intensely personal decision.

Take-home messages

- *About one in three U.S. women will have an abortion*
- *One abortion occurs for every five births nationwide*
- *Each year, about one in 50 women of reproductive age has an abortion*
- *Women disadvantaged by poverty, illness, and violence disproportionately depend on abortion*

Chapter 8
Pro-natal, prenatal diagnosis

When the technician stopped chatting during the routine obstetrical ultrasound examination, Mary had her first indication that something was amiss. In the darkened room, as the technician quietly moved the slippery probe over her abdomen, Mary's fear grew by the second. When the technician left the room to bring in the obstetrician, terror replaced concern. Mary could feel her heart pounding. After exchanging some pleasantries, the obstetrician repeated the scan of the fetus, again in silence and with no facial expression to provide a clue. After a moment of reflection and a deep breath, the doctor reported, "Mary, I'm afraid we have some very bad news for you......"

A small (1%) but important proportion of abortions in the U.S. is done for identified fetal abnormalities.[1] However, 13% to 14% of women report *possible* health problems with their fetus as a reason for abortion.[2] In England and Wales, more than 2,000 abortions are done each year for fetal anomalies.[3] Antenatal (prenatal) diagnosis of fetal abnormalities in early pregnancy is a routine component of contemporary obstetrical practice. Its provision is predicated upon the availability of safe, legal abortion to allow choices in the event of an adverse diagnosis. Knowledge of fetal defects in early pregnancy enables women to make important decisions for themselves and their families.

Depending on the seriousness of the diagnosis, many women opt for abortion; others can prepare for the birth of a handicapped child, which has important benefits for parents. Among couples who have had a child handicapped or killed by a genetic or other anomaly, such as Tay-Sachs syndrome, many would not chance another pregnancy. Others would abort any pregnancy that occurred. With the possibility of prenatal diagnosis, many of these couples can have children that they desperately want, but would not have dared to have without abortion as a back-up for prenatal diagnosis.[4,5] Ironically, in this regard, prenatal diagnosis and selective abortion are themselves pro-natalist: 98% of women who undergo genetic screening receive reassuring news.[6]

> *"... Reasons favouring legal abortion in such circumstances rest on the potential to do greater good than harm in the community, and reveal the positive, life-affirming aspects of legally available abortion services."*
>
> *Cook*[7]

Candidates for prenatal diagnosis

Maternal age is a frequent indication for prenatal diagnosis. Women are increasingly delaying childbearing to accommodate educational and professional goals. A by-product of this delay is an increased proportion of women with an elevated risk of chromosomally abnormal pregnancies solely because of their age.

The definition of "advanced maternal age" (35 years and older) is arbitrary; the risk of chromosomal defects increases in a smooth, exponential fashion with advancing age,[8] and no clear cut-off is evident (Figure 8-1). Many women in this age group are apprehensive about the potential for fetal chromosome disorders, and diagnostic testing can be reassuring.

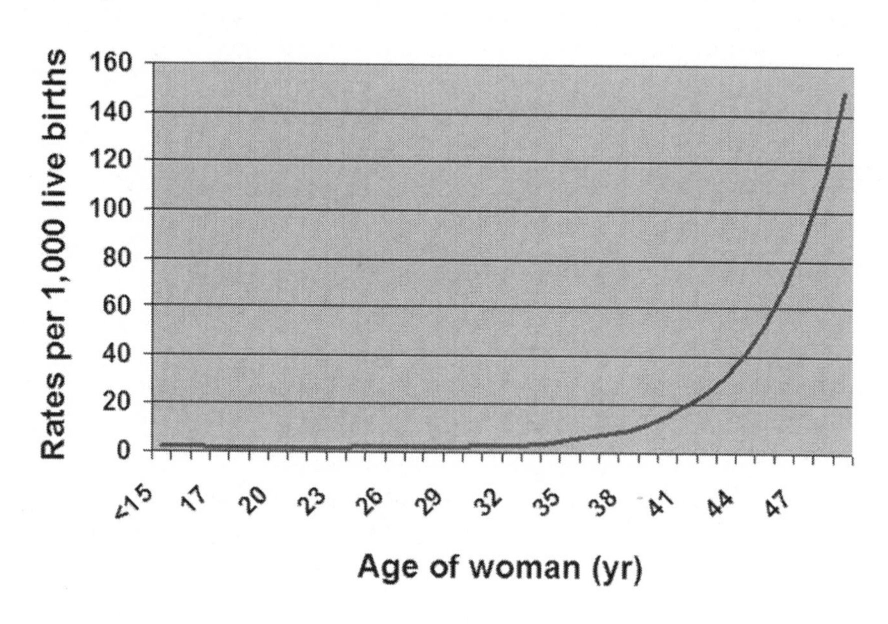

Figure 8-1
Estimates of chromosome abnormalities in live births by maternal age
Source: Hook[8]

Family history is a second important indication. Another group of candidates includes women who have already had a child with defects or those with a family history of a genetic disorder with risk of recurrence. Some aneuploidies (an abnormal number of chromosomes that are not a multiple of the usual number; Down syndrome is a common example) have a low risk of recurrence in later pregnancies (e.g., 1%); in contrast, autosomal dominant diseases have risks of 50%. Examples include

Huntington disease (a progressive neurological disorder) and Von Willebrand disease (a blood clotting disorder).

A third indication is an abnormal screening test. An example would be an abnormal maternal serum alpha-fetoprotein level. This blood test is routinely done in pregnant women to screen for neural tube defects (such as spina bifida and anencephaly) and for some chromosome disorders as well. Abnormalities in this test suggest an increased risk of neural tube defect or chromosomal anomalies. These possibilities commonly trigger further testing to establish the diagnosis.[5] New, less invasive screening tests such as cell-free fetal DNA (which can be found in maternal blood) may soon revolutionize screening for fetal chromosomal defects.[9]

Women's acceptance of prenatal diagnosis

Prenatal diagnosis has high acceptability around the world. For example, in public health screening programs in Europe, the uptake has been 60% to 90%. The most common reason for *not* having this testing is starting obstetrical care too late in the pregnancy to have an elective abortion.[10] In the U.K., only 7% of women declined prenatal diagnosis on moral grounds.[11] In South Africa, acceptance of both amniocentesis and abortion were high among women of differing ethnicities and religions.[12] Among women in the U.S. with an abnormal biochemical screening test for aneuploidy, about 80% chose to proceed with amniocentesis, in which a thin needle is inserted into the amniotic sac surrounding the fetus. This enables a small amount of amniotic fluid to be withdrawn for analysis of fetal chromosomes and for other tests.[13] More recently, a San Francisco study found that 30% of women with a positive screening test for Down syndrome declined amniocentesis for confirmation.[14] Given a diagnosis of a fetus with Down syndrome, the decision-making process about abortion or continuation of the pregnancy is complex.[15]

Similarly, in a Scottish study of 2207 women eligible to be screened for cystic fibrosis carrier status during pregnancy, 85% consented to be tested.[16] Cystic fibrosis stems from an enzyme deficiency that leads to chronic illness of lungs and digestive system; death usually ensues in the twenties or thirties after a lifetime of progressive disability. When a man and a woman, each of whom is a healthy carrier of the gene, have children, each child has a one-in-four chance of the disease.

Those directly affected by genetic diseases provide unique insight into the acceptability of prenatal diagnosis followed by selective abortion.

Choroideremia is an X-chromosome-linked degenerative eye disease that ultimately leads to blindness. Retinitis pigmentosa is another inherited eye disease leading to blindness. In a Finnish study of patients with these diseases, their relatives, and choroideremia carriers, most respondents had a favorable opinion of prenatal diagnosis. Only a minority, however, would seek an abortion based on this fetal diagnosis.[17]

In a similar study in the U.K., investigators queried adult cystic fibrosis patients and their parents about their views of screening programs for this disease. Both the patients and their parents supported prenatal screening (88% and 90%, respectively). The proportions that would opt for abortion given the diagnosis were 68% and 84%, respectively. The authors concluded that patients and parents whose lives have been affected by cystic fibrosis support genetic screening to enable women to make reproductive choices.[18]

Some critics label prenatal diagnosis a "search and destroy" mission, the prime intent being to abort abnormal fetuses. In contrast, prenatal diagnosis is designed to inform choices. Some women undergo prenatal diagnosis knowing that they would not choose abortion regardless of the outcome. A recent U.S. study followed women who received a prenatal diagnosis of aneuploidy (mostly Down syndrome) and who continued their pregnancies. These women were compared to others who discovered the chromosome problem only at the time of birth. Prenatal diagnosis of the defect was associated with a better experience for the woman during pregnancy and birth. Most reported that they would have done nothing differently in the pregnancy. In contrast, among women who did not learn of the abnormality before birth, more than 70% reported that they would have asked for more fetal diagnostic tests, such as serum screening or ultrasound examination. Women with a prenatal diagnosis of fetal aneuploidy found meeting with specialists, learning about Down syndrome, and making preparations for caring for a child with special needs to be valuable.[19]

Another recent study surveyed 141 women who received a prenatal diagnosis of fetal Down syndrome and continued the pregnancy.[20] Religious or personal reasons were the primary reasons for continuing the pregnancy rather than aborting. The author concluded that many women approach prenatal diagnosis either with ambivalence or no intention to abort regardless of the diagnosis. *Information, not action, is their goal.*

The impact of a child's disability

The burden of caring for a disabled child routinely falls disproportionately on the mother. The financial costs can be huge over a lifetime, both to the family and to society. Often, the mother must quit her outside employment and provide full-time care for the disabled child. The attention directed to the one child often results in real or perceived neglect of the husband and other children; many marriages do not survive the stress. The economic and social burdens may overwhelm the family. The husband may leave and take the older children with him, leaving the mother alone with limited means. With longer life spans of disabled persons, parents in their eighties may still be caring for middle-aged handicapped children in the home.[10]

Some families welcome and love handicapped children.[21] Stories of couples who have adopted large numbers of handicapped children are heart-warming and inspiring. Regrettably, American society has been consistently unwilling to assume responsibility for the care of the disabled; indeed, an estimated 80% of all disabled persons are cared for by their families at home. Some parents in this setting describe "a never-ending sense of loss."[10] Those who care for children with mental retardation or behavioral difficulties complain of "the immense difficulty of daily life." Especially for single parents and families of limited means, a severely disabled child can be a crushing burden. Hence, choice is important in this regard.

Factors leading to the abortion decision

The type of chromosome anomaly and, to a lesser extent, visualization of fetal anomalies on ultrasound examination, both influence women's decisions to proceed with abortion. As might be anticipated, the more severe the defect, the more likely is an abortion to be requested.[22-26] In contrast, whether the diagnosis was made by chorionic villus sampling (a placental biopsy done through the cervix in early pregnancy) or by amniocentesis had little influence on the abortion decision, unlike earlier studies.[27]

An exception may be the diagnosis of Down syndrome. In a large case-series from a single center, 87% of 145 women with a prenatal diagnosis of trisomy 21 chose abortion, and 13% elected to continue the pregnancy.[28] In a logistic regression model which controls for the potentially distorting effect of bias, the duration of pregnancy at the time of diagnosis emerged as an important predictor of the decision to abort. Women who elected to end the pregnancy were a mean of 15 weeks from last menstrual period, in contrast to those who chose to continue (19 weeks). Although women who chose abortion were older than those who continued their pregnancies, this difference was not significant. The same held true for number of prior births, race, religion, or type of insurance. The authors speculated that the higher abortion frequency with earlier prenatal diagnosis may reflect less bonding with the fetus than occurs in later gestation.

Abortion frequency by type of defect

The probability that an abnormal diagnosis will lead a woman to choose abortion varies widely by severity. A recent U.S. report found an overall abortion rate of 83% with fetal chromosomal abnormalities.[29] Another review indicated abortion rates for a range of fetal disorders. With trisomies (such as Trisomy 18) associated with early death or life-long mental retardation, the proportion of pregnancies electively aborted has ranged from 73% to 100% (Figure 8-2). For metabolic diseases, such as Tay-Sachs, characterized by profound mental retardation and agonizing death by age 5 years, 100% rates have been reported in several studies. For spina bifida (an open spine often associated with paralysis of the legs and loss of bladder and/or bowel control), the proportion has ranged from 74% to 100%.

The frequency of abortion varies by severity of blood diseases as well. Thalassemia is an uncommon defect in blood hemoglobin that causes pain, disability, and usually death by age 20 years. Sickle cell anemia is another serious hemoglobin defect causing painful crises and shortened life span. Abortion percentages for the former range from 90% to 100%, and the corresponding figures for the latter are 39% to 54%.[10]

Surveys of women with affected family members are instructive. In these studies, respondents are parents of ill children or adult children of affected parents.[10] In these families, the proportion of women favoring abortion with prenatal diagnosis of cystic fibrosis ranges from 20% to 95% (Figure 8-3). Hemophilia A is a severe bleeding disease caused by an inherited deficiency of a blood clotting factor. Victims require numerous transfusions of clotting factors throughout life and have painful bleeds into their joints and tissues. Given this prenatal diagnosis, 40% to 100% of women in North America, Europe, and Australia report that they would choose abortion. Huntington disease is an inherited neurological disease that first presents in mid-life with progressive mental and nervous system debility, leading inexorably to death. Given this prenatal diagnosis, from 30% to 71% of women in the U.S. and England would abort the pregnancy.[10]

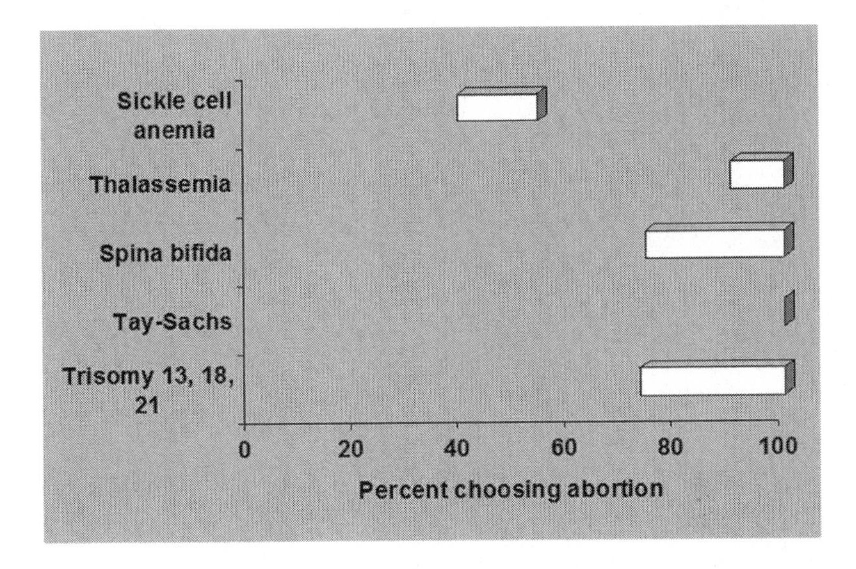

Figure 8-2
Percent of women who chose abortion for selected fetal defects
Bar indicates range reported in published studies
Source: Wertz[10]

A recent systematic review of the published literature provides the most detailed assessment of what women do in the face of an abnormal prenatal diagnosis. The authors compiled 37 data sets from 11 different countries and calculated confidence intervals around the proportions choosing abortion. Confidence intervals depict the precision in the calculated results: narrow confidence intervals imply good confidence in the result, and vice versa. Stated alternatively, larger numbers yield better precision.[23]

Abortion frequency varied by diagnosis. Abortion was chosen most commonly after a diagnosis of Down syndrome. Among 5035 women in 5 countries, 92% chose to abort the pregnancy for this indication (Figure 8-4). The narrow confidence interval (92% to 93%) indicates great precision; one can be reasonably sure that the true value for all women falls within this range. The high frequency of abortion after a diagnosis of Down syndrome reflected the broad reluctance to deliver a child with serious mental impairment.[30,31]

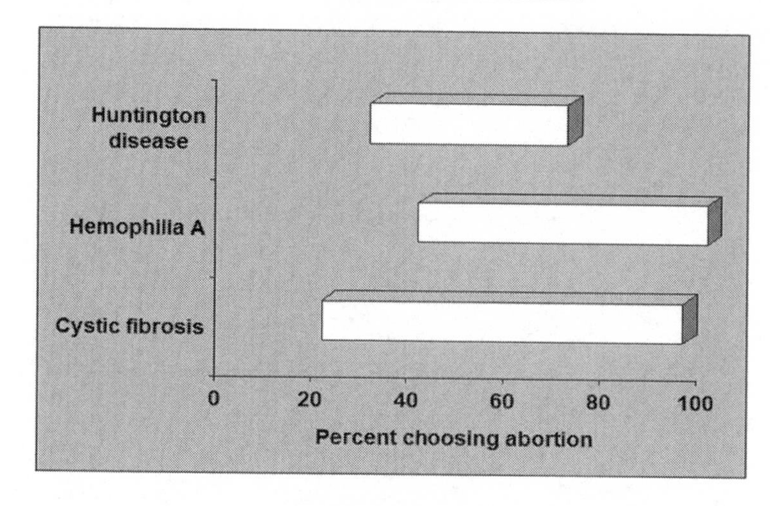

Figure 8-3
Percent of women with affected family member who would choose abortion, by fetal defect
Bar indicates range reported in published studies
Source: Wertz[10]

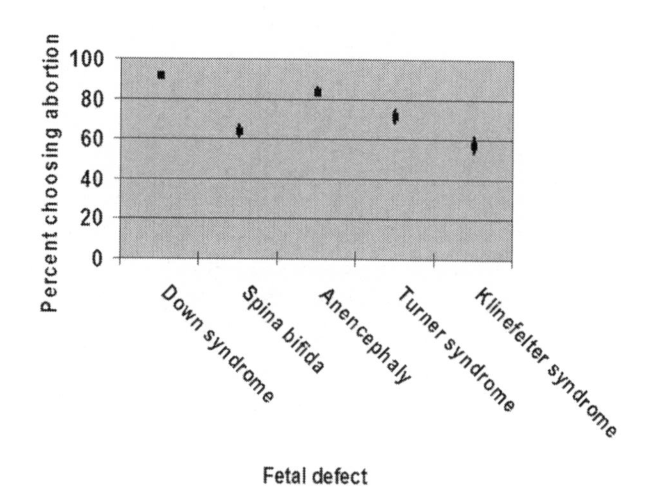

Figure 8-4
Percent of women in five countries who chose abortion for selected fetal defects
Square indicates the aggregate percent; vertical line represents the 95% confidence interval around the aggregate percent
Source: Mansfield[23]

Neural tube defects (failure of the developing nervous system to close during embryonic development) can be lethal. Failure to close at the lower end of the spine causes spina bifida, also known as meningomyelocele. Failure at the top leads to anencephaly or acrania. In a recent review of women's decisions, spina bifida, which differs in degree of handicap, had an overall abortion proportion of 63% (range 31% to 97%).[32] In contrast, anencephaly, a severe variant of neural tube defect, in which most of the brain is missing and which is nearly always fatal shortly after birth, was associated with a higher abortion frequency (83%; range 59% to 100%).[32] Abortion rates for these defects were higher in Europe than in North America.

Sex-chromosome abnormalities are less severe.[22,29] Turner syndrome in women (45X chromosome complement) causes short stature, webbed neck, susceptibility to several diseases, and failure of the ovaries to develop. Klinefelter syndrome in men (47XXY chromosome complement) is characterized by tall stature, obesity, lack of facial and body hair, language difficulties, and infertility. In a recent review of the literature, the average abortion rate for a fetus with Turner syndrome was 76% (range 33% to 100%), while that with Klinefelter syndrome was 61% (range 44% to 85%).[22]

The impact of prenatal diagnosis and abortion

Access to prenatal diagnosis and abortion has dramatically improved the lives of families around the world. Many studies have chronicled important declines in the prevalence of several important conditions.

Down syndrome

For example, the incidence of Down syndrome in Hawaii fell between 1963-1969 and 1971-1977. An estimated 43% of the decline was attributable to induced abortion.[33] While declining pregnancy rates and lower age-specific risk of Down syndrome also contributed to the overall improvement, legal abortion accounted for nearly half of this benefit.

Neural tube defect (open spine)

The incidence of neural tube defects has declined in many nations. In the U.S., the CDC has been conducting surveillance of spina bifida and anencephaly for decades. From 1985 to 1994, the prevalence of anencephaly was reduced by 60% to 70%, while that of spina bifida was lowered by about 20% to 30%.[34] Reports from the U.K. have corroborated major declines in both spina bifida and anencephaly.[35,36]

In Quebec, Canada, the prevalence of neural tube defects declined three-fold over the past three decades, with cranial defects more likely to be detected during pregnancy than were spinal defects.[37] In Ontario, Canada, the incidence of birth with neural tube defects declined from 11 per 10,000 in 1986 to 5 in 1999. The proportion of abortions for neural tube defect or hydrocephalus (excess fluid in the brain) rose correspondingly during the interval. The authors concluded that prenatal screening and selective abortion accounted for most of the decline prior to 1995, while folate (a B vitamin) supplementation of food after 1995 contributed to a lower risk of neural tube defect.[38] The prevalence of a neural tube defect among stillbirths in Ontario was 32 times higher than that in live births, indicating a continued winnowing of abnormal fetuses into late pregnancy.

Other birth defects

Use of ultrasound during pregnancy has led to similar reductions in other congenital anomalies. Birth defects occur in 2% to 3% of all live births, and these anomalies are important causes of suffering and death in infants. In one hospital in Haifa, Israel, with ready access to prenatal ultrasound examination, investigators tracked the frequency of newborns with birth defects from 1989 to 1993.[39] Although the incidence of congenital anomalies remained stable during the interval, the incidence of affected *newborns* dropped from 1.95% to 1.34%, a difference unlikely to be due to chance. Correspondingly, the proportion of abortions done for congenital anomalies increased. The range of fetal defects found by ultrasound examination was broad: hydrocephalus, spina bifida, Tetralogy of Fallot (a serious heart defect), cystic kidney disease, intestinal obstruction, and abdominal wall defect.

Financial support

Large racial discrepancies persist in the use of prenatal diagnosis. African-American women are 40% less likely than white women to undergo amniocentesis.[40] Some of this disparity may relate to cultural or religious differences. However, financial barriers may play a large role as well. As of 1988, all State Medicaid programs paid for amniocentesis, and most States provided coverage for new prenatal diagnostic procedures for eligible women.[41] However, only 13 States paid for women to have an abortion after diagnosis of an anomalous fetus. Hence, the lack of financial ability to end a pregnancy found to be defective may have dissuaded many poor women from seeking prenatal diagnosis.

A recent sophisticated analysis lends support to this possibility. Using logistic regression to control for potential bias, epidemiologists found important socioeconomic differences in women's use of amniocentesis.[42] They also confirmed substantial differences in women's risk of having a child born with Down syndrome. Specifically, women of African-American or Mexican descent had both a lower likelihood of receiving amniocentesis and a higher risk of Down syndrome at birth. After controlling for the potentially distorting effect of maternal age, socioeconomic factors, and prenatal care, the researchers found an inverse relationship between State support of abortion services and the frequency of Down syndrome at birth. States that supported abortion services had a lower incidence of Down syndrome.

A bumper sticker popular among teachers notes, "If you think education is expensive, consider the alternative." The same holds true for prenatal diagnostic services. California has been a leader in screening for neural tube defects and aneuploidy. Mandatory offering of maternal serum screening began in 1986 and expanded to include other tests in 1995 (the triple-marker screen). About two-thirds of women offered the test accepted. More than 90% of women with positive screening results were seen at a prenatal diagnostic center. Analysis of the costs of the program in averting neural tube defects and Down syndrome found a savings of nearly $3 for every $1 invested in the program.[43]

The emotional impact of abortion for anomalies

The tragic discovery of serious fetal problems is often compounded by the need for rapid decision-making regarding abortion (because the diagnosis is usually made in the second trimester). The decision to proceed with abortion is often the most difficult decision parents ever have to make - one with which they struggle. Many couples never envisioned finding themselves in this situation, and some openly share their opposition to abortion—in principle. However, pregnancies do not occur in the abstract. Couples often conclude that this is the best—and most loving course—for the fetus, their current and future children, and their marriage.

Studies of couples who have been through this agonizing process suggest that most fare well emotionally. For example, in one survey, 70%

of couples reported that their marriages were made stronger through the process.[44] Most coped with the loss by relying on friends, families, or professional counselors. A home-visit study after abortion for neural tube defect in the U.K. found most women satisfied with their care during screening and prenatal diagnosis (82% of 166 women). Nearly as many were satisfied with their care during the abortion, performed in hospital (72%).[45] A Canadian study found that those couples deemed at increased risk and who received prenatal counseling before diagnosis did better emotionally than did couples who had a fetal defect found incidentally by ultrasound examination.[46]

Afterthoughts of patients

While the medical and public health benefits of prenatal diagnosis with abortion as a back-up are clear, the emotional impact of abortion in this setting is harder to quantify. Excerpts from recent thank-you notes I have received from such patients, however, may provide some insight (Table 8-1).

Table 8-1
Excerpts from thank-you notes from patients after abortions for fetal indications

"…thank you for the care you gave to me and my husband as we experienced the painful loss of our baby to anencephaly. We are so very thankful for your kindness and expertise. For me, this experience has made much more personal a conviction that has always been philosophical. I am so thankful that I was able to choose how to handle the very tragic news of our baby's future…."

"Thank you so much for your kindness during our recent procedure. Working with someone so compassionate made our difficult time a little easier."

"Thank you for your kindness, empathy and fine skills. I am doing well and feel very lucky to have had you as my doctor."

"Just a quick note to thank you for helping us through this very sad time in our lives. We felt very safe during this scary time with your care….When you look in the dictionary under doctor it should read your name. Warmest regards."

"This note is hard for me to write but I thought you deserved a thank-you for the kindness and compassion you showed me last week. I was the patient you fit into your schedule on Wednesday and Thursday for a D&E. In spite of the difficult and emotional situation, I felt your care to be exceptional. We are still recovering from our loss emotionally, but I am much better physically. Thank you for helping us and for making things go as smoothly as possible…."

"I am proud of you…for providing comforting bedside care and painless medical procedures for women in need of such. I will never forget you, nor the entire staff. Physically I am doing great; awaiting my first cycle in hopes of trying again soon. Emotionally both my husband and I are doing well. We planted a tree and plan to put the ashes there as well. I was told that grieving would be harder post D&E versus L&D and I think, for us, this theory is wrong."

"Thank you so much for fitting me into your busy schedule last week. Also thank you for making me feel like I was your only patient. You and your staff made a very difficult time for J. and me very comforting."

"We wanted to thank you for being a kind, knowledgeable and caring doctor. You were able to make a difficult situation easier to handle."

"We just wanted to send a quick note to sincerely thank you for the wonderful care you and your team provided to us. At our most vulnerable moments, we felt comfortable and reassured….we are moving ahead with our sights set on brighter days."

"I would first like to thank you for your guidance during a difficult time during our lives last February when we chose to terminate a pregnancy due to a major congenital heart malformation. I am very grateful that I had the right to choose and the resources to be able to have this option at a respected institution well into my second trimester. I am disheartened to know that not all women have this opportunity…."

"We just wanted you to know how much we appreciated your compassion and gentle method of applying your obvious skill and expertise in such a delicate area of medicine. Your presence was truly a gift from God as we journeyed through a very difficult end to a very challenging pregnancy. Thank you for taking such good care of us."

Take-home messages

- *Prenatal diagnosis is predicated upon the availability of abortion as a back-up*
- *The availability of abortion enables many families to have children that they would not have had otherwise*
- *The more serious a fetal abnormality diagnosed during pregnancy, the greater the likelihood that abortion will be chosen*
- *Prenatal diagnosis and abortion as a backup have been associated with declines in Down syndrome, anencephaly, and spina bifida*

Chapter 9
The economics of abortion: Pay now or pay later

Paying for abortion: women's realities

By 2008, more than 40% of women having abortions were poor. Compared to women better off financially, poor women are five times more likely to have an unplanned pregnancy and abortion; they are six times more likely to have an unplanned birth.[1]

Even among those with insurance, most women pay out-of-pocket for abortion care. In 2011, half of all women had to turn to others to help pay for the procedure. Non-medical expenses included transportation (mean $44) and for some, lost wages, childcare, and hotel costs. Making ends meet is a challenge for many women: one-third had to defer paying basic living expenses such as rent, food, and utilities to pay for their abortion.[2] Clearly, paying for care is a hardship for many.

Who bears the burden?

The funding of abortions for poor women has been as contentious as other aspects of the national debate. Women's health advocates argue that abortion is a fundamental part of reproductive health care and that it should be covered by Medicaid; those opposed to abortion disagree with both assertions. Some take an intermediate view: abortion is a woman's choice, but she should pay for it herself. While these philosophical differences are irreconcilable, the provision of abortion as a public health program is amenable to objective economic analysis. These dispassionate analyses shed important light on the role of abortion in American society. For a third-party payer or society, an unwanted birth and its consequences are expensive.

> *"Why should your tax dollars pay the bill for those who choose to have an abortion?...This is an unfair burden imposed on Coloradans, who each year see their taxes grow higher and higher."*
>
> *Citizens for Responsible Government*[3]

As noted in the prior chapter, many view abortion as an abstract notion. This holds true in abortion economics as well. Abortion occurs in the context of an unwanted or a wanted but abnormal pregnancy. Refusing to pay for abortions for poor women saves tax dollars in the short run. However, the later costs to taxpayers are large, both in direct and indirect costs. In New York, insurance companies often refuse to pay $150 for a diabetic patient to see a podiatrist, yet they readily cover foot amputations, which cost more than $30,000.[4] This chapter explores the subsequent costs to society, describes the public-health consequences, and contrasts the costs of abortion with a common public-health program, vaccination.

Levels of prevention

Preventive medicine has three tiers (Figure 9-1).[5] Primary prevention avoids illness. Prototypes include immunization against polio and addition of fluoride to municipal water supplies to prevent tooth decay. Secondary prevention involves prompt diagnosis and treatment of disease once present. A prototype is Pap smear and treatment of precancerous growths of the cervix. Tertiary prevention can be considered rehabilitation and mitigation of long-term consequences. In the case of polio, this would include providing "iron lungs" to those permanently paralyzed and incapable of breathing independently.

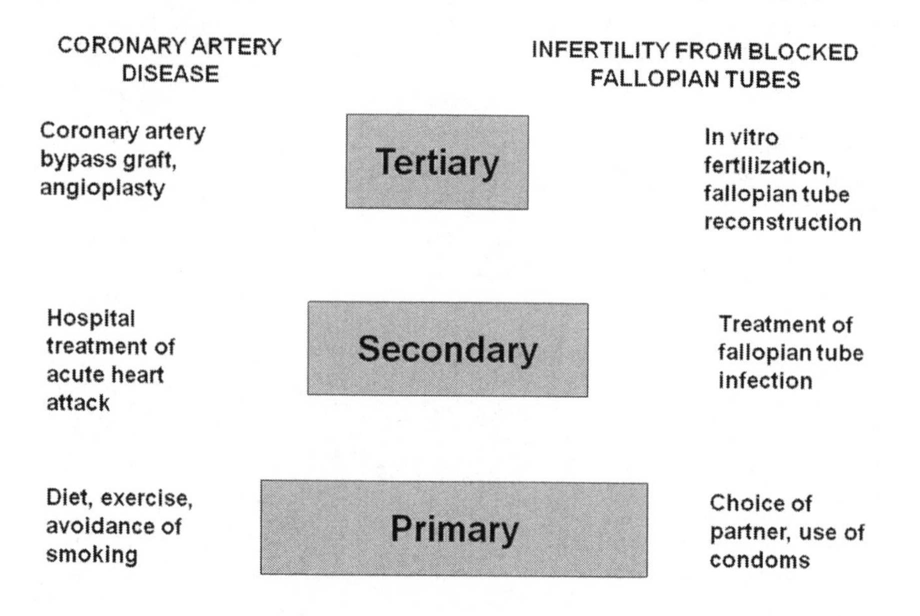

Figure 9-1
Levels of prevention with examples
Source: Grimes[5]

Prevention examples

Heart disease and infertility are important conditions affecting millions of U.S. women. On the left side of Figure 9-1, coronary artery disease is a leading cause of death among both women and men. On the right side of the Figure is infertility from blocked fallopian tubes. Scarring resulting from sexually transmitted infection of the tubes is an important cause of obstruction.

Primary prevention of coronary artery disease includes diet, exercise, and avoidance of smoking. Primary prevention of tubal infertility avoids infection through prudent choice of sexual partner and use of condoms as appropriate.[6,7]

Secondary prevention of coronary artery disease includes prompt, effective care of an acute heart attack. Drugs given during the acute episode can minimize damage to the heart muscle. Secondary prevention of tubal infertility would encompass prompt diagnosis and treatment of fallopian tube infection with two or more antibiotics; the sooner therapy is begun, the lower the risk of infertility.[8]

Tertiary prevention of coronary artery disease may entail coronary artery bypass grafts (designed to bring new blood supply to the heart) or angioplasty, during which the clogged arteries are dilated and cleared of obstructing material. For tubal infertility, tertiary prevention involves surgical reconstruction of the tubes or bypassing the tubes altogether and proceeding directly to *in vitro* fertilization.

As suggested by these examples, primary prevention strategies tend to be simple, non-technical, widely available, and fairly inexpensive. Tertiary prevention is the opposite; secondary prevention is intermediate in complexity and cost. In general, primary prevention is more cost-effective than higher levels of prevention. Moreover, preventing illness or disability is clearly more compassionate than treating illness after it has developed.[5]

Prevention of unwanted fertility and resulting pregnancy can be viewed in the same schema. Primary prevention would include sex education and contraception, including emergency contraception. While behavioral strategies to prevent unintended pregnancy appear ineffective,[9] contraception has a dramatic benefit.[10-12] Induced abortion constitutes secondary prevention of an unintended birth. Abortion also has a favorable cost-benefit equation. Tertiary prevention of unwanted births includes adoption and public assistance for health care, food, and other needs. As with coronary artery disease and tubal infertility, tertiary care here is more expensive than primary or secondary prevention.

Primary prevention: contraception

Contraception saves money. In recent decades, numerous analyses, both cost-benefit and cost-effectiveness, have reached the same conclusion. An early report from Iowa showed that publicly funded family planning saved the State money, especially by preventing births to teenagers, who would then become eligible for public assistance.[13] For example, fiscal year 1989 public-sector expenditures for contraception in California ($46 million) saved an estimated $232 to $509 million in public costs for abortion, obstetrical and newborn care, and welfare and nutritional programs during the first two years after an unintended birth.[14] Thus, for every public dollar spent on contraception, around $8 was saved as a result. More recent evaluations of the Family Planning, Access, Care, and Treatment (Family PACT) program in California found the savings to range from $3 to $4 for every dollar invested.[15,16]

On a national basis, the same analysis for fiscal year 1987 yielded savings of more than $4 for every public dollar spent on contraception.[17] An update to that analysis provided a range of savings from a national average of $3 to as high as $8 per dollar invested.[18] More recently, every dollar spent on taxpayer funding of family planning was estimated to save $4.[19]

Family planning methods varied widely in their cost-effectiveness. A recent analysis of nine methods found that the hormonal and copper intrauterine devices (IUDs) were the most cost-effective methods of reversible contraception.[20] Another recent study carried this assessment further by using a cost-utility analysis.[21] This study included the net impact of contraception on women's health as well as on economic outcomes. The authors compared 13 different contraceptive methods according to health care costs and quality-adjusted years of life. Compared to not using contraception, all methods saved money over two years.

The economic benefits of contraception are not unique to the United States. Countries as diverse as the United Kingdom[22] and Turkey[23] have documented large savings from contraceptive expenditures. In Turkey, every dollar invested yielded a savings of nearly $18.

The Hyde Amendment: cruel and costly

In the early years after the *Roe v. Wade* decision, poor women who qualified for Medicaid had coverage for abortions along with other routine obstetrical services. Medicaid is the largest public provider of funds for family planning in the United States today. It has eclipsed the Title X national family planning program, Maternal and Child Health block grants, and other programs.[24]

In 1976, the Hyde Amendment was enacted by Congress, limiting Medicaid coverage to only those abortions necessary to save the life of the woman. Because of a federal court's restraining order, implementation was delayed until August of 1977. Since then, except for a seven-month interval in 1980 during which the Supreme Court reviewed the constitutionality of the Amendment, Medicaid funding for abortions has been severely restricted.[25] For example, in 1994, 19 States spent $464,000 in federal Medicaid money to pay for 282 abortions.

At present, States are required to pay for abortions that meet three federal exceptions: life endangerment, rape, and incest. Thirty-two States and the District of Columbia follow these guidelines (Table 9-1), but South Dakota is in violation of federal law by paying only for abortions for life endangerment.[26] In addition, three of these States pay for abortions performed because of fetal abnormality, and three fund abortions necessary to prevent serious damage to the woman's health. In contrast, 17 States use their funds to pay for most or all abortions. Four do so voluntarily, and the remainder because they were directed to do so by a court. Six States have additional exemptions, such as fetal impairment or threats to the woman's health.

Secondary prevention: abortion

Refusing to fund Medicaid abortions defers – but then ultimately increases – costs to taxpayers. Studies in four geographic areas have examined this question over the past three decades, and each has reached the same conclusion (Figure 9-2). The only substantive differences relate to how much bigger is the deferred tax burden.

New York had been one of the earliest States to fund Medicaid abortions, starting when State law was liberalized in July 1970. In response to early threats to Medicaid coverage, the New York City Department of Health tallied just the first-year costs for staff, supplies, and facilities should the $7.75 million in payment for Medicaid abortions (29,362 in 1973) to city residents be cut off.[27] Of note, the time horizon was limited to only the first year, and costs other than those borne by governments (federal, State, and city) were not considered.

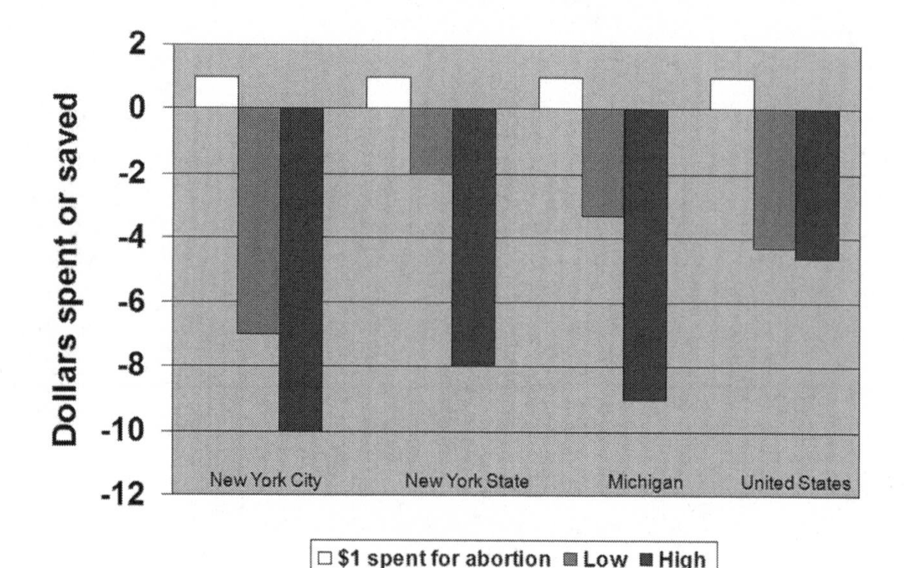

Figure 9-2
Economic impact of Medicaid funding of abortions, by locale
For every tax dollar spent for abortion (positive axis), subsequent tax savings range from about $2 to $10, depending on the date of study, duration of costs, and components of cost
Sources: Torres[25], Robinson[27], Guttmacher Institute[28], Evans[29]

Four types of services were considered: septic, incomplete abortions; medical care for full-term pregnancy; social welfare programs; and well-baby services. Medical care costs included hospitalization during pregnancy, normal newborn hospitalization, and hospitalization of premature babies. Social welfare programs included foster and institutional care, Aid to Dependent Children and Home Relief (both old and new cases), and infant day care. Well-baby costs encompassed child health visits and supplemental feeding for infants. An estimated $3 million would be spent on care for septic, incomplete abortions. The medical care for an estimated additional 10,000 to 15,000 births to poor women would range from $17 million to $26 million. Social welfare costs would add another $27 million to $41 million, without counting the administrative costs of these programs. The well-baby costs would contribute another $3 million to $4 million, for a total of $51million to

$74 million dollars (in 1974 dollars) versus $7.75 million for Medicaid abortions.

The benefit-cost calculation was simple:

$51 million/$7.75 million = 7 (low estimate of the savings for every dollar spent on Medicaid abortions)

$75 million/$7.75 million = 10 (high estimate of the savings)

In other words, for every dollar spent for Medicaid-funded abortions, the savings in the next year alone ranged from $7 to $10. The New York City health officials cautioned that these estimates probably were low.

A second analysis from New York, which examined the entire State, reached a similar conclusion.[28] In this analysis, researchers assumed that only 25% of pregnancies that would have ended through Medicaid abortions would continue to delivery. Stated alternatively, 75% of Medicaid-eligible women would still obtain an abortion without Medicaid funding. In this scenario, for every Medicaid dollar spent for abortions, two were saved in public health and welfare costs. If 100% of pregnancies continued to delivery (improbable), the savings would rise to $8.

> *"...our estimate of a cost-to-benefit ratio of from seven-to-one to ten-to-one must be considered as extremely conservative...The most important costs, however, are human ones."*
>
> *Robinson*[27]

A more recent State analysis from Michigan confirmed the New York conclusions.[29] Although Michigan initially had paid for Medicaid abortions from its State funds, a referendum in November, 1988, ended that practice the following month. Several years into the Medicaid-restriction era, researchers at Wayne State University and the University of Michigan assessed whether the funding restriction had, indeed, saved the State money.

Adjusted for inflation, the annual Medicaid abortion tab for Michigan would have been about $7 million by 1991. They estimated that only 20% of pregnancies that would have ended through induced abortion, were it not for the Medicaid cut-off, would continue to delivery.[30] Using different sources of information led to an estimated additional 2,120 to 5,800 births each year in the absence of Medicaid funding.

Cost estimates were made for a five-year interval. These included obstetrical care, delivery and newborn costs; public assistance costs related to the pregnancy; and continuing assistance (an average of five years receiving Aid to Families with Dependent Children [AFDC]). The estimates of these federal and State costs ranged from $51 million to $139 million. The portion paid by Michigan alone varied from $23 million to $63 million. If one divides these costs by the $7 million saved by not paying for Medicaid abortions, the ratio ranges from 3.4 to 9.4. Stated alternatively, for every dollar saved by Michigan taxpayers by ending abortion funding, an extra three to nine were spent over the next five years – a poor trade-off.

Finally, State-by-State and national estimates corroborate the findings of these three studies.[25] Nationwide, in the mid-1980s an estimated 303,000 to 394,000 abortions would have been paid for by Medicaid in the absence of the Hyde Amendment. Assuming that 20% of women without Medicaid funding would continue their pregnancies to delivery,[30] these Medicaid-funded abortions would have averted 61,000 to 79,000 Medicaid-funded births. The costs considered included medical services for pregnancy, delivery, and infant care and for social welfare benefits during only the first two years of the infant's life. Social welfare benefits comprised Aid to Families with Dependent Children, food stamps, and the WIC (Women, Infants, and Children) program designed for those at risk of poor nutrition.

If publicly funded abortions had been available to indigent women nationwide, the federal and State savings in the next two years would range from $435 to $540 million. In contrast, the savings from not funding such abortions would range from $95 to $125 million. Thus, the ratio between immediate savings and deferred costs ranges from 4.3 to 4.6. Stated alternatively, for every dollar spent for a Medicaid abortion, more than $4 would be saved in the next two years.

The savings varied widely from State-to-State. Massachusetts had the greatest savings (about 9:1), and Hawaii and Pennsylvania the least (about 2:1, reflecting the higher abortion costs in those States). Importantly, in every State, funding Medicaid abortions saved tax dollars.

These estimates are conservative. While abortion incurs a modest one-time charge, unintended births continue to cost society—often for decades. This is especially true for women giving birth in their teens. As of 1992, every birth to a teenage mother on public assistance was determined to cost the federal government $54,000 over the next 20 years.[29] These costs include Aid to Families with Dependent Children, food stamps, and Medicaid health benefits.

Prohibiting second-trimester abortions for fetal abnormalities diagnosed during pregnancy would have similar negative financial consequences.[31] Birth defects are the leading cause of infant death and the second leading cause of death among children aged one to four years. Extrapolating from an eight-year database of prenatal diagnosis and second-trimester abortion, researchers examined the lifetime costs for selective birth defects. If a legislative ban prevented such abortions, the lifetime costs for an average year of abnormal fetuses diagnosed at their Detroit hospital was $8 million. Such a ban would have an annual net cost to the State of Michigan of $74 million, and to the United States, $2 billion.

The aftermath of the Hyde Amendment

Several studies reported on the public-health impact of restricting abortion funding. North Carolina's on-again, off-again State abortion funding enabled investigators to measure the impact of restricted funding on pregnancy outcomes.[32] About one-third of pregnancies among poor women that would have ended in publicly funded abortions continued to birth. African-American women aged 18 to 29 years, many of whom lived in desperate poverty, were most affected. Another report from North Carolina, Colorado, and Pennsylvania found an inverse relationship between public funding of abortion and live-birth rates.[33] Before the funding restriction, each State had a declining birth rate and a rising abortion rate. When the States imposed restrictions in 1985, the trends reversed.

The irony becomes even more remarkable in light of the U.S. Department of Agriculture's estimate that rearing a child to age 18 years costs about $240,000 for a middle-class family.[34]

The natural experiment provided by the Hyde Amendment and the patchwork-quilt of State funding thereafter enables assessment of the health (and thus economic) impact of abortion funding.[35] During the interval of 1982-88, States without public funding had 136,500 more births to teenagers (associated with worse outcomes for mother and infant), 18,500 more low-birth weight babies (stunted infants have a range of medical problems), and 20,800 more premature births. Prematurity can lead to weeks of expensive nursery care, respiratory disease, and often life-long disability, including lung and brain problems, such as cerebral palsy. In addition, 201,900 more births occurred after late or no prenatal care, a marker for poorer outcomes.

Hurdles in accessing Medicaid funding

Even in States paying for most or all abortions with State funds, women must often run an administrative gauntlet to get payment for justified claims.[36-39] In Massachusetts, the current administrative structure does not provide timely abortion coverage.[38] Young women in particular find navigating the administrative process difficult. State Medicaid personnel may provide inconsistent information and generally discourage women from seeking justified coverage.[36] In contrast, in other States nearly all submitted claims are funded, due to electronic billing procedures and responsive Medicaid staff.[37]

> *"It is rather remarkable that the necessity of paying a couple-of-hundred-dollar fee for an abortion is sufficient to persuade (or compel) some women to incur the much larger financial and personal costs of bearing an unwanted child."*
>
> Cook[32]

The vaccination example

How does Medicaid funding of abortion for poor women compare with other contemporary public-health programs? Vaccination (immunization against diseases such as polio, tetanus, and measles) is the most revered, effective, and widely used public-health program today. Since experience with vaccination is ubiquitous, it offers a useful economic benchmark against which Medicaid abortion can be compared.

Several recent studies have examined the return on investment of vaccination; all saved society money (Figure 9-3). The strategy of annually immunizing healthy, working adults against influenza is on average cost saving, although the savings are modest.[40] Using the societal perspective for direct and indirect costs, a study estimated that vaccination saved a mean of $14 per person vaccinated. For every dollar invested in influenza vaccination for working adults, $1.82 was saved as a result.

Routine vaccination of children against chickenpox (varicella) pays longer-range and thus bigger dividends.[41] From the societal perspective, every dollar spent saves more than $5 in medical and work-loss costs. From the health-care payer perspective (considering medical costs only), the savings were smaller, about $2 per case of chickenpox averted.

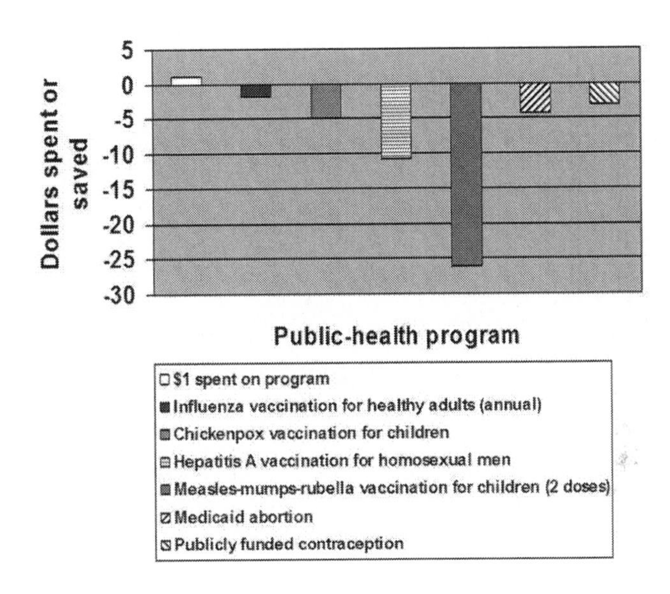

Figure 9-3
Cost savings from selected vaccination programs, United States
Savings per dollar invested range from about $2 to $26;
by comparison, Medicaid funding of abortion is intermediate
(low estimate $4.30)
Sources: Forrest[18], Torres[25], Nichol[40], Lieu[41], Rosenthal[42], Zhou[43]

The return on investment with immunization against hepatitis A varied greatly according to the population in question.[42] In homosexual men, a group at increased risk of this liver infection, vaccination reaped large savings. Each dollar spent on the program saved nearly $11: $2.31 in treatment costs, $5.46 in reduced absenteeism, and $2.96 in losses related to early death.

Immunization of children with two doses of measles-mumps-rubella (German measles) vaccine saves money.[43] According to the CDC, the societal costs of administering the vaccine to all U.S. infants in 2001 were $0.30 billion. The subsequent cost savings were $7.88 billion, or $26 for every dollar spent on the program. This high yield reflects, in part, the appeal of preventing three different diseases with a single combined vaccine.

The Medicaid abortion example

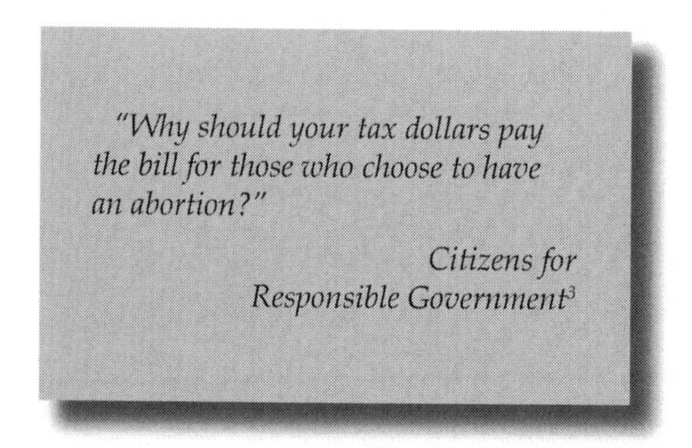

How does Medicaid abortion funding compare to these popular programs? Vaccination saved from $2 to $26 per dollar invested in these four examples above. The lower estimate for the savings from Medicaid abortion ($4.30 per dollar invested)[25] falls in the middle of the vaccination range (Figure 9-3). Similarly, the lower estimate of savings from contraception is similar ($3).[18] Medicaid abortion approximates the yield on investment of chickenpox vaccination for children.[41]

Prevention pays

Primary and secondary prevention of unwanted births saves taxpayers money, as do immunization programs. Ostensibly designed to save taxpayers money,[3] State and federal restrictions of abortion funding have a paradoxical, clearly unintended, effect: they cost taxpayers more. Regardless of one's views on abortion, the arithmetic is incontrovertible: public funding of abortion is good fiscal policy.

Compassion and social justice aside, the answer is pragmatic: paying now is far better than paying later. For unwanted pregnancy, an ounce of prevention is worth a *ton* of cure.

Table 9-1
Medicaid funding of abortion by State, 2013

Pays for all or most medically necessary abortions (N=17)	Pays for life-threatening pregnancy, rape, and incest (N= 32 plus District of Columbia)
Alaska	Alabama
Arizona	Arkansas
California	Colorado
Connecticut	Delaware
Hawaii	District of Columbia
Illinois	Florida
Maryland	Georgia
Massachusetts	Idaho
Minnesota	Indiana*
Montana	Iowa*
New Jersey	Kansas
New Mexico	Kentucky
New York	Louisiana
Oregon	Maine
Vermont	Michigan
Washington	Mississippi*
West Virginia	Missouri
	Nebraska
	Nevada
	New Hampshire
	North Carolina
	North Dakota
	Ohio
	Oklahoma
	Pennsylvania
	Rhode Island
	South Carolina
	Tennessee
	Texas
	Utah*
	Virginia*
	Wisconsin*
	Wyoming

*Other exceptions exist, such as fetal impairment or physical health of the woman

Source: Guttmacher Institute[26]

Take-home messages

- *Denying Medicaid coverage of abortion is "penny wise and pound foolish"*
- *Immediate cost savings to taxpayers are offset by much larger economic burdens over several decades*
- *As a public health prevention program, abortion rivals vaccination programs in economic benefit to society*
- *The Hyde Amendment has been a costly social experiment for taxpayers*

Chapter 10
Stronger families, healthier babies

> "...Roe v. Wade *has had a destructive effect upon family order and relationships*....Roe *has emasculated the man by preventing the exercise of his authority."*
>
> **Patriarch** *Magazine*[1]

Critics allege that abortion has shredded the fabric of American society. Their attacks extend to such sinister groups as the Girl Scouts[2] and doll manufacturers.[3] In reality, the legalization of abortion in the U.S. benefited marriage, family formation, and newborn babies. Legal abortion provides an alternative to mandatory motherhood (with the potential for adoption) or unsafe abortion – the only options in the "bad old days." The public health benefits of abortion extend beyond the pregnant woman to her family; this chapter summarizes some of these salutary effects.

"Shotgun marriages"

Until the 1960s, premarital pregnancy often led to illegal abortion (discussed in Chapter 1) or to precipitous, often ill-advised, marriage to legitimize the impending birth (Figure 10-1).[4] The term "shotgun" reflected the threat of violence from the pregnant woman's father should the couple not marry. These marriages, often created under duress, were at high risk for failure through early divorce.[5,6] The "shotgun marriage" rate can be calculated as the proportion of births conceived out-of-wedlock with subsequent marriage during that pregnancy.[6] Several studies have addressed the temporal relationship between access to abortion and marriage rates; they indicate that abortion provided an important alternative to forced marriage.

An early study found an inverse relationship between access to abortion and crude marriage rates.[7] Until 1969, marriage rates were similar among the States. However, after abortion became increasingly available, States with larger numbers of abortions experienced declines in overall marriage rates, unlike States with fewer abortions.

A follow-up study lent further support to the hypothesis. In the latter report, the researchers examined marriage rates by age group, since they felt that premarital pregnancy would be a stronger predictor of "shotgun marriage" among younger women as compared with older women. A significant inverse relationship was evident between abortion provision and marriage rates among younger women, but not among those over 30 years of age.[8] That the legalization of abortion preceded the changes in marriage rates strengthened the notion that the two events were causally related and not due to chance or other concurrent trend, such as better use of family planning methods. They repeated the analysis using several different measures of abortion frequency: numbers of abortions per 1,000 single women, per 1,000 women aged 15-44 years, and per 1,000 live births. Each iteration of the analysis confirmed the association. Moreover, when restricted to just first marriages, the same relationship held. The findings were robust.

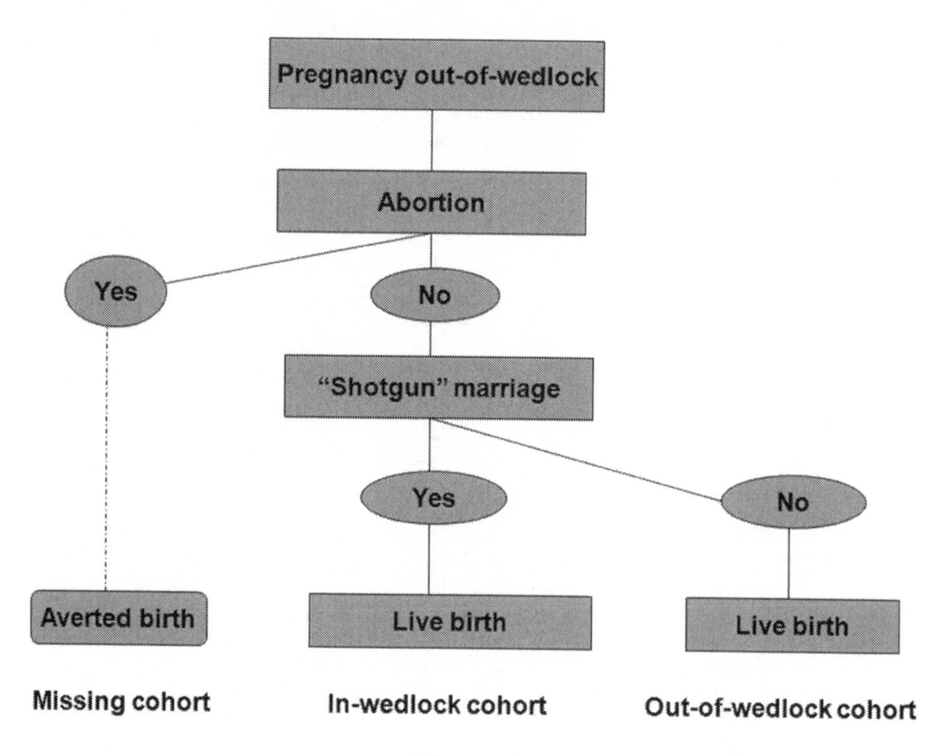

Figure 10-1
Several options for unmarried pregnant women
Source: Lundberg[15]

More recent analyses have confirmed the dampening effect of abortion on "shotgun marriages." The erosion of the custom of "shotgun marriages" was related to better contraception, improved access to legal abortion, and more acceptance of out-of-wedlock childbearing. Economists concluded that trying to "turn back the clock" on access to contraception and abortion would be "both undesirable and counterproductive."[6] These economists predicted that such a policy shift would lead to greater poverty.

Born out-of-wedlock

> *"...denial of access would probably increase the number of children born out of wedlock and reared in impoverished single-parent families."*
>
> *Akerlof*[6]

Children born out-of-wedlock face special challenges. Numerous studies have documented that such children are disadvantaged—both medically and psychologically—as compared to children born to married couples. Birth out-of-wedlock has long been linked with both low birth weight and with infant death.[9] The adverse effect of unmarried status appears greatest for white women 20 years and older. In the U.K., where health care is provided by the government and barriers to pediatric care should not be a factor, infant death rates for children born out-of-wedlock remain higher than for other children.[10]

Traditionally, children born out-of-wedlock have been impoverished both financially and socially. This picture may be changing, however, with an increasing proportion of out-of-wedlock births being planned and wanted. About 80% of all children in the U.S. receiving public assistance live in families headed by mothers who have never married.[11] Even when differences in housing, income, use of social services, and pediatric care are accounted for, the association between birth out-of-wedlock and children's deprivation persists. This relates to the high frequency of single parenthood and the large proportion of young, single, poor women giving birth for the first time.[12] Growing up in a single-parent household has been linked with decreased verbal capacities among children and an increased risk for their abuse and neglect. In one study of boys in trouble with the authorities, those born out-of-wedlock and reared in single-parent homes had the largest discrepancies between verbal and performance IQ (intelligence quotient) scores and the most frequent abuse and neglect.[13]

Legalization of abortion was temporally linked with a dramatic drop in out-of-wedlock births. Rates of out-of-wedlock births in California had been climbing inexorably until 1971. With the advent of wide availability of legal abortion in the State, rates of out-of-wedlock births plummeted.[4] This reflected, in part, the much greater use of abortion by unwed as opposed to married women, as noted in Chapter 7. Researchers estimated that had the 1966-1970 trends in births out-of-wedlock continued in 1972 and had abortion not been accessible, an additional 5,383 children would have been born to unmarried teenagers in California. Abortion averted an estimated 4,243 such births among older unmarried women.

The same phenomenon occurred simultaneously on the East Coast. In New York City, the number of out-of-wedlock births had been climbing steadily since records were first kept in the 1950s. In 1970, 32,000 such births were recorded in the city; the number promptly fell to 28,000 in 1973.[14] The authors noted that reducing dependence upon welfare for these women "has provided an important measure of dignity and hope to many individual women."[14] Many of the pregnant women were teenagers. In 1969, the fertility rate was 42 births per 1,000 women aged 15 to 19 years in New York City. By 1973, the rate had plummeted to 31. The number of births to teens dropped by one-fifth and the rate by one-fourth in a span of four years.[14]

Abortion legalization had the same salutary effect on out-of-wedlock births nationwide. In the pivotal years of 1970-1971, 15 States had liberalized abortion laws, which facilitated access to care. While all States had declining rates of out-of-wedlock births, the decline in States with easy access to abortion was six times higher than that in States without easy access (12% vs. 2%). Abortion lowered fertility in general— but out-of-wedlock fertility in particular— during the early years after legalization. Had legal abortion not been available, an additional 39,000 children would have been born out-of-wedlock, about one-tenth of all delivered out-of-wedlock in that year. More than three decades ago, demographers noted that if liberal abortion laws were overturned, "not only would there be an upturn in illegal abortions and pregnancy-related marriages, but also a marked rise in illegitimacy, particularly among women who do not have the means to obtain an illegal abortion."[5]

> *"...legal abortion was of pivotal importance in the nationwide declines in illegitimacy between 1970 and 1971."*
>
> *Sklar*[5]

Women as heads of households

When young, unmarried women become pregnant, policies and access to abortion services influence their decisions about the pregnancy. As might be anticipated, liberal abortion policies in States are associated with a higher abortion rate among pregnant unwed young women, and vice versa.[15] Policies hindering access to abortion have been linked with several worrisome trends. The nationwide decline in access to abortion services in the 1980s was associated with a small but significant rise in the proportion of women heading families (about 2% increase). After the Hyde Amendment was enacted, the cut-off in Medicaid funding for abortion accounted for about half of the increase in African-American women heading households. In contrast, new State laws requiring parental notification or consent for minors was linked with a modest increase in white women heading households.[16]

The *Roe v. Wade* decision led to nationwide declines in fertility, especially among the women at highest medical and social risk. The prompt decline in fertility (about 4%) in States that liberalized laws early provides indirect evidence of a causal association.[17] When corrected for the influence of interstate travel to obtain abortions, the estimated decline in fertility rates becomes 11%. Of importance, the youngest and oldest women, women of minority races, and unmarried women had the greatest declines, suggesting greater use by those most threatened by an unintended pregnancy. Access to legal abortion disproportionately benefits the most vulnerable women and their families.

> *"A cautious interpretation of our results is that cutbacks in abortion access have instead contributed modestly to the upswing in female headship [of households]...."*
>
> *Lichter*[16]

Healthier babies

Legalization of abortion dramatically improved children's health as well. While many clinicians and laypersons are aware of abortion's health benefits for women, fewer are aware of the benefits to babies born wanted. By all the customary indices of infant health (Figure 10-2), abortion has been linked with enduring health benefits.

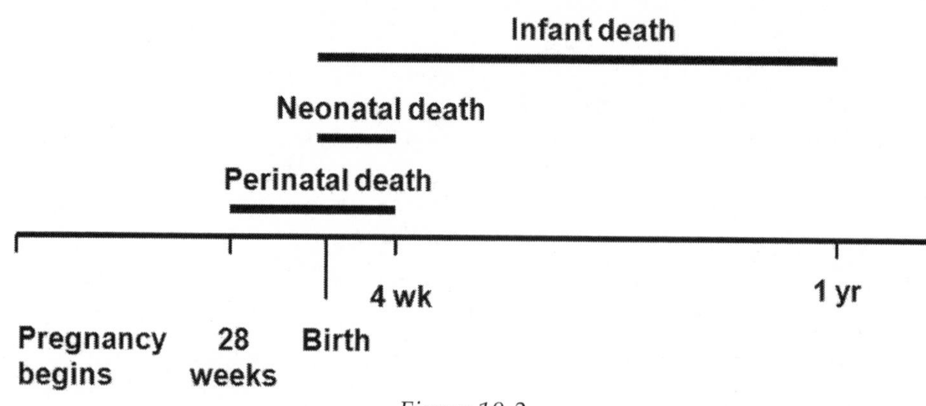

Figure 10-2
Schematic of common measures of infant health
Source: Barfield[18]

Several indices of health are used to measure the health status of infants (Figure 10-2).[18] Neonatal ("newly born") deaths are those that occur within the first 28 days of life; these are sometimes further divided into "early" (first 7 days) and "late" (days 8-28) neonatal deaths. The neonatal mortality rate is the number of deaths within the first 28 days of life per 1,000 live births. The term "perinatal" ("around birth") includes both neonatal deaths and fetal deaths during late pregnancy. The perinatal mortality rate is defined as the number of stillbirths plus neonatal deaths per 1,000 total births. Infant deaths include all deaths of live-born infants from birth through 12 months of life. Some further divide infant deaths into the neonatal and post-neonatal (days 29 through 365). Hence, the infant mortality rate is the number of infant deaths per 1,000 live births.

Neonatal deaths dominate infant mortality. Deaths in the first four weeks of life are about three times more frequent than are deaths in the next 48 weeks. As might be anticipated, the causes of deaths at different ages differ as well. Neonatal deaths are commonly due to birth defects and complications of prematurity. Later deaths are often due to infections and accidents.[19]

An early report from Oregon, which liberalized abortion law in 1969, associated legal abortion with declines in rates of prematurity (born weighing less than 2500 grams, or about 5.5 pounds), "spontaneous" fetal deaths, and infant mortality.[20] Another early report, using States as the unit of study, found no association between legalization of abortion and trends in fetal and infant deaths in the U.S. in the early 1970s.[21]

Experience in New York City was similar. In 1969, the year before legalization of abortion in New York State, the infant death rate in New York City was 24 deaths per 1,000 live births. By 1973, it had fallen to 20. The decline in numbers of deaths and in the rate of infant deaths was unprecedented. While not all this decline was due to abortion, that fewer babies were born into adverse conditions played an important role.[14]

Three other reports from demographers and health economists document a powerful influence of legal abortion on infant deaths. Beginning around 1963, the infant mortality rate in the U.S. began to decline sharply. In a county-by-county analysis for the U.S., the advent of legal abortion emerged as the single most powerful factor in reducing neonatal deaths among both white and African-American babies from 1964 to 1977. The expansion in abortion services dwarfed the influence of other public policies affecting health, but, interestingly, it also overshadowed the impact of the mother's education and poverty.[19] The authors predicted in 1981 that a severe restriction in abortion access (to 1969 levels) would translate into a 19% increase in the neonatal mortality rate. The use of counties as the unit of study, rather than larger heterogeneous States, probably accounted for the different findings than a prior negative study.[21]

> *"According to the estimates from this study, legislative attempts to ban abortion would have a negative impact on birth outcomes."*
>
> *Joyce*[23]

A later race-specific analysis by the same investigators corroborated the health benefits of legalization of abortion. When the authors disentangled the effects of abortion availability, newborn intensive care, women's educational attainment, Medicaid funding, poverty, organized family planning programs, maternal nutrition programs, and Bureau of Community Health Services projects, the answer was the same. Abortion was the single most important determinant of lowered neonatal mortality rates among African-Americans.[22] The drop in neonatal deaths attributable to abortion was about one death per 1,000 live births, or 9% of the observed decline. Among white populations, the effect of abortion was muted. Abortion availability ranked as the sixth leading contributor to the lowering of neonatal deaths. Among whites, schooling was the dominant influence on neonatal mortality rates.

A third sophisticated demographic analysis supported the infant-health benefits. Using counties as the unit of measurement, the author compared the impact of legal abortion with that of prenatal and newborn care, smoking, and organized family planning use. The study focused on three key outcomes: neonatal death, low birth weight, and preterm birth. Among white women, the impact of abortion was beneficial and statistically significant (unlikely to be due to chance) for each newborn outcome. Findings for African-American babies were similar, although not as strong.[23]

Several mechanisms may have accounted for the observed benefits, and these may vary by race. First, abortion averts an unwanted birth, thus altering the distribution of births among women at higher risk of poor infant outcomes. Second, those pregnancies continued to delivery may have been more wanted and thus have received ancillary benefits that accrue with being desired. As the report noted, "....especially among women of low socioeconomic status, abortion will remain an important option for many pregnant women."[23]

A fourth study from New York City confirmed the inverse relationship between abortion use and low birth weight. Using monthly data beginning in 1972, the investigators used a sophisticated technique (vector autoregression) to explore the dynamic relationship between abortion and low birth weight. Among African-American women, they found that as the proportion of pregnancies aborted declined, the proportion of live births with low birth weight rose. This was not seen among white women. The inverse relationship observed between abortion and low birth weight among African-Americans confirmed this finding in other studies with different research designs. Simulations based on the model predicted that any unanticipated reduction in abortions among African-Americans in New York City would have "a substantial impact on the rate of low birth weight among blacks."[24]

"Wantedness"

Being wanted may carry important—yet poorly characterized—benefits for babies.[23] Several groups of investigators have examined the effect of wantedness on pregnancy outcomes and infant well-being. These rigorous studies confirm what one would expect.

Unwanted pregnancies tend to deliver prematurely. A Belgian study of preterm labor (spontaneous onset of labor before the fetus is due) found that a poor emotional investment in the pregnancy strongly predicted preterm birth.[25] Indeed, women who scored highest on the scale of poor investment in the pregnancy had a risk of preterm labor seven times higher than that of women who were emotionally invested.

A more recent study in Baltimore corroborated the adverse effects of an unplanned pregnancy on birth outcomes. Poor African-American women completed a questionnaire regarding the "intendedness" of their current pregnancies. These responses were divided into intended (wanted either now or sooner) or unintended (mistimed, unwanted, or not sure). In a sophisticated analysis that controlled for the potential influence of many other factors, unintended pregnancy emerged as a powerful risk factor for preterm birth. Women whose pregnancy was unintended were 80% more likely to deliver early than were women who had planned the pregnancy.[26]

The planning status of a pregnancy is associated with other important infant outcomes as well. Researchers at the Guttmacher Institute analyzed large national databases.[27] Overall, unintended pregnancies were associated with worse outcomes. Pregnancies that were unwanted or mistimed had higher proportions of infants who were premature, low birth weight, or small for gestational age (stunted growth). These infants were also less likely to be breast fed than were wanted children. When potentially distorting factors were accounted for, the effect of an unintended pregnancy was no longer statistically significant. However, a simple query about whether the pregnancy was intended can identify women at increased risk of poor obstetrical outcomes.[27]

Child abuse and neglect

More than 3 million cases of child maltreatment are reported to State authorities each year, and about three children die from abuse or neglect every day in the U.S.[28] Unplanned births are associated with child abuse, and large family size is related to both abuse and neglect.[29] A child in a family with two unplanned births is about three times more likely to be abused than a child in a family with no unplanned births. In a family with three unplanned births, the risk increases five-fold.

Abortion availability is inversely related to reported child maltreatment in the U.S.[28] States with Medicaid funding restrictions (and thus less access for poor women) have significantly higher rates of child maltreatment than do States without restrictions.[28] Similarly, availability of abortion is inversely related to the proportion of children receiving social services in a given State.[30] However, no relation to child deaths or murders, the most extreme forms of maltreatment, is evident.[30]

The "missing cohort" of fetuses

One study attempted to characterize the "missing cohort" (Figure 10-1) of fetuses terminated through legal abortion. What would their lives have been like had their pregnancies continued? Using the natural experiment provided by the legalization of abortion in the early 1970s, the researchers compared the living standards of children born versus the standards that would have prevailed for the aborted fetuses. The study examined all non-institutional children born between 1965 and 1979 and identified in the 1980 U.S. census, about 2.4 million children. The principal outcomes of interest were living in poverty, living in a single-parent home, and receiving welfare.

The typical living conditions of children born in the U.S. soon after legalization of abortion improved dramatically, relative to children born in preceding years. The same improvement was observed when the comparison entailed States without early legalization of abortion.[31] Had the fetuses not been aborted, the "missing cohort" would have endured substantial hardships. They would have been 60% more likely to live in single-parent households and 50% more likely to live below the poverty level. Moreover, this "missing cohort" would have been 45% more likely to live in a household receiving public assistance and 40% more likely to die in the first year of life.[31] The "missing cohort" would also have suffered higher rates of low birth weight, although this difference was not statistically significant.

The investigators concluded that induced abortion plays a screening role analogous to that of spontaneous abortion (Chapter 4). The fetuses aborted through induced abortion would have grown up in more disadvantaged circumstances than their peers. This indicated a powerful positive selection process during pregnancy: those pregnancies that continue to delivery have brighter prospects for not only surviving, but also thriving.[31] The selection

> *"In other words, this evidence strongly suggests that abortion is used by women to avoid bearing children who would grow up in adverse circumstances."*
>
> *Gruber*[31]

process of induced abortion is associated with benefits lasting into adulthood; these include an increased likelihood of college graduation, lower risk of being a single parent, and lower use of welfare.[32]

Take-Home Messages

- *The availability of abortion benefits children and their families*
- *"Shotgun marriages" and out-of-wedlock births declined after* Roe v. Wade
- *Infant deaths dropped quickly after the legalization of abortion*
- *Rates of child abuse are inversely related to abortion availability*

Section IV

Hot-button issues: the collision of politics and science

Barbara

Barbara, the wife of a surgical resident at a large university hospital, was thrilled to discover that she was pregnant. She and Bill eagerly anticipated starting their family. What began as routine prenatal care soon took a grim turn at 16 weeks. The pregnancy had almost no amniotic fluid surrounding the fetus, and the fetal kidneys could not be seen on fetal ultrasound examination. Without amniotic fluid, the alveoli (air sacs) in the lungs do not develop, and survival after birth is not possible. Fetuses that grow in the absence of amniotic fluid often have orthopedic defects as well from compression in the uterus. When born, these babies suffocate in the first few minutes of life. She requested an abortion at her husband's hospital, but she was told that the hospital board of directors opposed abortion. She was referred to our hospital, where she had genetics consultation, grief counseling, and a dilation and evacuation abortion. She told us that she felt betrayed by the hospital where her husband was training.

She was furious about the State-required recitation read to her over the telephone, especially section (d.) about options. She felt she had no other option. The pregnancy could end now, or she could continue the pregnancy, deliver the child, and watch it suffocate in her arms. Pursuant to the Woman's Right to Know Act of 2011,[1] women in North Carolina, regardless of the medical circumstances, must be informed of this information at least 24 hr in advance of the abortion:

a. That medical assistance benefits may be available for prenatal care, childbirth, and neonatal care.

b. That public assistance programs under Chapter 108A of the General Statutes may or may not be available as benefits under federal and State assistance programs.

c. That the father is liable to assist in the support of the child, even if the father has offered to pay for the abortion.

d. That the woman has other alternatives to abortion, including keeping the baby or placing the baby for adoption.

e. That the woman has the right to review the printed materials described in G.S. 90-21.83, that these materials are available on a State-sponsored Web site, as well as the address of the State-sponsored Web site. The physician or a qualified professional shall orally inform the woman that the materials have been provided by the Department and that they describe the unborn child and list agencies that offer alternatives to abortion. If the woman chooses to view the materials other than on the Web site, they shall either be given to her at least 24 hours before the abortion or be mailed to her at least 72 hours before the abortion by certified mail, restricted delivery to addressee.

f. That the woman is free to withhold or withdraw her consent to the abortion at any time before or during the abortion without affecting her right to future care or treatment and without the loss of any State or federally funded benefits to which she might otherwise be entitled.

1. General Assembly of North Carolina. Session Law 2011-405 House Bill 854. http://www.ncleg.net/Sessions/2011/Bills/House/PDF/H854v6.pdf, accessed December 26, 2013.

Chapter 11
Missing criminals?

> *"Statistics can be used to support anything, especially statisticians."*
>
> *Anonymous*

Few could have anticipated the abortion firestorm ignited by a couple of economists studying criminology. Economists from Stanford University and the University of Chicago examined the potential association between the legalization of abortion in the U.S and crime in the aftermath of that Supreme Court ruling.[1] They concluded that abortion led to important declines in crime two decades later. Donohue and Levitt were not the first to suggest this; a Rockefeller Commission on Population and the American Future had advanced the theory two decades earlier.[2] Donohue and Levitt postulated two explanations for this striking and unexpected phenomenon: fewer unwanted births overall and better-timed births. Both, they reasoned, would translate into fewer individuals bent on crime.

> *"... one need not oppose abortion on moral or religious grounds to feel shaken by the notion of a private sadness being converted into a public good."*
>
> *Levitt*[5]

In part because of its clear eugenic ("social cleansing") implications,[3] skeptics on both sides of the abortion debate quickly attacked. According to the New York Times, "it provoked angry op-ed columns, tirades on radio talk shows and expression of indignation by groups on both sides of the abortion divide."[4] Several papers reaching the opposite conclusion have appeared since then. This chapter will point out the fundamental problem with correlation studies of this type, and then summarize recent key studies.

Research credibility

Human research has several basic designs, and these vary widely in validity (the ability to find the truth). A useful hierarchy (Table 11-1) was developed by the U.S. Preventive Services Task Force.[6]

Table 11-1
A hierarchy of clinical research

I: Randomized controlled trials

Researchers assign participants to treatments by chance alone, which avoids bias

II-1: Non-randomized controlled trials

Researchers assign participants to treatments, but not by chance. Examples include assignment by hospital chart number, date of birth, or alternate assignment

II-2: Comparative studies (cohort and case-control studies)

Researchers do not assign treatments. Instead, they observe what happens in usual clinical practice

II-3: Multiple time-series reports (before-after studies)*

Researchers do not assign treatments. Instead, they look for changes before and after an event

III: Descriptive studies

Researchers do not assign treatments. These studies lack comparison groups and thus do not allow inferences about causal effects.

*Studies of abortion and crime
Source: U. S. Preventive Services Task Force[6]

Randomized controlled trials (Level I evidence) stand at the pinnacle of research design. In these studies, investigators assign participants to treatments by random allocation, which ensures that comparison groups are similar at the start. Bias is minimized in such studies, and results tend to be more trustworthy than with other types of research. An example is the Women's Health Initiative,[7] which randomized thousands of menopausal women to receive either hormone therapy or placebo (inactive pill). The trial found that hormone therapy increased, not lowered, the risk of heart disease.

Level II research has comparison groups. However, these comparative studies have a built-in bias. Non-randomized controlled trials would include studies in which participants are assigned to treatment in alternate fashion (e.g., treatment A, treatment B, treatment A, etc.) Cohort studies follow participants forward in time after an exposure (e.g., abortion performed under local versus general anesthesia) to study outcomes. Case-control studies work in reverse: they start with an outcome, such as infection, and look back in time to see what exposure preceded it.

The weakest kind of comparative study (Level II-3) is the time-series, or before-after study. If a profound difference occurs in the population soon after an event, this suggests that that event may have caused it. Prototypes in medicine include death rates from pneumonia before and after the introduction of penicillin in the 1940s. Death rates from tuberculosis before and after the launch of streptomycin (an antibiotic effective against the tubercle bacillus) provide compelling indirect evidence of cause and effect. In abortion epidemiology, prototypes include deaths from abortion nationwide before and after legalization of abortion and maternal mortality in Romania before and after abortion restrictions by Ceausescu (Chapter 19). In these examples, the population effects were prompt and large, lending support to a causal link.

Because they lack concurrent controls, before-after studies are weaker than other types of comparative studies. Many factors influence most medical outcomes, and teasing out the contributions can be difficult—or impossible. For example, newborn outcomes improved in the U.S. between the 1960s and 1970s, during which time electronic fetal monitoring was introduced. Using before-after studies, some obstetricians claimed that electronic fetal monitoring caused the improvements. Decades later, randomized controlled trials showed such monitoring to be ineffective.[8] Other factors, such as better nutrition and general health, child-spacing, and improved economic conditions were likely the explanation.

Research Pitfalls

Ecological fallacy

Investigators sometimes draw incorrect causal inferences from population characteristics, an error termed "ecological fallacy."[9] For example, a strong, statistically significant association exists between the number of telephone poles in a community and the risk of heart disease. Similarly, a strong relationship exists between listening to rap music and acne. Few would suggest, however, that wooden poles damage the heart or that music (no matter how violent or misogynistic) causes acne. Telephone poles are simply a marker for socioeconomic status (related to heart disease), and listening to rap music is common among adolescents (oily skin leads to acne).

Post hoc ergo propter hoc reasoning

Post hoc ergo propter hoc (after the thing, therefore on account of the thing) is another common error in logic. That one event follows another does not imply that they are causally related; the relationship may be only temporal.[9] For example, assume that a woman has corn flakes for breakfast each morning. One day, prompted by a cheery television commercial, she has oatmeal instead. On her commute to work that morning, she is involved in a fender-bender. What could have caused this rare outcome, she wonders? Ah, oatmeal! Oatmeal must be the culprit, since this was the only difference in her usual morning routine.

Donohue and Levitt: the first volley

The drop in crime in the U.S. in the 1990s was precipitous and unexpected. Not since the end of Prohibition had murder rates fallen so much (40%), and violent and property crimes declined nearly as much (30%).[1] Many theories were advanced to account for this phenomenon, none as inflammatory as that of two economists, Donohue and Levitt. They proposed that the decline in crime in the 1990s was a late effect of *Roe v. Wade* two decades earlier. Their theory got a boost with the publication of Levitt and Dubner's popular book, **Freakonomics**.[5]

Two mechanisms were advanced: a smaller cohort of persons at risk of committing crime, a lower likelihood of crime in the cohort, or a combination of the two. They argued that because of legal abortion in the 1970s, fewer men would reach their late teens and early twenties, the peak years for crime. The second explanation was more controversial: women who are teenage mothers, unmarried mothers, and economically disadvantaged simply give birth to children who are more likely to become criminals.[10] These groups were disproportionately the women obtaining abortions after *Roe v. Wade*. Children born in the era of legal abortion were "better born" and thus less likely to resort to crime.

Several lines of evidence support the notion that better families lead to better personal development. For example, an analysis of the Longitudinal Survey of Youth from Ohio State University examined the impact of family structure on delinquent behavior.[10] Having a father present in the home at the child's age of 14 years was related to a significantly lower risk of the child being stopped by the police. The same held true for being charged with or convicted of a crime by age 22 years. Having the father present in the home overwhelmed all other factors associated with delinquency, including family income.

Donohue and Levitt offered several lines of indirect evidence to support their hypothesis. First, the timing was right: individuals ages 18-24 years are statistically most likely to commit crime, and the children born soon after *Roe v. Wade* (January 1973) would have reached that age in the early 1990s. Thus, the lag in the observed effect was consistent with a delayed effect of the legalization of abortion. Second, the five states that liberalized their abortion laws in 1970 had earlier declines in crime than did the other states, which had abortion liberalized by *Roe v. Wade* in 1973. Third, crime rates from 1985 to 1997 were inversely related to abortion rates in the 1970s and early 1980s. Higher abortion rates were associated with lower crime rates in subsequent years (Figure 11-1).

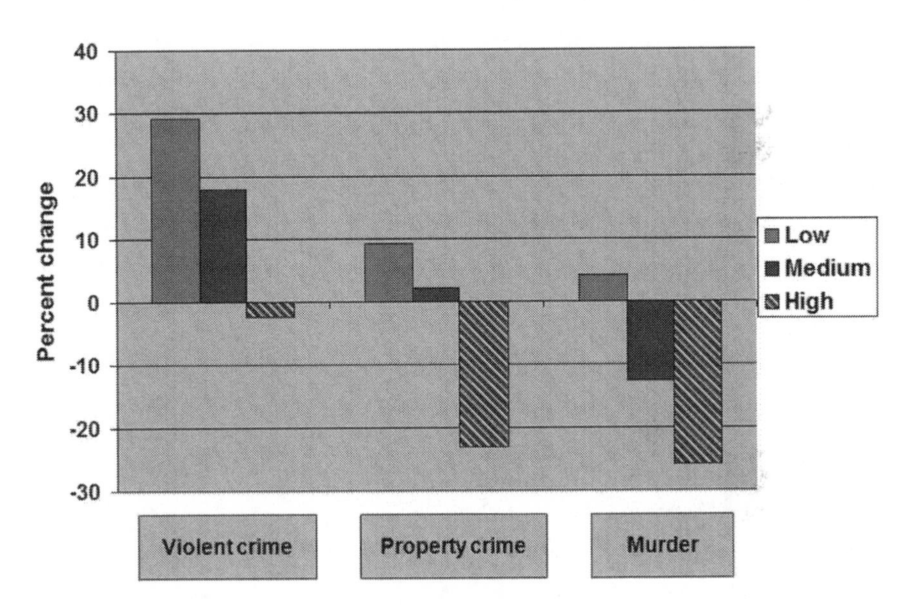

Figure 11-1
Changes in crime rates between 1985 and 1997 as a function of abortion frequency (low, medium, high) between 1973 and 1976, selected states, United States
Source: Donohue[1]

Fewer young men meant fewer criminals. Donohue and Levitt noted that legalization of abortion was associated with a 5% drop in birth rates; because of this decline, groups of people reaching the peak crime years would be commensurately smaller. However, they argued that the social

> *"Consequently, the life chances of children who are born only because their mothers could not have an abortion are considerably dampened related to babies who were wanted at the time of conception."*
>
> *Donohue*[1]

winnowing of abortion leads to a smaller proportion of criminals in the cohort. As noted in Chapter 10, abortion has been associated with better prospects for children, including better family environments. Among the social factors related to crime, African-American race, having a teenage mother, and growing up in a single-parent home are all associated with crime.[11] Similarly, being an unwanted child is also related to criminal activity.[11]

Not so: crack cocaine confounding

Soon after appearance on the Internet of the Donohue and Levitt analysis, another economist attempted to corroborate the association.[12] Joyce criticized the approach of Donohue and Levitt[1] for failing to take into account the rise and fall of the crack cocaine epidemic, which many observers have linked to the proliferation of guns and violent crime among youth. Joyce argued that the mathematical models used by Donohue and Levitt omitted this key variable, which biased the results. Joyce's detailed mathematical analysis led him to a different conclusion, "In sum, legalized abortion has improved the lives of many women by allowing them to avoid an unwanted birth. I found little evidence to suggest, however, that the legalization of abortion had an appreciable effect on the criminality of subsequent cohorts."[12]

The retort: we underestimated!

Donohue and Levitt quickly rebutted the criticisms of Joyce, noting that his analysis actually shored up their thesis.[13] Moreover, Donohue and Levitt reported that, if anything, they underestimated the effect of abortion on later crime. They note that in an earlier publication, Joyce[14] observed that by reducing unwanted births, abortion improves the chances that children born will be healthy and wanted by their mothers. Indeed, attempts to ban abortion would hurt birth outcomes. Regarding the failure to account for the crack cocaine epidemic, Donohue and Levitt noted that Joyce's focus on only the years of the epidemic biased his analysis, because the crack epidemic differentially affected the states that liberalized abortion before *Roe v. Wade*. When the men's criminal experience over their lifetimes, not just their teenage years, was analyzed, a different conclusion resulted. Indeed, they argue that Joyce's entire analysis was doomed because it focused on the six years when the crack epidemic was raging.

Joyce rebuts: more negative evidence

Undaunted by these criticisms, Joyce conducted other analyses of the putative association.[15] He conducted separate analyses by race, since African-American women disproportionately relied on abortion to control their fertility. In addition, young African-American men have higher murder rates that those of whites. Joyce found little difference in the reduction in crime by race. An examination of murder rates before and after legalization, within states, by year of age, also provided no support for an effect of abortion.

Illicit drugs

Other authors have examined the potential relationship between abortion, family formation, and crime, with mixed results. One of the most rigorous analyses studied the association between legalized abortion and substance abuse. Charles and Stephens[16] advanced the two central themes of Donohue and Levitt: a smaller cohort at risk of illegal behavior and a cohort with a lower intrinsic risk.[1] Charles and Stephens had unique data that avoided many of the problems with aggregate statistics used by Donohue and Levitt[1] and by Joyce.[12] In studying substance abuse, these authors could examine the proportions of persons from specific birth cohorts who used illegal substances in a specific year at a known age. Using data from the Monitoring the Future Survey, they examined substance use by adolescents born in the states that liberalized abortion before *Roe v Wade*. These young persons were significantly less likely to use controlled substances, especially illegal narcotics, than were adolescents born at the same time but in other states. After abortion became legalized nationwide, the differential effect disappeared. Charles and Stephens concluded that children born in an environment when abortion was legal were less likely to have used controlled substances than were other 12th grade students. Availability of legal abortion at the time of birth was associated with a 15% reduction in use of illegal drugs.

Murder and Mathematical Machinations

The legalization of abortion was associated with declines in murders of young persons and toddlers, but not infants, according to researchers at UCLA. Sorenson and associates postulated that children born into families with insufficient means might be at peril for homicide.[17] For example, children are at increased risk of murder if the mother is unmarried, a teenager, and not a high school graduate. The investigators studied murder statistics for the U.S. from 1960 to 1998 for children under five years. They found no *sudden* decrease in child murder rates immediately after legalization. However, legalization *was* associated with a steady decline in murders of children aged one to four in later years. They postulated three explanations: less stress on parents, less financial strain, and fewer unwanted children who might have become both a cause and the target of parental frustration.

> *"In short, the balance of evidence suggests there is a decline in growth of homicide victimization that is roughly consistent with abortion legalization."*
>
> *Berk[18]*

A subsequent analysis found a link between legalization of abortion and murders of young men. Using the national mortality statistics from 1970 to 1998, Berk and associates examined the relationship between *Roe v. Wade* and homicides among young men 15 to 24 years of age.[18] Instead of using regression models of earlier authors,[1] they used interrupted time-series analyses and examined national rather than State data. Overall, the authors found a strong association between *Roe v. Wade* and homicide rates, but for males only. They considered and rejected alternative explanations, such as changes in police and incarceration practices or the rise and fall of the crack epidemic. The abrupt reduction in male homicides in the early 1990s, the weak associations observed for females, the tendency for longer lagged effect in older-age cohorts of men, and larger effects for men during the peak crime years all fit with their hypothesis.

More dissent

Lott and Whitley reached the opposite conclusion: legalization of abortion was associated with a paradoxical increase in murder.[19] These investigators, from Yale University and the University of Adelaide (Australia), began by noting flaws in the assumptions of Donohue and Levitt.[1] For example, Lott and Whitley correctly reported that substantial migration of women across State borders occurred, especially in the early 1970s, and this would contaminate State-to-State comparisons. In addition, they used a different information source, the Supplemental Homicide Report, which directly links the age of the murderer with the year of the crime. Their results suggested a small to modest increase in murder rates in response to the legalization of abortion. They concluded that while abortion may benefit many women, it can make women unable to bring themselves to have an abortion worse off, specifically with regard to out-of-wedlock birth.

> *"There are many factors that reduce murder rates, but the legalization of abortion is not one of them."*
>
> *Lott and Whitley*[19]

Evidence from other countries

Researchers in other countries have found beneficial associations between legal abortion and crime. A Finnish cohort study followed 11,017 children from their mothers' pregnancies until the child was 28 year old.[11] Information from the mothers and children was then linked with crime data from the Ministry of Justice. Since crime was infrequent among women, the analysis was restricted to men. If the mother reported during the pregnancy that it was unwanted, the risk of violent crime in the child was doubled. Other risk factors were significantly associated with violent crime, including maternal smoking during the pregnancy, a single-parent family, and the mother being less than 20 years old at the child's birth.

A recent analysis of the Romanian experience identified other adverse social consequences of the restriction of abortion (Chapter 19).[20] Pop-Eleches of Columbia University used a 15% sample of the Romanian 1992 census, focusing on 55,000 children born between January and October, 1967 (the sharp increase in the birth rate occurred in the latter half of that year). Pop-Eleches found some suggestive evidence that cohorts born in a time without access to abortion may have higher crime rates in adulthood. They also had poorer educational attainment and worse labor-market outcomes.

Investigators in Australia provided some support from Down Under for the Donohue and Levitt theory.[21] They noted that homicide rates in Australia fell in the 1990s, but limited State-specific data precluded more detailed analyses for the nation.

Further support came from Canada as well. Sen of the University of Waterloo found that the drop in teenage childbearing in the 1960s and 1970s had the strongest association with the drop in crime.[22] Sen attempted to disentangle the contributions of smaller cohorts of potential criminals from the lower likelihood of criminality in the cohort (the two hypotheses advanced by Donohue and Levitt[1]). The only significant determinants of violent-crime trends were abortion and teenage fertility rates. The effect was limited to teenagers, not the general population. The analysis suggested that lower crime rates in Canada were the result of "better timing of births." Thus, "better births," rather than smaller numbers of births, were judged responsible. In Canada, with more refined abortion data and with no crack cocaine epidemic and proliferation of handguns, Sen corroborated the initial findings of Donohue and Levitt.

> *"...persuasive – although not conclusive – demonstration of the commonsensical point that unwanted children are quite likely not to turn out to be the best citizens."*
>
> *Richard Posner*
> *Chief Judge*
> *Seventh US Circuit Court of Appeals*[21]

Levitt attempted to put the debate into context.[23] He argued that four factors were probably involved in the decline in crimes, while six were not. Among those related to crime reductions are the following: increases in numbers of police officers, more incarcerated citizens, the ebb of the crack epidemic, and the legalization of abortion. The number of police per capita has grown in recent years, and this may have played a role in deterring crime. The number of people behind bars had grown to two million by the year 2000; these individuals were off the streets and unable to commit crime. The crack cocaine epidemic in the mid-1980s spawned a steep rise in murder rates among young African-American men that diminished as the epidemic ebbed.

Those explanations that he discounted include the following: the strong economy during the Clinton Presidency, aging of the baby boomers, better police tactics, gun control legislation, laws concerning carrying concealed weapons, and increased reliance on capital punishment.

An exercise in futility

The debate continued among economists. Donohue and Levitt did further calculations in response and again found evidence supporting their theory.[24] Joyce rebutted with further analyses that again showed no association between abortion and crime rates.[25] In a comprehensive review article, Joyce aptly noted that the controversy had done little to enlighten.

> *"Of the four factors that I believe to account for much of the recent crime decline, only rising numbers of police officers and legalized abortion are likely to be continuing contributors to future crime declines."*
>
> *Levitt*[23]

Joyce[26] provided an example of the model's complexity:

$$(2)\ C_t = \alpha_0 + \alpha_1 UW_{t-a}$$
$$(3)\ UW_{t-a} = \beta_o + \beta_1 AR_{t-a-1} + \beta_2 Pill_{t-a-1} + \beta_3 S_t$$
$$(4)\ AR_{t-a-1} = \lambda_0 + \lambda_1 Pill_{t-a-1} + \lambda_2 S_{t-a-1}$$
$$(5)\ Pill_{t\text{-}a\text{-}1} = \phi_0 + \phi_1 AR_{t-a-1} + \phi_2 S_{t-a-1}$$
$$(6)\ S_{t-a-1} = \kappa_0 + \kappa_1 AR_{t-a-1} + \kappa_2 Pill$$
$$(7)\ C_t = \gamma_0 + \gamma_1 PA_{t-a-1} + PP_{t-a-1}$$
$$(8) AR_{t-a-1} = c_0 + c_1 PA_{t-a-1} + PP_{t-a-1}$$

Judging causation

The fatal flaw with all of these complex mathematical analyses of large databases is that they simply do not allow causal inferences to be made. While economists and statisticians may find numerical *associations,* these exercises do not allow *causal* inferences.

A temporal association does not imply a causal association. Oatmeal does not cause car wrecks. These large correlation analyses lack sufficient detail to identify the potential effect of a single possible factor (abortion) among the many factors

> *"The debate often reduces to dueling regressions, which can be numbing to observers not enmeshed in the issues."*
>
> *Joyce*[26]

related to crime rates. Unlike the physical sciences, which have absolute laws (such as thermodynamics), assessing causal associations is more difficult in the biological and social sciences. For example, strong statistical associations may be completely *bogus* (due to bias), *indirect* associations (due to the presence of a confounding or distorting factor), or *causal*, the real McCoy.

In 1965, Sir A. Bradford Hill proposed a set of criteria for judging whether statistical associations (such as those found in these crime studies) might be bogus, indirect, or causal.[27] When evaluated by these widely-accepted principles,[28-30] these economic analyses fall short. Unlike the prompt decline in mortality from pneumonia after the introduction of penicillin in the U.S., the putative effects here lagged two decades after the event (*Roe v. Wade*), so drawing inferences from before-after studies is tenuous at best. These authors can report associations, but the level of evidence mustered here (level II-3)[6] does not allow inferences about causality.

> *"What the link between abortion and crime does say is this: when the government gives a woman the opportunity to make her own decision about abortion, she generally does a good job of figuring out if she is in a position to raise the baby well. If she decides she can't, she often chooses abortion."*
>
> *Levitt*[5]

Take-home messages

- *Economists have proposed that abortion helped to reduce crime in the U.S.*
- *Others were unable to corroborate these findings*
- *Many factors, such as the cocaine epidemic of the 1980s, may have played a role in crime rates*
- *Correlation does not imply causation*
- *Studies like these are simply incapable of drawing cause-and-effect conclusions*

Chapter 12
Abortion on the Internet: Penile amputation and other dreaded complications

> *"Oh, what a tangled [World Wide] web we weave*
> *When first we practice to deceive"*
> *Walter Scott*[1]

As I counseled the couple before the abortion, my patient's husband asked me if the planned suction curettage procedure would harm his wife's ability to have a baby in the future. They had conceived accidentally while using oral contraceptives and they planned on starting their family in a few years when their financial situation would be better. I answered his question—and then inquired about the source of his concern. I got the usual response: the Internet. As noted by others, "The best thing about the Internet is that you can find anything there; the worst thing about the Internet is that you can find anything there."

Medical information on the Internet

The Internet has emerged as a major source of medical information for Americans. Indeed, in 2002 the Pew Internet Project estimated that 73 million persons living in the U.S. have searched the web for medical information.[2] Most of these looked for information on a specific illness or condition. In a 2003 report, the Project found that half of all American adults surveyed had searched online for health information. Analysis of health topics searched revealed that "reproductive health" and "pregnancy/obstetrics" were among the five most common.[2]

Others have confirmed the wide use of the internet for medical information.[3] A 2006 survey of Americans found that 80% had searched online for medical information; another found that 70% were influenced by the material that they read.[4] Search engines remain the most frequent means of access to medical topics, suggesting that many persons are unaware of comprehensive, high-quality sites, such as that of the Centers for Disease Control and Prevention (www.cdc.gov).

Adolescents are among the heaviest users of the Internet for information on health topics.[5] A Kaiser Family Foundation study found that 55% of teens in grades seven to 12 had used the Internet to find health information of relevance to them or others.[6] An estimated 75% of today's teenagers have searched online for medical information at least once. Among topics searched, "sex" and "fitness and exercise" were the most frequent.

Health advice on the Internet poses a double-edged sword. Unfettered access to information has appeal, but the lack of quality control is a huge deficiency.[3,7] While documented medical harm from misinformation on the web appears to be either limited or underreported,[5,8] the impact of incorrect or misleading information remains a concern to health care providers and patients alike.

Accuracy

Audits of medical information on the Internet have yielded conflicting results. Some sites provide high-quality information on topics ranging from female urinary incontinence[9] to smoking cessation.[10] In contrast, poor sites abound. Incorrect advice or commercial bias is evident on topics as diverse as plantar fasciitis (a cause of foot pain),[11] "alternative" cures for depression,[12] fever in a child,[13,14] the safety of intrauterine devices for contraception,[15] and prevention of malaria for international travelers.[16] Internet information on obsessive-compulsive disorder is of good quality,[17] while that on treating anxiety disorders is fair to poor.[18] In contrast, information about dementia[19] and treating inflammatory bowel disease[20] varied widely in quality. The leitmotif throughout audits of Internet medical information has been the urgent need for quality assurance.

For less than $100, anyone with the desire can dispense health information over the Internet to millions of potential users. While healthcare providers must complete approved training and then pass certifying examinations before being allowed to provide medical advice, no such prerequisites exist on the Internet. For a modest fee, interested persons can dole out health information to the unsuspecting. Consumers often cannot determine the credentials of the web site author, how current is the information, and whether the information has received peer review for quality assurance.[12] Assessments of quality have found that commercial websites have inaccuracies more often than do those of professional groups or organizations, like the American Congress of Obstetricians and Gynecologists.[4]

> *"Only a few web sites provided complete and accurate information for this common and widely discussed condition [fever in a child]. This suggests an urgent need to check public oriented healthcare information on the internet for accuracy, completeness, and consistency."*
>
> *Impicciatore*[13]

Readability

Readability poses yet another challenge to those surfing the Web. Health literacy is the "capacity to obtain, interpret, and understand basic health information and services and the competence to use such information and service to enhance health."[21] Health literacy is the single best predictor of one's health, yet many Internet users in the U.S. are handicapped. According to the National Adult Literacy Surveys, nearly half of the U.S. population is either functionally illiterate or marginally literate, meaning eighth-grade reading skills or less. One in five U.S. citizens reads at or below the fifth-grade level.[21] Much of the material on the Internet is written at a reading level out of reach to millions of users. For example, an assessment of the readability of web sites about warfarin (a common blood thinning medicine) found that seven of 11 web sites had poor readability.[22]

Quick searches

Another danger of Internet browsing is superficiality. Few Internet users venture beyond the first or second page of websites retrieved in a search.[2] Hence, the sequence of hits largely determines what information users will find. Clever Internet marketing, which can of course be learned on the Internet, allows those who post content to drive traffic to their sites very effectively, regardless of the validity of their information.

"....on the World Wide Web there is often no verification of validity – let alone peer review – of the information submitted. Difficulty in judging the validity of the information thus poses a problem for people using the internet."

Impicciatore[13]

Subversive web sites

On a darker note, some sites intentionally mislead the public. For example, an audit of "alternative" medicine sites found some that overtly discouraged patients from using established, mainstream medicines that have been shown to be effective.[12] Other advice was potentially harmful. Women in greatest need may disproportionately turn to the Internet for information. The volume of Internet searches for abortion is inversely related to local abortion rates and directly related to anti-abortion restrictions.[23] This suggests that women living in restrictive areas turn to the Internet to find providers elsewhere.

Although credible information about abortion is available on the Internet,[24] it can be hard to find. Anti-abortion websites are much more likely to feature medically incorrect information than sites that are neutral or supportive of mifepristone abortion.[25] In a formal study, university students evaluated the credibility of various abortion web sites. In terms of accuracy, honesty, promotional nature, and overall quality, "pro-life" sites were ranked the lowest.[26] Another evaluation of the accuracy of health information sites young people visit found that 35% of the abortion sites had at least one error.[27]

Crisis Pregnancy Centers

Crisis pregnancy centers, also known as pregnancy resource centers, are non-profit organizations designed to discourage women from having abortions. Anti-abortion Christian activists run most of these centers, which outnumber abortion clinics in the U.S. by a substantial margin.[28] An important outreach of these centers is disseminating medically incorrect and misleading information about abortion via the Internet. By buying advertisements through Google's *AdWords* program, they can steer searches in their direction.[29]

According to Brian Fisher, the CEO of Media Revolution Ministries, "There are over 6 million Internet searches each and every month in the United States alone for the word 'abortion' and other related keywords.... We intercept them when they're searching for abortion information, and instead we direct them to pro-life pregnancy centers in their area where they're counseled for life."[29]

Because both State and federal money has subsidized crisis pregnancy centers (including funds from "choose life" license plates), Congressman Henry Waxman requested a "mystery shopper" approach to study the advice being provided by telephone. Of the 23 centers contacted, 87% provided false and misleading information about abortion. Released in 2006, the report concluded that, "This tactic may be effective in frightening pregnant teenagers and women and discouraging abortion. But it denies the teenagers and women vital health information, prevents them from making an informed decision, and is not an accepted public health practice."[30]

More recently, a "secret shopper" study of these centers in North Carolina examined both in-person and Internet advice about abortion.[31] Thirty-six web sites were scrutinized for accuracy. Most (86%) had at least one error. Common (but incorrect) claims included infertility, breast cancer (Chapter 13), mental health problems (Chapter 14) and preterm birth (Chapter 15).

Non-medical medical advice

One organization, the Elliot Institute in Springfield, Missouri, sponsors numerous web sites featuring junk science. David C. Reardon, whose undergraduate degree is in electrical engineering, founded this anti-abortion organization.[32] Reardon holds a "Ph.D." from Pacific Western University (Hawaii), a non-accredited diploma mill that required no course work for the diploma. The State of Hawaii sued the school for misrepresentation and other violations, and it was forced to close and pay a civil penalty of half a million dollars.[33]

A self-proclaimed expert on alleged abortion complications, Reardon startled the medical world by concluding that penile amputation with a kitchen knife is a late consequence of abortion.[34-36] He determined that Lorena Bobbitt cut off her husband's penis because of an abortion many years earlier. Perhaps because of its rarity, penile amputation had eluded comprehensive inventories of abortion complications.[37,38] However, in the interest of full disclosure, this unpleasant outcome should likely be added to State-mandated counseling of abortion risks.[39]

The medical advice provided by The Elliot Institute illustrates its scholarship and attention to detail. A web page reportedly authored by Reardon[40] claims that in one (unidentified) hospital study, 12.5% of first trimester abortions required suturing to treat cervical lacerations. (The citation provided for this claim is a book by a former President of National Right-to-Life rather than a published article in the peer-reviewed medical literature.) The web site warns that in another study, the frequency of cervical laceration was an astounding 22%. In reality, the cited article[41] clearly states that 22.5% (about 38) of 167 *complications* were cervical lacerations. These 38 cervical injuries occurred among 3,643 abortions studied, for a frequency of 1 per 100 cases, typical of reports of that era.[37] The Elliot Institute web site exaggerated the true risk by more than 20-fold.

A quick Internet search reveals that most abortion information accessible to the lay public is scripted by political organizations, not health care providers. These exaggerated or false claims have been refuted by impartial medical and public health organizations, such as the American Medical Association[42] and the Centers for Disease Control and Prevention.[43] Needless to say, many of these anti-abortion sites are inconsistent with recommended scientific and ethical standards for health information on the web.[44,45]

Stacking the deck

A Google web search of "abortion and prematurity" on December 29, 2013 revealed how "hijacking" the Internet misleads the lay public (Figure 12-1). Since most web users do not venture deeply,[2] consideration was limited to the 13 "hits" on the first page of results; three ads identified at the bottom were not considered further. (The Google search engine frequently places ads physically above search results on the screen, dramatically increasing the possibility that individuals will mistakenly click on an advertisement rather than an actual search result.)

Figure 12-1

First page of Google search for "abortion and prematurity"

https://www.google.com/#q=abortion+and+prematurity, accessed December 29, 2013
About 5,240,000 results (0.25 seconds)

Scholarly articles for abortion and prematurity

1... of spontaneous and induced abortion on prematurity ... - Papaevangelou - Cited by 43

2 Prevention of habitual abortion and prematurity by ... - Saling - Cited by 30

3 Chlamydia trachomatis and mycoplasmal infections in ... - Harrison - Cited by 261

Search Results

4 Researchers admit link between abortion and premature births ... https://www.lifesitenews.com/news/researchers-admit-link-between-abortion-and-premature-births Jul 18, 2013 - The new study analyzes medical records from women in Scotland between 1980 and 2008.

5 Abortion and premature birth – new Finnish study raises serious https://www.lifesitenews.com/blogs/abortion-and-premature-birth-new-finnish-study-raises-serious-questions Sep 5, 2012 - The link between abortion and premature birth is already well established but largely underplayed or denied by authorities in Britain. However ...

6 Abortion and preterm births studied - Los Angeles Times articles. latimes.com/2013/jul/09/.../la-sci-abortion-preterm-birth-201307... Jul 9, 2013 - In a finding likely to reignite debate over proposed new limits on abortion, British researchers have found that years ago, women who ...

7 New Study Finds "Strong" Link Between Abortion and Premature Birth www.lifenews.com/.../new-study-finds-strong-link-between-abortion-and. Jul 16, 2013 - A new study found a "strong independent relationship" between a history of abortion and the risk of a subsequent preterm birth. For the pro-life ...

8 Modern methods of abortion are not linked with an increased risk of www.sciencedaily.com/releases/2013/07/130710062437.htm□ Jul 10, 2013 - The link between previous termination of pregnancy (abortion) and preterm delivery in a subsequent pregnancy has disappeared over the last ...

9 Changes in Association between Previous Therapeutic Abortion and http://www.plosmedicine.org/article/info%3Adoi%2F10.1371%2Fjournal.pmed.1001481 Jul 9, 2013 - Changes in Association between Previous Therapeutic Abortion and Preterm Birth in Scotland, 1980 to 2008: A Historical Cohort Study.

10 Induced Abortion and Pre-Term Birth - American Association of Pro www.aaplog.org/...of...abortion/induced-abortion-and-pre-term-birth/□ AFRICAN-AMERICAN WOMEN, PRETERM BIRTH, AND ABORTION WHAT IS THE ASSOCIATION? The huge increase in preterm birth in the United States [PDF]

11 Abortion's Impact on Prematurity - North Carolina Family Policy http://www.ncfamily.org/magazine/abortions-impact-prematurity/by M McCaffrey - □Related articles: Family North Carolina. 2 devastating is the financial impact. The estimated annual cost for care attributable to preterm birth in the United States is $26 billion.1.

12 Induced abortion and prematurity in a subsequent pregnancy: a www.ncbi.nlm.nih.gov/pubmed/12521858 by Y Che - □2001 - □Cited by 9 - □Related articles: To evaluate the impact of a first trimester induced abortion on the risks of low birth weight (LBW) and preterm birth in a subsequent pregnancy we conducted a ...

13 122 Studies from 1960s to Present: Relationship Between Abortion med.studentsforlife.org/122-studies-from-1960s-to-present-relationship-... +10 Papaevangelou G, Vrettos AS, Papadatos D, Alexiou C. The Effect of Spontaneous and Induced Abortion on Prematurity and Birthweight. The J Obstetrics ...

Ads related to abortion and prematurity

The first three hits came from Google Scholar. None was relevant, since the only one to examine abortion studied illegal abortion in Greece more than four decades ago.[46]

Number 7, posted on July 16, 2013 misrepresented the conclusions of a study from Scotland: "New study finds 'strong' link between abortion and premature birth." The posting refers to a recently published study from Scotland.[47] This database study[48] examined the potential relationship between abortion and preterm birth from 1980 to 2008. The association progressively decreased from weak (odds ratio of 1.32, a 32% increase) to nil by 2000. Odds ratios less that 2.0 are below the discriminatory limits of such studies, because of inadequate control for confounding and other biases.[49] Weak associations of this size are not credible.[50] In contrast, "strong" relationships in research have odds ratios on the order of 14 to 32.[51]

Number 4 repeated the misrepresentation. This posting just two days later (July 18) from the same author (Rebecca Oas, Ph.D.) concluded that "Researchers admit link between abortion and premature births…" The Scottish study found a weak association that disappeared over time; bias and confounding are more likely explanations than causality.

Number 5 commented on a Finnish study published in 2012.[52] Overall, abortion was not related to preterm birth. This, too, was a database study incapable of controlling for bias and confounding. Preterm birth (<37 weeks) was increased *only* for women with three or more abortions (odds ratio 1.35), again below the discriminatory ability of a database study. As the authors of the Klemetti report noted, "Observational studies like ours, however large and well-controlled, will not prove causality."[52]

Numbers 6 and 8 were a newspaper and science web site, respectively, commenting on the Scottish study.[47]

Number 10 was a listing of anti-abortion documents from the American Association of Pro-Life Obstetricians and Gynecologists, including a reference to hit number 11.

Number 11 was an essay supporting passage of NC Senate Bill 132, which would require schools to teach that abortion increases the risk of prematurity. The document omitted mention of the conclusions of any mainstream medical or public health organizations. All have concluded that no causal association exists. The essay appears on the web site of the North Carolina Family Policy Council, whose mission includes "a battle to retain the Judeo-Christian values that are the foundation of western civilization."[53]

Number 12 was a large cohort study from Shanghai, China, that compared pregnancy outcomes of women with and without a history of abortion. No association was found between abortion and preterm birth. The site was sponsored by the National Library of Medicine of the National Institutes of Health.

Number 13 was a compilation of studies of this question. The web site was that of Med Students for Life, whose mission is "abolish abortion in our lifetime."

In summary, the first page of this Google search cited three published studies, none of which found a strong or statistically significant relationship between abortion and prematurity in contemporary practice. However, the first page of the Google search results yielded six hits suggesting otherwise. Three of these, from www.lifesitenews.com, intentionally misrepresented published results.[47,52] Lay readers easily can be—and undoubtedly are—confused by this obfuscation of the evidence.

The world is flat

To test the hypothesis that Internet disinformation is more acute in the area of abortion than in other common human activities, I searched with Google for "international travel." Scores of travel agents, airlines, government agencies, and public health sites popped up in the first 100 hits. Not surprisingly, the Flat Earth Society[54] and similar skeptical sites were missing from my search results. If most Internet sites for international travel warned that the world is indeed flat, prospective travelers would be understandably confused. Cruisers would be particularly worried about falling off the edge or being eaten by the dragons that inhabit the perimeter. Should high-school students looking for Internet help with their chemistry homework find numerous intentionally misleading alchemy sites,[55] grades would likely suffer.

The risk of complications associated with induced abortion is low (Chapter 5). These risks are substantially lower than with continuing the pregnancy (Chapter 6).[56-58] Few women choose the course of their pregnancy based on these medical risks. Nevertheless, women deserve the truth. To mislead women is not only dishonest but also paternalistic; this implies that women are not capable of making independent decisions based on accurate information. Like my couple choosing an abortion, most people are reassured to learn that the alleged epidemic of abortion complications[34,36,40] exists only in cyberspace. And perhaps Springfield, Illinois.[32]

Lying to advance one's political agenda is inconsistent with the ethical principles of beneficence (doing good for others) and autonomy (being able to make free choices based on accurate information).[59] Internet sites that subvert searches and mislead users are blatantly unethical. For abortion opponents (who claim the "moral high ground") to use these tactics is telling.

Take Home Messages

- *The Internet is a common source of information about abortion*
- *Much Internet information about abortion is posted by abortion opponents with no medical expertise*
- *Crisis pregnancy centers spread disinformation about abortion on the Internet*
- *Misleading consumers is unethical*

Chapter 13
Breast cancer: the jury is in

When a Mississippi physician plans to perform an abortion, he or she must counsel the woman about "the particular medical risks... including, when medically accurate, the risks of...breast cancer..."[1] The implication (Miss. Code Ann. § 41-41-33 (2005)) that abortion increases the risk of breast cancer is, however, wrong.[2] Other states, such as Alaska, Kansas, Texas, and West Virginia have similar language, revealing that State legislatures are intentionally misleading the public.[3,4] The Texas Department of State Health Services' required booklet is more forceful: "The risk [of breast cancer] may be higher if your first pregnancy is aborted...If you have a family history of breast cancer or clinical findings of breast disease, you should seek medical advice from your physician before deciding whether to remain pregnant or have an abortion."[5] This novel cancer prevention strategy has been suggested by others: "...a young woman facing an unwanted or crisis pregnancy can and should be informed of the loss of that protection that would derive from a decision to terminate her pregnancy and delay having a baby."[6] One of the fundamental principles of medical ethics is beneficence: the physician must act in the patient's best interests.[7] Providing frightening, medically incorrect information is unethical,[4] especially for young teenagers.This chapter describes the scope of breast cancer in the U.S., explains the types of studies used to study it, and portrays measures of association that scientists use. A summary of the extensive literature follows in a few simple graphs. Finally, the conclusions of national and international medical and public health organizations are summarized.

Why it matters

That abortion might influence the risk of breast cancer is a reasonable biological question. Doctors and scientists have known for decades that pregnancies and hormones influence a woman's risk of developing this common cancer. For example, giving birth, especially at an early age, lowers the risk. Surgical removal of the ovaries, which make female hormones, also reduces the risk.

Any association between abortion and breast cancer could be of public health importance, because of the frequency of both. In past decades, over a million women per year have had abortions in the U.S., and breast cancer has long been one of the most common female cancers. In 2013, about 230,000 women developed this cancer, and about 40,000 died from it.[8] Assume that abortion doubled a woman's risk of breast cancer in later life: because of the millions of women who have had abortions, this would translate into tens of thousands of extra cases of cancer attributable to earlier abortion.

Research tools

The essence of the scientific method is the use of a control group. The simplest study design with a control group is a cohort study (Figure 13-1).[9] In this type of prospective study, those with and without the exposure of interest (here, abortion) are followed forward in time to see how many develop breast cancer (the outcome). This provides Level II-2 evidence.[10] If a higher percentage of abortion patients develop breast cancer than control women who have not had an abortion, then abortion is associated with breast cancer. Cohort studies enable investigators to confirm the abortion history at the start of the observation period, a key advantage, as will be discussed later.

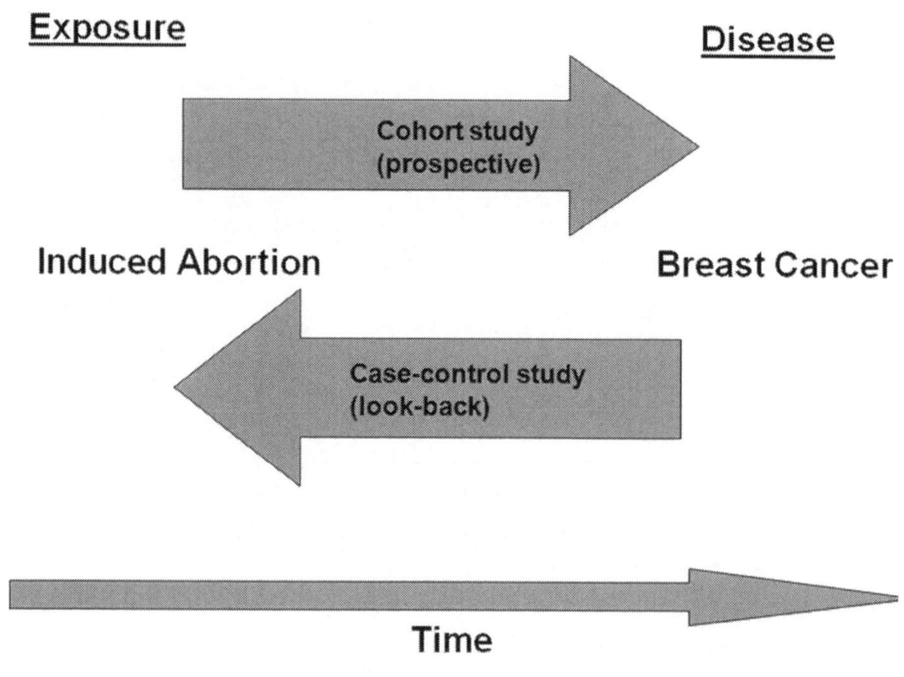

Figure 13-1
Schematic of cohort and case-control studies of abortion and breast cancer
Source: Schulz [11]

Because cancers take a long time to develop, cohort studies are impractical. Large numbers of women need to be followed for many years, which can be a huge logistical and financial challenge. Hence, the more common way that researchers study cancer causes is the case-control study.[11] This is research in reverse: the researchers start at the outcome and then look backwards (Figure 13-1).

A case-control study starts with women who already have developed breast cancer (cases). The tricky part is then finding controls (women without breast cancer) who are representative of women in the community at risk of breast cancer.[12] Once the researchers have a group of cases and controls identified, they then look back in time to determine how many in each group had an abortion (Figure 13-2). If a higher proportion of cases (women with cancer) have this history than controls (women free of this cancer), then an association exists between abortion and breast cancer.

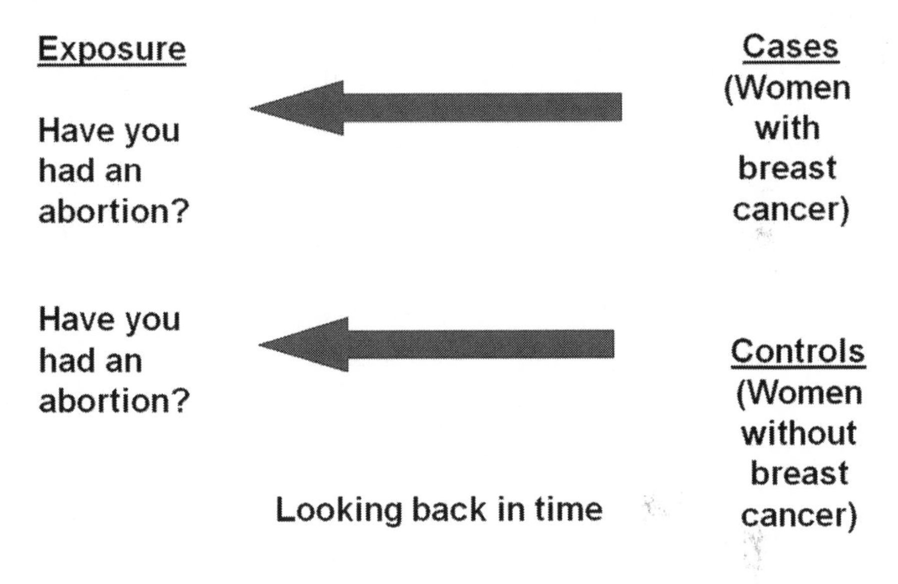

Figure 13-2
Design of a case-control study of abortion and breast cancer
Source: Schulz [11]

Fuzzy memories

Accurate determination of abortion history has been the Achilles heel of case-control studies of this question.[13] Because abortion is a private, often painful, part of a woman's life, and because of the controversy and potential stigma, many women are reluctant to acknowledge this history. Women's abortion histories as reported in the National Survey of Family Growth, a large federal health survey of reproduction in the U.S., are revealing. The percentage of women acknowledging an abortion falls far short of the predicted proportion based on numbers of abortions performed; about half of abortions are not reported to interviewers.[14]

However, this under-reporting of abortion history is not uniform. Women with breast cancer, whose lives are in jeopardy, are more likely to search their memories for any past experience that might have put them at risk. Healthy women have no such motivation. This "recall bias" is rampant in case-control studies of abortion that rely on women to provide abortion histories after the diagnosis of cancer is made.[15,16] If women are queried about abortion history at the start of the observation period or if medical records are used to obtain this information, this type of bias can be minimized.

This differential determination of abortion history for cases and controls can be a powerful and pervasive bias. In research, "bias" means a systematic distortion of the truth (not "prejudice" in common parlance). An example shows how under-reporting of abortion among healthy women leads to a bogus (false) increase in the risk of breast cancer after abortion (Figure 13-3). Assume that 100 women have breast cancer (cases) and 100 comparable controls are free of disease. In reality, 40 women in each group had an abortion earlier in life. Prompted by medical concern, all breast-cancer cases correctly recount their abortion history when interviewed by a researcher. Only 30 controls (less inclined to disclose this information) relate the same history. In reality, equal proportions of cases and controls had an abortion. However, selective recall among the controls (but not cases) yields a spurious association.

Figure 13-3 shows how inaccurate recall by controls produces a bogus association. In the left panel, all 100 cases and 100 controls are correctly classified by prior abortion. The odds ratio (or relative risk) in this case-control study is calculated by multiplying the number of women in cell "a" times those in "d" and dividing that product by "b" times "c." As noted above, the true relative risk here is 1.0, meaning no association. In the right panel, all of the cases recount their abortion history accurately, but ten controls do not. Now the calculation gives a relative risk of 1.6, or a 60% increase in the apparent risk of breast cancer associated with abortion - due to faulty recall by just 10% of the control women.

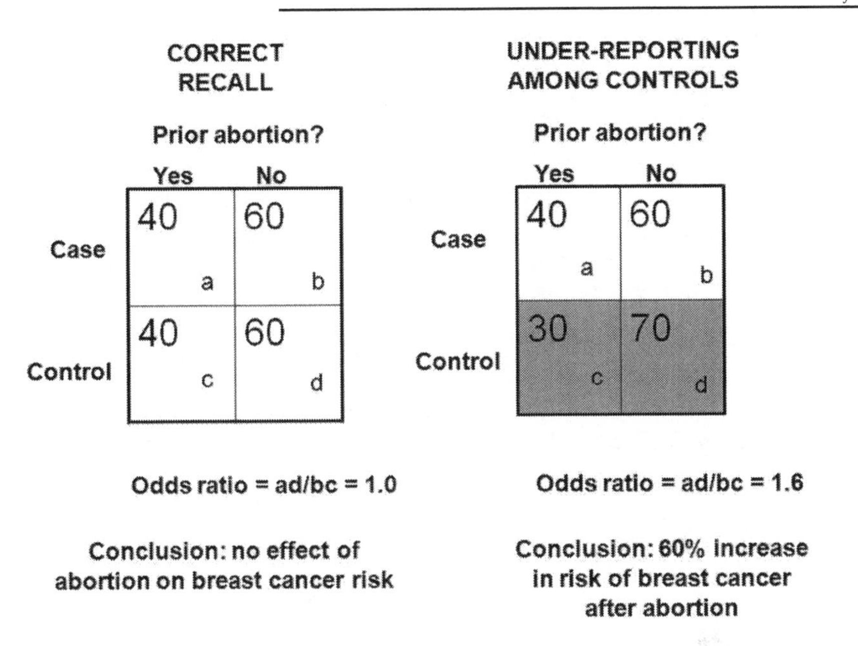

Figure 13-3
Recall bias in a case-control study of abortion and breast cancer
Source: Grimes[15]

Recall bias in action

Several researchers have documented the amount of recall bias in studies of abortion and breast cancer. In Sweden, as in other Scandinavian countries, every citizen is assigned a unique number at birth. This number enables Swedes to be tracked throughout life. Since most of the health care is provided by the national government, linkages are possible for research studies. A team in Sweden assessed the amount of recall bias among controls by performing the same case-control study two different ways. In the traditional way, investigators interviewed cases and controls personally and asked about prior abortions. In the second approach, they used government health care data to provide the history of abortion.[17]

The disparities were striking. According to the national register, 59 healthy controls had abortions provided in Sweden. When the same controls were interviewed in person in the usual way, only 44 of them acknowledged having had an abortion. Thus, 15 of 59 known abortions (25%) were denied by the respondents. The statistical probability of getting a disparity this large due to chance alone is less than 1%.[17] In other words, the odds are better than 99 to 1 that systematic under-reporting of abortions occurred among controls.

In a Dutch case-control study, the authors explored the influence of religion on recall.[18] The risk of breast cancer after abortion in Roman Catholic areas of the country was large (15-fold increase in risk). In contrast, in the more liberal areas, the risk was modest and likely due to chance (30% increase). The authors concluded that the controversial nature of abortion led to selective underreporting of prior abortions by controls (healthy women) in the Catholic areas of the country. The net effect of underreporting is to produce a spurious association between abortion and breast cancer.

Measuring associations

Simple ratios measure associations in epidemiological research. In a cohort study moving forward in time, the measure is the "relative risk."[19] This term expresses the risk of disease in one group relative to another. Assume that 4% of women with a family history of ovarian cancer (the exposure) develop ovarian cancer themselves. In contrast, only 1% of women without this family history (unexposed) develop the cancer. The rate of cancer in those with the family history is four times that of women without the history (4%/1% = 4.0, the relative risk). If the risk of ovarian cancer in vegetarian women (the exposed) were 1% and that in non-vegetarians were also 1%, then the relative risk of ovarian cancer among vegetarians would be 1.0 (1%/1% = 1.0, indicating no increase in risk related to a vegetarian diet). To complete this example, assume that women who ever used birth control pills had a rate of ovarian cancer of 0.5% and those without pills, 1%. Here, the relative risk of ovarian cancer would be 0.5 for those exposed to oral contraceptives (0.5%/1% = 0.5). This implies that the risk in those exposed to birth control pills is reduced by one-half.

> In case-control studies, the measure of association is the odds ratio. It represents the odds of the exposure among cases compared with the odds of exposure among the controls. For diseases that occur in less than 5% of the population (like breast cancer), the odds ratio becomes an excellent proxy for the relative risk. Hence, I will refer to both as "relative risk" for simplicity.

A simple graph facilitates the interpretation of these measures of associations reported in breast cancer studies (Figure 13-4). If the rates of disease are the same in the exposed (e.g., past abortion) and unexposed (no prior abortion), then the relative risk is unity. Thus, relative risks start at 1.0 and go up to infinity and down to zero. Relative risks greater than 1.0 imply an increased risk of disease among those exposed compared to those not exposed (the zone of harm in Figure 13-4). Conversely, relative risks less than 1.0 imply protection (the zone of protection).

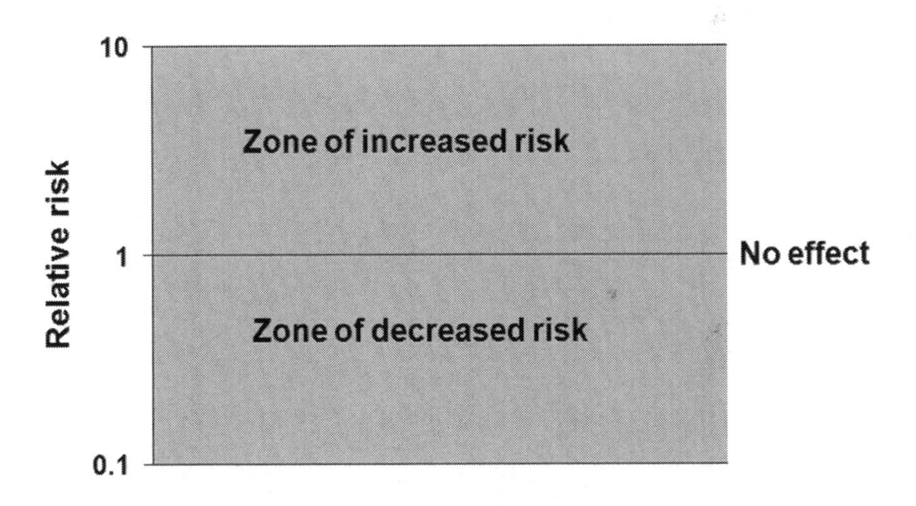

Figure 13-4
Schematic of relative risks
Source: Grimes[19]

How sure are the results?

In any study, researchers get only an estimate of what is happening in the entire population. Both small numbers of participants in studies and few events, such as cancer, lead to imprecise estimates. On the other hand, with more persons under observation, precision improves. This precision is usually portrayed in "confidence intervals." These confidence intervals depict a range of plausible values for the relative risk; the true value for the entire population of women is more likely to be at the middle of the range than at its extremes (the middle of the bell-shaped curve, rather than in the tails). When interpreting research results, narrow confidence intervals indicate good precision, and vice versa. When the 95% confidence interval does not cross 1.0, the difference found in a study is statistically significant at the usual 0.05 level. This means that the likelihood of a difference this large being due to chance alone is less than one in 20.

Relative risks and confidence intervals are easier to understand in graphic format. In Figure 13-5, study A had an increase in risk of 2-fold. The 95% confidence interval runs from 1.1 to 3.6, meaning one can be fairly sure that the true risk for all women lies within that narrow range. All these values fall in the zone of increased risk. Study B found a protective effect: the relative risk was 0.6, and the 95% confidence interval extended from 0.4 to 0.8. Again, all these values fall in the zone of protection. In contrast, study C found no association at all: the relative risk sits squarely on 1.0.

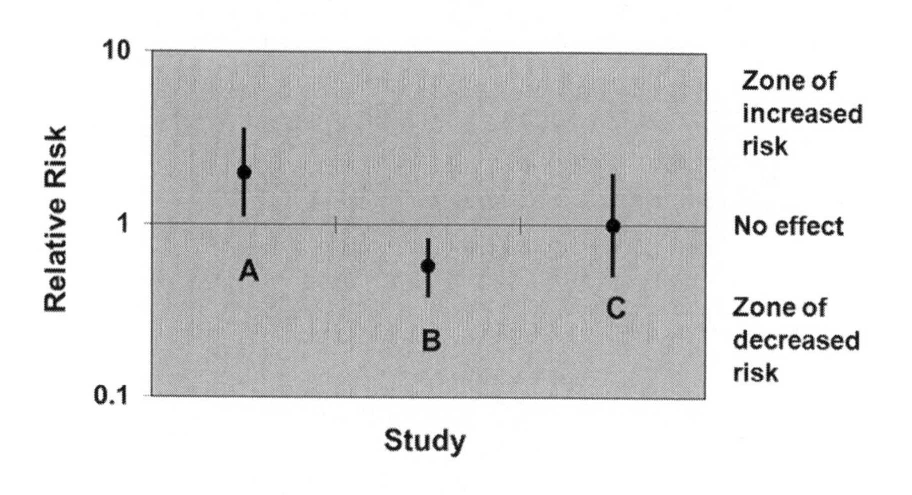

Figure 13-5
Examples of relative risks with 95% confidence intervals
Source: Grimes[19]

Avoiding recall bias

Because of recall bias, studies in which the history of abortion is obtained after the diagnosis of cancer cannot be trusted.[20] Hence, only those studies that determined abortion exposure before the outcome (breast cancer) deserve consideration. A comprehensive review of the entire world's literature on this question was published in 2004. Conducted by several international luminaries in epidemiology, the Collaborative Group on Hormonal Factors in Breast Cancer encompassed 53 studies from 16 countries and 83,000 women with breast cancer.[20] Twelve of these studies had abortion history obtained prospectively, such as the Danish study.[21] After controlling for potentially distorting biases, the overall relative risk for induced abortion was 0.93, indicating a slight reduction in risk (Figure 13-6). The 95% confidence interval was narrow: 0.89 to 0.96, meaning that the true value for the universe of all women probably lies within that range. The probability that this result was due to chance was remote, less than one in a thousand.

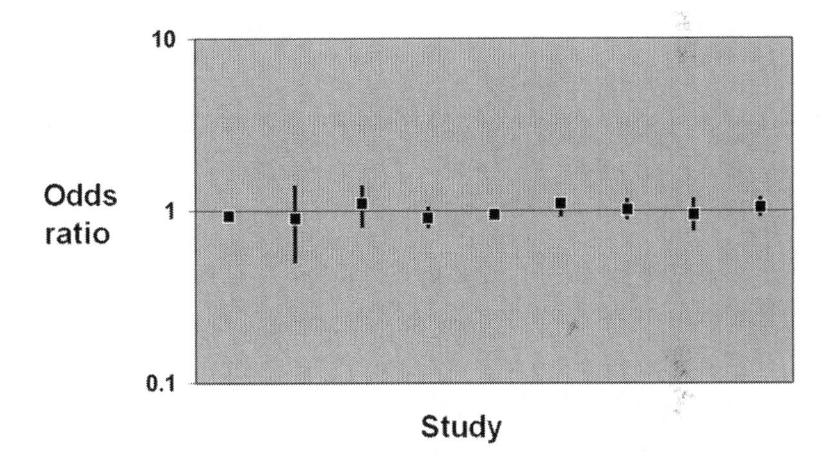

Figure 13-6
Results of recent studies of abortion and breast cancer, from left to right, Beral[20], Palmer without children[22], Palmer with children[22], Lash[23], Reeves[24], Rosenblatt[25], Michels[26], Henderson without children[27], Henderson with children[27]
Sources: References 20-27

Since publication of this major review, other rigorous studies have determined exposure to abortion before the diagnosis of breast cancer.[22,23] In the Black Women's Health Study, history of abortion was determined at the outset.[16] Over the course of 205,000 woman-years of follow-up, 348 new cases of breast cancer occurred. Among women who had not given birth before, the relative risk of breast cancer associated with a prior abortion was 0.9 (near unity, meaning no effect). Among those who had delivered a child, the corresponding relative risk was 1.1, again near unity. For both relative risks, the 95% confidence intervals crossed 1.0 and were not statistically significant (Figure 13-6).

A Massachusetts study derived information about prior termination from birth certificates.[23] In this case-control study, the relative risk of breast cancer associated with any pregnancy termination was 0.91. Again, the 95% confidence interval portrayed in Figure 13-6 is narrow, indicating a precise result.

In 2006, Reeves and others published results from the European Prospective Investigation into Cancer (EPIC) study; this multicenter study enrolled 4800 women with breast cancer. Among women who reported ever having an induced abortion, the relative risk of breast cancer was 1.0 (95% CI 0.9-1.0), showing no effect.[24]

In the same year a large cohort study was published from China; it followed over a quarter million women forward in time. Women who reported an induced abortion had no increase in the risk of breast cancer (relative risk 1.0 [95% CI 0.9-1.1]).[25]

Results from a large U.S. prospective cohort study, the Nurses' Health Study, provided further evidence of no effect.[26] More than 100,000 young women were followed forward from 1993, for a total of nearly a million woman-years of observation. The breast cancer hazard ratio (interpreted like a relative risk) for one or more induced abortions was 1.0 (95% CI 0.9-1.2).

In 2008, investigators reported the experience of California teachers followed forward in time from 1995.[27] Women who had ever had an induced abortion had no increased risk of breast cancer; this held true for women who had never given birth and for those who had delivered a child. Seldom in epidemiology is such consistency in results seen.

Other systematic reviews

Three other teams of investigators published overviews of this question. Epidemiologists at Harvard reviewed the published literature and concluded any relationship "is likely to be small or nonexistent."[28]

In the same year, Brind and associates reached a different conclusion, noting a relative risk of 1.3 (a 30% increase in risk) associated with prior abortion.[29] Of note, the authors did not critically evaluate the quality and credibility of the reports they included. Finally, a review by epidemiologists at the Centers for Disease Control and Prevention and American Cancer Society concluded that abortion was unrelated to later breast cancer.[30] Of note, these reviews did not go back to investigators and gather individual patient data as was done in the Collaborative Group report.[20]

Views of major medical organizations

Several important national and international medical organizations have independently summarized the literature and made recommendations. The World Health Organization reviewed the evidence with a panel of experts and concluded that no association exists (Table 13-1).[31] The American Cancer Society concurs.[32] The American College of Obstetricians and Gynecologists, which includes over 50,000 members, critically reviewed the evidence on two occasions; the College has concluded that no association exists.[33] The Royal College of Obstetricians and Gynaecologists, the U.K. counterpart of the American College, has reached a similar conclusion about safety.[34]

Table 13-1
Conclusions of disinterested medical organizations regarding abortion and breast cancer

Organization	Conclusion
American Cancer Society	"Still, the public is not well-served by false alarms. At this time, the scientific evidence does not support the notion that abortion of any kind raises the risk of breast cancer or any other type of cancer."[35]
Canadian Cancer Society	"The body of scientific evidence does not support an association between abortion and increased breast cancer risk."[36]
National Cancer Institute	"Induced abortion and miscarriage have not been shown to increase a woman's chance of developing breast cancer."[37]

World Health Organization	"...results from epidemiological studies are reassuring in that they show no consistent effect of first trimester induced abortion upon a woman's risk of breast cancer in later life."[31]
American College of Obstetricians and Gynecologists	"Early studies of the relationship between prior induced abortion and breast cancer risk were methodologically flawed. More rigorous recent studies demonstrate no causal relationship between induced abortion and a subsequent increase in breast cancer risk."[38]
Society of Obstetricians and Gynaecologists of Canada/Society of Gynecologic Oncologists of Canada	"A history of induced or spontaneous abortions is not associated with an increased risk of breast cancer."[39]
Royal College of Obstetricians and Gynaecologists	"From the literature we reviewed there is no established link between induced abortion or miscarriage and development of breast cancer."[40]

On-again, off-again website

The National Cancer Institute (NCI), a component of the National Institutes of Health, also concluded that abortion and breast cancer were not causally linked. However, during the first administration of George W. Bush, the web site mysteriously disappeared. In November, 2002, the NCI quietly replaced its "no association" conclusion with a statement that, "Some studies have reported statistically significant evidence of an increased risk of breast cancer in women who have had abortions, while others have merely suggested an increased risk. Other studies have found no increase in risk among women who have had an interrupted pregnancy." Interestingly, no major new studies prompted the switch. Around the same time, the Centers for Disease Control and Prevention had to bowdlerize its website's statement that condoms protect against sexually transmitted diseases. Since neither condoms nor abortion sat well with some members of the Administration, scientists and physicians suspected politics was at work.[41]

The outcry from Congress and from the medical and public health communities was prompt and strong.[42] To its credit, the NCI quickly empanelled a blue-ribbon commission of distinguished scholars to review the totality of the evidence. On March 21, 2003, a new NCI website appeared, concluding that no causal association existed between abortion and breast cancer.[43] The progression from "no association" to "inconclusive" to "no association with strong evidence" all in the course of four months (without new evidence) suggested political censorship, according to the Union of Concerned Scientists[41] and editorial writers across the nation.[44]

Mandated misinformation

State legislatures have entered the fray. In Mississippi and Texas, this improper warning is now required by law. Health officials in Kansas and Louisiana "voluntarily" issue the warning as well.[1] Montana law required the same warning, but the State Supreme Court struck it down. Other states are lining up to require similar incorrect warnings: Georgia, Hawaii, Illinois, Iowa, Massachusetts, Minnesota, New Hampshire, New Jersey, New York, North Carolina, Oklahoma, Vermont, Washington and West Virginia.[1]

To frighten a woman needlessly with medically inaccurate information is unethical. For women faced with the difficult decision about an unwanted pregnancy, it is cruel as well. State-mandated misinformation violates the basic ethical principal of beneficence.[7] Moreover, it threatens another fundamental ethical precept: autonomy. When provided incorrect information about risks and benefits, women cannot exercise truly informed consent.

The jury is in

The potential association between abortion and breast cancer was a legitimate, important public health question to address. Early case-control studies were sabotaged by recall bias. In recent years, more than a dozen studies — free of recall bias — have tackled this issue.[20,22,23] The consistency of findings has been striking: relative risks all cluster tightly around 1.0, indicating no effect. Women who have had an abortion and those considering abortions in the future can be confident that abortion does not influence the risk of breast cancer.

Take-home messages

- *Hormones and pregnancy are involved in the development of breast cancer*
- *Hence, the hypothesis that abortion might influence the risk is reasonable*
- *Early reports finding an association between abortion and breast cancer were handicapped by recall bias (underreporting of abortion among healthy controls)*
- *Well-done studies have consistently found no association*
- *All major medical and public health organizations concur that abortion does not influence a woman's risk of breast cancer.*

Chapter 14
Abortion and mental health: apples and oranges, chickens and eggs

In a Marx brothers' movie, a brother asks Groucho, "Say, Groucho, how's your wife?" Without hesitation, Groucho retorts, "Compared to what?" While sexist by contemporary standards, Groucho's response highlights the challenge of studying emotional responses to abortion.[1] Like medical effects, the emotional consequences of abortion cannot be viewed in a vacuum. As noted by Stotland, "Flagrant and almost universal...is the tendency to come to consider the outcomes of abortion without the recognition that the only alternative to abortion is the continuation of pregnancy and the delivery of a baby."[2]

This chapter reviews the psychological effects of abortion and compares them when possible to those of childbirth, miscarriage, and adoption. Context is critical: the emotional effects of abortion can never be disentangled from the effects of an unwanted pregnancy and the events that led to the pregnancy being unwanted.[2,3] Moreover, comparing women seeking abortions (an unhappy choice) to those choosing childbirth (often a joyous, eagerly anticipated event) is fundamentally biased: these women differ at baseline, resulting in an "apples and oranges" comparison.[4]

Flaws in the literature

Overall, research has found few harmful effects. However, the literature on this topic is extensive, emotionally charged, and generally of meager quality. Hence, detailed consideration of individual studies is unwarranted.[5-11]

A recurring theme of literature reviews is the poor quality and subjective nature of most studies. For example, use of comparison groups was rare in early reports (fewer than 10% of articles).[5] Without a comparison (or control) group, researchers cannot deduce whether emotional outcomes are more or less common than after other conclusions of pregnancy. For example, from 5% to 15% of women in general may suffer from a psychiatric disorder at any point in time.[9] Postpartum depression occurs in at least 10% of women giving birth in the U.S.[2] The emotional status of abortion patients before the unwanted pregnancy was described in fewer than 10% as well.[5] Since pre-existing emotional state is a powerful predictor of response to abortion and other pregnancy events, this is a fatal flaw.[11]

Selection bias was common. Participants who took part in studies often were not representative of all women having abortions. This is a special concern for reports from single centers, whose referral patterns

> *"Finding that women in these studies had a higher rate of mental health problems after abortion... would be like finding out that people who took hangover remedies had an increased risk of headache."*
>
> Kendall[12]

may have led to atypical patients.[9] For example, those with emotional problems are more likely to be interested in psychological assessments than are well-adjusted women.[6] Attrition rates (losses to follow-up) were often high, undermining the validity of any conclusions drawn.

Information bias was frequent as well. Researchers commonly developed novel measures for their studies, without having validated the instruments.[8] Early studies that reflected Freudian psychoanalytic theory focused on anticipated negative responses, thereby introducing bias.[8] The duration of follow-up was often limited to weeks or months, thus precluding any assessment of potential long-term consequences. Large numbers of representative women need to be studied to provide stable, precise estimates of psychological effects; small sample sizes in early reports led to a lack of precision in conclusions drawn.[6]

Chicken or egg: which came first?

In studying the potential relationship between abortion and mental health, a fundamental challenge is establishing the order of events. Which came first, the abortion or the reported psychological problems? If the psychological problems preceded the abortion (and may themselves have been an indication for the abortion), any claim of causation evaporates. Cause must precede effect.[13]

As noted previously in this book, women choosing abortion tend to be disadvantaged in many ways. Psychological problems are often among these challenges. This has been repeatedly documented in the medical literature: women choosing abortion have more difficult lives than do women choosing to continue their pregnancies. A recent Dutch study documented major differences in the psychological well-being of women who had abortions vs. those who did not.[14] Women in the abortion sample had three times the incidence of psychiatric disorders as did the general population. Abortion patients had statistically significant increases in mood disorders (e.g., major depression or bipolar disorder), anxiety, substance abuse, impulse control, and antisocial behavior. A secondary lifetime-minus-last-year psychiatric history analysis eliminated effects from a recent abortion or delivery. The results were unchanged. Several explanations were offered for the findings: 1) women with a history of psychological problems may be at increased risk of unintended pregnancy, 2) given an unwanted pregnancy, women with such a history are more likely to choose abortion, or 3) both.

However, a better comparison group is pregnant women. (The optimal comparison group would be women with an unwanted pregnancy who continued to childbirth.) A national database study from Denmark confirmed that women choosing abortion had higher use of psychiatric services both before and after abortion than did women who continued their pregnancies.[15] Administrative databases lack sufficient detail to enable sophisticated analyses,[16] but they can identify large relationships. In Denmark, health care is funded by the government, so both abortions and psychiatric care should be documented. Between 1997 and 2005, women had much higher use of psychiatric services (either inpatient or outpatient) before a first abortion than before a first birth (15 contacts per 1,000 person-years vs. 4) (Figure 14-1). After the first abortion, use of psychiatric services remained fairly stable at 15 contacts per 1,000 person-years. In contrast, after first birth, psychiatric contact increased significantly from 4 to 7 per 1,000 person-years, presumably reflecting the demands of motherhood and sleep deprivation.[2] Two inferences emerge: no temporal association was seen between abortion and use of psychiatric services, and women who chose abortion had higher use of psychiatrists before the abortion than did women continuing pregnancies.

Assessing the evidence

By 1981, more than 1,000 studies of varying credibility had addressed the psychological impact of abortion.[5] The number has grown since then. Much of the literature consists of case-series reports, which, lacking a control or comparison group, allow no valid conclusions. Before-after studies attempt to use each woman as her own "control." Stated alternatively, each woman provides a baseline assessment for later comparison.

An example of a before-after study would be administering a new blood pressure medication to treat hypertension (high blood pressure.) A study participant would have her baseline blood pressure determined, receive the experimental drug for several weeks, and then have a repeat blood pressure check. If the latter measurement were lower, this would provide indirect evidence that the medicine lowered the blood pressure. Without a contemporaneous control group, however, this conclusion may not be warranted. For example, a program of weight loss and exercise prompted by the alarming initial blood pressure might have lowered the pressure…and not the medicine. Alternatively, the abnormally high pressure might have improved on its own due to the phenomenon of "regression to the mean:" the more aberrant an initial observation, the more likely it is to be closer to the center of the population distribution on repeat observation.[17]

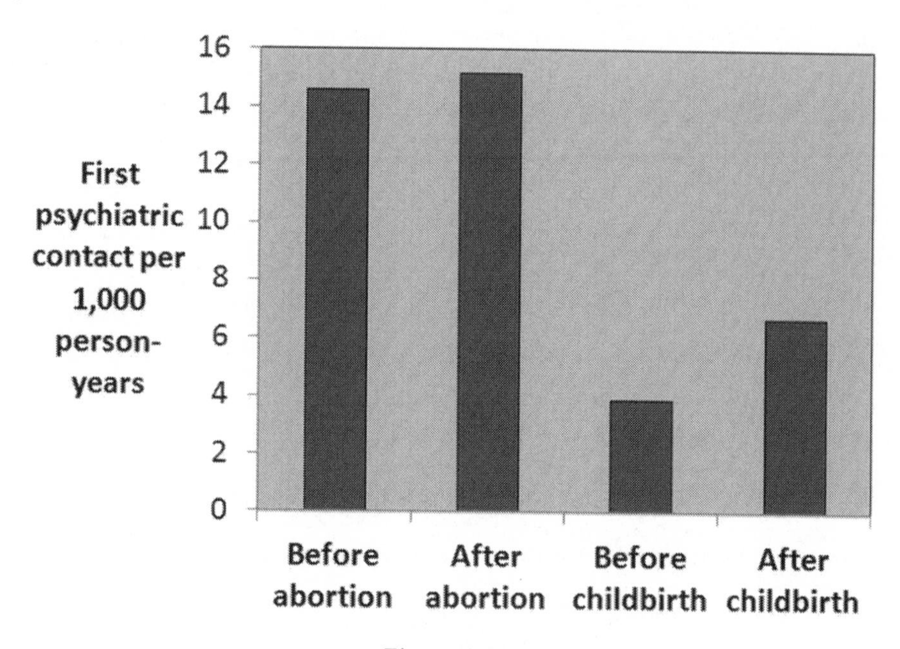

Figure 14-1
Rate of first psychiatric contact before and after abortion and childbirth, Denmark, 1995-2007
Source: Munk-Olsen[15]

Before-after studies

Several before-after studies have examined the potential emotional impact of abortion. An early study from the U.S. compared the responses of women having a first abortion with those having a repeat abortion.[18] Distress was evident in both groups before the abortion. Afterwards, those having repeat abortions continued to manifest distress in dimensions related to interpersonal relationships. Which came first: difficult relationships or repeat unwanted pregnancies? A smaller study at a private clinic in North Carolina also showed emotional distress before abortion.[19] Given the small sample size, few conclusions could be drawn regarding hypothetical coping mechanisms. A larger before-after study of suction curettage patients at a single facility found attribution to play a role.[20] Women who blamed the unwanted pregnancy on their *personal* failings coped less well than did other women. As anticipated, those who expected to cope well with the process fared better than less optimistic women.

A large before-after study attempted to track more than 800 women for two years after abortion.[21] Unfortunately, only 50% completed two-year evaluations. Of the 418 with follow-up, 72% of women were satisfied with their abortion decision. Six women (1%) self-reported symptoms consistent with post-traumatic stress disorder, although these reports were not confirmed by formal psychiatric evaluation.

While most women reported no psychological problems or regret after abortion, those who did tended to have been depressed before the operation. This common-sense finding echoes throughout the abortion literature. As is true with hysterectomy, the best predictor of a woman's emotional response after the operation is her emotional status before it.

Abortion vs. childbirth: single centers

Cohort studies provide the most rigorous assessment of the emotional impact of abortion. In these studies, investigators identify women with unplanned pregnancies, commonly at the time of a request for a pregnancy test. Once the pregnancy is confirmed, women then segregate into two comparison groups: those opting for abortion and those planning to continue the pregnancy. An advantage of this research design is use of a contemporaneous control group. An insurmountable deficiency is that women who choose these options are fundamentally different (apples vs. oranges). Hence, differences in psychological status at follow-up may be due to baseline differences and not to the abortion or delivery.

Single institutions have sponsored several such studies. Additional limitations of single-center studies include small sample sizes (and thus little ability to study rare outcomes, such as serious depression) and questions as to how representative women are of other women having abortions. In an early and important cohort study,[22] investigators at

Johns Hopkins identified 373 women in early pregnancy; 268 planned to undergo abortions, and 105 wanted to continue the pregnancy. Women in both groups were interviewed and administered attitude scales, the Minnesota Multiphasic Personality Inventory, and the Symptom Check List. Few differences emerged among those who had suction curettage abortion, labor-induction abortion with hypertonic saline, or childbirth. If anything, those who had suction curettage had fewer bodily complaints than did women in the other two groups. As the authors concluded, "Induced abortion appears to be a benign procedure compared to term birth, psychologically and physically."

A second key study from Johns Hopkins focused on poor, urban African-American teenagers aged 17 years and younger.[23] Of 334 teenagers who requested a pregnancy test in the mid-1980s, 141 aborted, 93 carried to delivery, and 100 were found not to be pregnant. Women in all three groups were interviewed and given a battery of psychological tests. The follow-up rate here was high (90%). Emotionally, women fared well with either abortion or childbirth over the two years of the study. Socioeconomically, those who opted for birth were significantly worse off. Those who became mothers fell behind in school or dropped out more often that did those who chose abortion. A similar study of teenagers seen at a county clinic corroborated the favorable outcomes after abortion.[24] Most women (more than 80%) who chose abortion or delivery were comfortable with their decision six months afterwards.

> *"If any conclusions were to be drawn from the few differences we did find, it would be that early abortion by suction curettage was possibly more therapeutic than carrying a pregnancy to term."*
>
> *Athanasiou*[22]

Another study from a California community-based health clinic found the distress related to abortion to be transient.[25] Women who requested a pregnancy test were enrolled in a longitudinal study with assessments planned at enrollment, a day after receipt of positive pregnancy test results, and four weeks later. Among the 98 women who agreed to participate, only 44 were actually pregnant. Thirty-three completed all measurements, with 21 having abortions and 12 continuing the pregnancy. Those opting for abortion had more distress initially, but this dissipated by the one-month follow-up. At one month, emotional responses of women who had abortions and those continuing their pregnancies were similar.

Abortion vs. childbirth: population-based studies

> *"...abortion is a relatively brief stressor relatively soon after which women return to a state of affect comparable to women not experiencing abortion."*
>
> Cohan[25]

Large, population-based studies have confirmed the lack of adverse psychological effects of abortion for most women. Such studies offer the power to study rare events and the ability to generalize findings, at a cost of less detail than studies from single centers. The earliest such report stemmed from the West Midlands Region of the U.K.[26] Cooperating psychiatrists in the region reported hospitalizations for psychosis within 3 months after abortion or childbirth. The denominators for both had to be estimated. The author estimated the risk of hospitalization for psychosis was significantly higher after birth than after abortion (1.7 vs. 0.3 per 1,000 events, respectively), a difference unlikely to be due to chance.

A U.S. report provided further support.[27] The National Longitudinal Surveys of Labor Market Experience is a national study conducted by Ohio State University in conjunction with the U.S. Bureau of the Census. Interviews with a national probability sample of men and women aged 14 to 21 years in 1979 provided detailed information about reproductive histories. This large national sample was followed for eight years, with a high follow-up rate (90%). The Rosenberg Self-Esteem Scale evaluated well-being. Of 5,295 women in the sample, 773 had abortions during the study, and 3,053 became mothers. Of the latter, 980 identified the births as wanted, in contrast to 173 who reported the birth as unwanted.

Women who had abortions had significantly higher global self-esteem than did women in the total sample, a difference unlikely to be due to chance. This finding did not change over time. On the other hand, women who had unwanted births had the lowest self-esteem. The authors concluded that the greater feelings of worth and capability associated with having had an abortion (despite the stress of an unwanted pregnancy) argued strongly against adverse psychological effects of abortion. In this sample, women did not use abortion to avoid childrearing responsibilities; indeed, two-thirds of women opting for abortion were already mothers (as is true nationwide).[28]

A collaborative study by the Royal College of General Practitioners and Royal College of Obstetricians and Gynaecologists in the U.K. has provided important population-based data on psychiatric disorders associated with pregnancy.[29] Beginning in 1976, these two medical organizations recruited women requesting abortion and a comparison group of women with unplanned pregnancies but not requesting abortion. Extensive information was gathered on 13,261 women who agreed to follow-up, which ended in 1987. The researchers attempted to control for previous psychiatric history and demographic factors among the four comparison groups: women who obtained an abortion, those who did not request abortion, those who requested abortion but were refused, and those who requested abortion then changed their minds.

> *"Total number of children, however, continues to have a negative relationship to women's well-being. This suggests that abortion's positive relationship to well-being may come through its contribution to reducing women's total number of children rather than through a psychological effect of feeling empowered by having an abortion experience."*
>
> *Russo*[27]

Women without serious mental disorders who had abortions were significantly less likely to develop a psychosis than were women who did not request an abortion (Figure 14-2). For women with no such history, abortion of an unplanned pregnancy was associated with a 70% lower risk of having a psychotic episode afterwards. For women with a history of mild psychiatric disorders, the reduction in risk of psychosis associated with abortion was 60%. Differences of this size were unlikely to be due to chance.

A different pattern emerged with deliberate self-harm. Among women with no history of psychiatric disease, the relative risk of self-harm (mainly drug overdoses, none fatal) was significantly increased (by 70%) after abortion. The relative risk of such harm was higher still among women denied abortion (a 190% increase). No temporal association was seen with anniversary dates of the abortion or estimated date of delivery.

The authors of this long-term, longitudinal study of thousands of women after abortion or childbirth found total rates of psychiatric illness to be similar. Given the power of this large study to find important differences if they existed, the findings are reassuring regarding the emotional consequences of abortion.

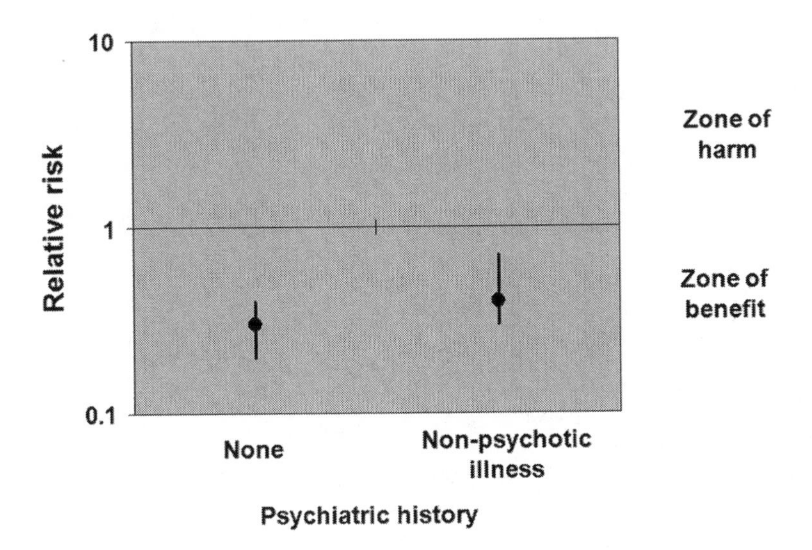

Figure 14-2
Relative risk of psychotic episode after abortion vs. birth of unplanned pregnancy, by psychiatric history, U.K., 1976-1987
Source: Gilchrist[29]

The Surgeon General's (almost) report

As noted previously, Surgeon General Koop was directed by President Reagan to write a white paper on abortion. A comprehensive report was written, then suppressed. In January, 1989, Dr. Koop met with Otis Bowen, Secretary of Health and Human Services, and decided not to release the report.[8] In March of that year, Congressman Ted Weiss (D-NY) held Congressional hearings to investigate the apparent discrepancies between the draft report and information made public, which suggested that data were inconclusive. During his testimony before the Congressional Committee, Dr. Koop acknowledged that the issue of psychological problems stemming from abortion is "miniscule from a public health perspective."[11,30]

Reviews of the literature since the Surgeon General's (almost) report

About 50 million U.S. women have had abortions; if this process substantially increased the risk of psychological illness, an epidemic should be evident by now. No such epidemic has occurred. Because of the continuing claims of abortion opponents, several medical organizations and individuals have published literature reviews of abortion and mental health in recent years. As with the underlying literature, the quality and rigor of the reviews varies widely.

Two reviews by professional organizations merit special attention. The Task Force on Mental Health and Abortion of the American Psychological Association published a comprehensive review and critique of the literature from 1989 to 2008, the full report of which is available on the Internet.[31] The updated findings also appeared in the peer-reviewed literature the following year, authored by members of the Task Force.[32]

The report reached six conclusions. A single abortion carries no greater risk of mental health problems than delivery of an unwanted child. The emotional burden of an abortion because of fetal anomaly is similar to that of a miscarriage, stillbirth, or newborn death. In several countries (Norway, Australia, and New Zealand) mental health problems are slightly but significantly more common among women who reported one or more abortions. Because the social context of abortions differs in those countries (e.g., requiring two physicians to document the need for abortion), the relevance to the U.S. may be limited (where abortion is available upon request). No causal relationship between abortion and mental health problems is established. Most women who have an abortion do not have emotional problems afterwards. However, some women do have regret or emotional problems after abortion; these responses deserve compassionate consideration.

The U.K. federal government commissioned a similar review in 2007.[33] The House of Commons asked the Royal College of Psychiatrists and Royal College of Obstetricians and Gynaecologists to update knowledge concerning the relationship, if any, between abortion and mental health. Funded by the Department of Health and coordinated by the Academy of Medical Royal Colleges, the comprehensive report was published in 2011. The 252-page document formally graded the quality of scientific evidence.

The Academy report echoed the problems with scientific quality of much of the published literature, especially inadequate control for mental health before abortion. Overall, the Academy found that any relationships between abortion and mental health problems were "unlikely to be meaningful."[33] The report largely confirmed the reviews of others.[31,34] "When a woman has an unwanted pregnancy, rates of mental health problems will be largely unaffected whether she has an abortion or goes on to give birth."[33]

In 2008, researchers at Johns Hopkins University published a systematic review of the literature; it reached a reassuring conclusion.[34] The report included 21 studies published between 1989 and 2008; each study was assigned a quality rating, similar to that used by the Academy of Medical Royal Colleges.[33] Overall, a common pattern (reported elsewhere in this book) emerged: better quality studies found no association between abortion and adverse emotional effects. In contrast, those with biased methods reported adverse emotional consequences (Figure 14-3).

In 2009, U.S. and Canadian collaborators published another synthesis of the literature.[35] As noted in the review above,[34] findings were related to study quality. Deficiencies among the studies claiming a harmful effect of abortion included poor choice of control groups, inadequate control of confounding (a blurring of effects), improper analysis, and unjustified inference of causal relationships. The report concluded that, "In studies that adequately control for wantedness of pregnancy or exposure to violence, a significant association between abortion and negative mental health outcomes is not to be found…."[35]

Because the American Psychological Association report focused on first-trimester abortion (by far the more common), less was known about the potential impact of abortions later in pregnancy. Hence, Steinberg at the University of California-San Francisco synthesized the literature on second-trimester abortion, which dealt with women having abortions because of fetal defects.[36]

The eleven studies shared common weaknesses. For example, none considered the woman's mental status before the abortion. Despite these limitations, the report concluded that, "The existing literature suggests that women who have later abortions for reasons of fetal anomaly have no worse mental health than women who give birth to infants with lethal or severe mental or physical conditions or experience other types of later perinatal loss (e.g., stillbirth or later miscarriage)."[36]

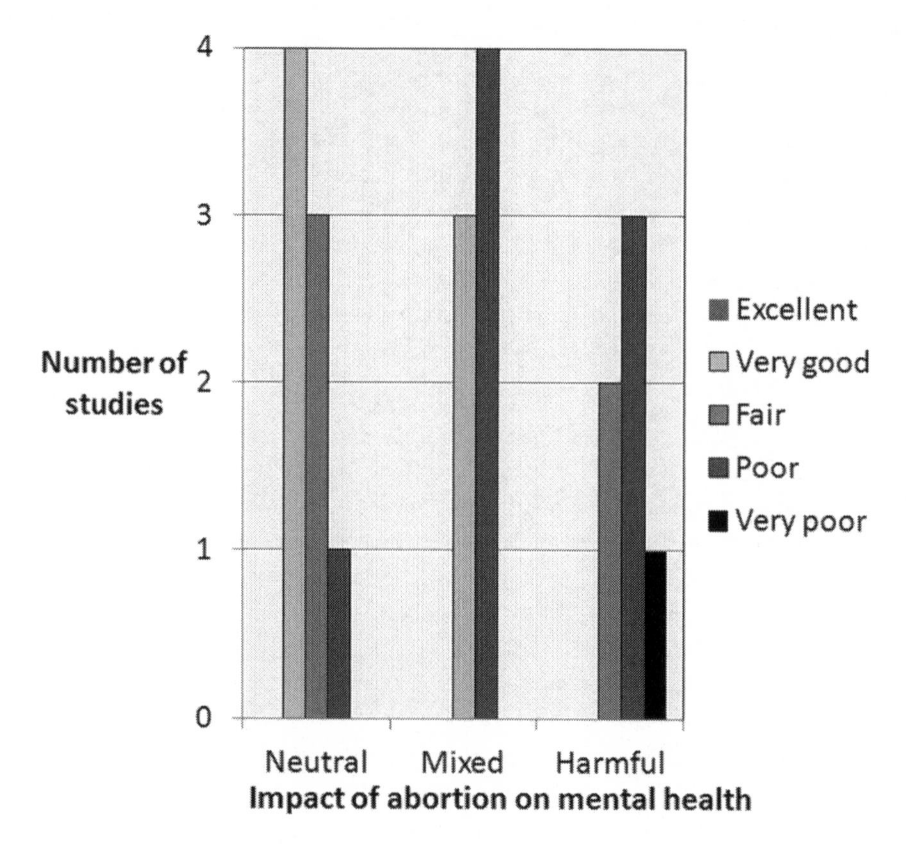

Figure 14-3
Inverse relationship between study quality
and impact of abortion on mental health
Higher quality studies generally find no effect;
lower quality studies report adverse effects
Source: Charles[34]

An outlier

One literature review was an anomaly. Coleman of Bowling Green State University reported a "meta-analysis" of 22 observational studies.[37] Published in 2011, this "meta-analysis" failed to follow (or cite) the internationally accepted rules of conduct, published over a decade earlier, for a synthesis of this type.[38] Whether Coleman was unaware of the rules of conduct or chose to disregard them is unclear. However, as noted previously,[34] reports with poor quality methods are more likely to claim adverse emotional effects.

Aggregating heterogeneous (dissimilar) results into a single summary number, as done by Coleman, is scientifically improper.[38-44] However, the published article skirted the troublesome issue of heterogeneity by not mentioning it.

The Coleman "meta-analysis" sparked immediate criticism internationally. Numerous letters to the editor documented its scientific shortcomings, and authors of the Academy of Medical Royal Colleges report noted that "it cannot be regarded as a formal review."[12]

Coleman had scientific problems in an earlier study.[45] This led to publication of a correction, which still failed to overcome the incorrect conclusions.[46] Coleman and her co-authors used lifetime mental disorders and assumed that all these events followed, not preceded, the abortion. This assumption was reviewed by the journal editor and the Principal Investigator of the National Comorbidity Study (which Coleman used); they concluded, "That some previous researchers used the same flawed method cannot be taken as evidence in its favor."[47]

Impartial scientists in the U.K evaluated the quality of Coleman's publications as part of their systematic review.[33] Coleman's reports ranged from fair[45] to poor[48] to very poor.[49,50] Coleman's co-author David Reardon,[48] a frequent contributor to this literature, also had his publications evaluated. His research was judged fair[51,52] to poor.[53,54]

Emotional consequences of miscarriage

While rigorous studies have compared the psychological impact of abortion vs. childbirth, little is known about its impact compared with miscarriage. Miscarriage can have a powerful but transient adverse effect on mental health. Over 90% of women have grief reactions after the unexpected loss of pregnancy.[55,56] From 48% to 51% of women experiencing miscarriage suffer from depression in the early months thereafter.[57]

Cohort studies comparing women having miscarriage to other women confirm the harmful emotional consequences of this pregnancy loss. In a study of anxiety disorders after miscarriage, women having losses were compared with nonpregnant women drawn from the same community.[58] The risk of obsessive-compulsive disorder was eight times higher among women having miscarriage than among other women. Another cohort study compared women having miscarriage to women having live births.[59] In the first six months after the loss, these women had more depression, anxiety, and other symptoms than did women who gave birth. By one year, these symptoms had abated. Thus, most women seem to recover from miscarriage by one year without psychiatric help.

The adoption option

Relinquishing a child for adoption is psychologically unacceptable for most women faced with an unwanted pregnancy.[60] National data confirm the near disappearance of child relinquishment. According to the National Survey of Family Growth, the proportion of babies given up for adoption by never-married women under age 45 years was about 9% before 1973. By 1989-95, the figure had plummeted to less than 1% (Figure 14-4).[61] Between 1973 and 2002, the proportion of ever-married women aged 18 to 44 years who had adopted varied from 1 to 2%.[62]

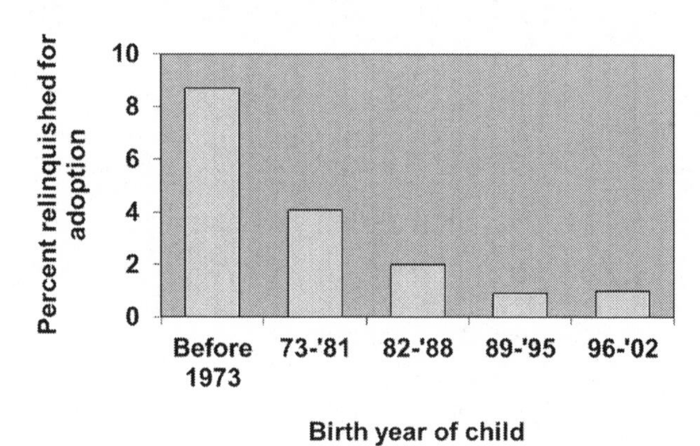

Figure 14-4
Percentage of children born to never-married women under 45 years of age, by year of birth relinquished for adoption, selected years, United States
Sources: Chandra,[61] Jones[62]

The proportion of non-marital births has grown commensurately. Recent declines in abortion rates suggest that the growing unacceptability of relinquishment is unrelated to abortion.[61] In a random sample of 5,295 U. S. women followed for eight years, only 33 were identified as having placed a child for adoption.[27] The number was considered too small to assess its emotional impact.

Because of its rarity, the psychological impact of adoption has not been well studied, and case-series reports are limited. However, the sparse literature supports the common perception that relinquishment is traumatic. Among 20 Australian women giving their babies up for adoption, pathological grief reactions were common.[63] Among 119 women hospitalized for serious psychiatric disorders, emotional consequences of relinquishment of custody were rated as significantly more severe than were those after abortion.[64] In another study, 95% of women felt grief and loss after signing relinquishment papers, and two-thirds harbored these feelings for five to 15 years after the adoption.[65]

The "abortion trauma syndrome:" an opportunity to evangelize

Since harmful effects of abortion have been so hard to document, critics of abortion had to invent one: the "abortion trauma syndrome."[66] A "syndrome" is the medical term for a cluster of signs and symptoms that characterize a disease or condition. "Trauma" is borrowed from "post-traumatic stress disorder," a recognized psychiatric illness featuring nightmares and flashbacks, related to a traumatic event outside the range of usual human experience.[67] Examples would include rape or war. The "abortion trauma syndrome" does not feature a recognized constellation of signs or symptoms, and abortion is an experience shared by one out of three U.S. women.[68] Abortion is thus well within the usual range of human experience.

Thorough searches through the psychology and psychiatry literature do not reveal an "abortion trauma syndrome." Moreover, the term is not included in the Diagnostic and Statistical Manual of Mental Disorders of the American Psychiatric Association.[69] Rather than being a legitimate psychiatric condition, this "syndrome" was viewed by its inventors to be "an opportunity to evangelize."[2]

Conclusions of national organizations

The American Psychological Association, which has no vested interest (and whose members do not provide abortions), has repeatedly reviewed and summarized the world's literature on the impact of abortion on women's psychological health.[8] Its briefing paper on the topic confirmed the safety of the procedure.[3] The Association concluded that the risk of psychological harm is low. Depression and post-traumatic stress disorder after abortion are uncommon – indeed, less common than in the general population. Scores on psychological tests for both minors and adults fall within the normal range. Stress is greatest *before* abortion, and positive emotions predominate afterward. Women with pre-existing emotional problems fare less well after abortion than do healthy women. The circumstances causing a pregnancy to be unwanted may be the source of distress, not the abortion itself. Unintended childbearing hurts the life prospects for both mothers and children.

> *"Access to legal abortion to terminate an unwanted pregnancy is vital to safeguard both the physical and mental health of women. Research indicates that abortion does not generally have a negative impact on either women's physical or mental health. A forced, unwanted pregnancy, on the other hand, could place women's health and well-being at risk."*
>
> *American Psychological Association*[3]

In 2008, the Association updated its review of the literature, confirming its earlier conclusions.[70] Its Task Force on Mental Health and Abortion considered all the English-language publications on the topic since 1989. The Association concluded that, "In general, however, the prevalence of mental health problems observed among women in the United States who had a single, legal, first-trimester abortion for nontherapeutic reasons was consistent with normative rates of comparable mental health problems in the general population of women in the Unites States." The strongest predictor of mental health after an abortion was the woman's mental health before it.

The American Psychiatric Association, representing U.S. psychiatrists, also found no cause for concern.[71] The organization stated that it "opposes all constitutional amendments, legislation, and regulations curtailing family planning and abortion services to any segment of the population." The psychiatrists noted that the "freedom to act to interrupt pregnancy must be considered a mental health imperative..." If abortion were emotionally harmful, why would the nation's psychiatrists be so supportive of it?

Obstetricians and gynecologists elsewhere have concurred: "The overwhelming indication is that legal and voluntary termination of pregnancy rarely causes immediate or lasting negative psychological consequences in healthy women."[72]

In summary, the American Psychological Association, the American Psychiatric Association, the Academy of Medical Royal Colleges, and the Royal Australian and New Zealand College of Obstetricians and Gynaecologists concur that abortion does not damage mental health.

The principal voices of dissent today are authors of suboptimal scientific reports.[33,73,74] The most common emotional response to abortion is a profound sense of relief, not grief.

Take-home messages

- *Studies of the potential impact of abortion on psychological health are generally of poor quality*
- *Many studies did not consider women's emotional state before the abortion*
- *Many studies did not have a comparison group*
- *Large, longitudinal studies have found no evidence of harmful effects*
- *The American Psychological Association has repeatedly concluded that abortion does not have adverse emotional effects*

Chapter 15
Prematurity and abortion (and gum disease)

Scope of the problem

Premature birth (before 37 completed weeks of pregnancy) remains the worst—and most stubborn—problem in childbearing today. Forty weeks from the last menstrual (or a couple of weeks on either side) are considered a completed pregnancy, i.e., "full term." In the U.S. many babies are born short of the mark. The national preterm birth rate hovers around 12%;[1] stated alternatively, one in eight U.S. births occurs too soon. The earlier the birth, the greater is the risk of death or complications for the baby. These risks include breathing problems, severe bowel disease, cerebral palsy, and bleeding into the brain.

About four million births and one million abortions occur each year in the U.S. Hence, if abortion were causally related to preterm birth in later pregnancies, the public health impact could be large. As with the abortion and breast cancer question (Chapter 13), many studies have addressed this question, and some have found a weak association. Poor research standards readily explain these findings: most studies have not been able to control for built-in bias. Women who are unhealthy and disadvantaged by poverty, abuse, poor nutrition, and low educational attainment are more likely to have abortions than are women without these handicaps. These same unhealthy and disadvantaged women also have worse pregnancy outcomes than do other women. Thus, disadvantaged women have worse pregnancy outcomes, with or without a history of abortion. Stated alternatively, women who have abortions are at higher risk for poor obstetrical outcomes.

Research challenges

Studying a potential relationship between abortion and prematurity is difficult.[2] A fundamental problem is choosing an appropriate comparison group. A woman pregnant for the second time and with one birth is different than a woman pregnant for the second time with one abortion. Similarly, a woman pregnant for the first time differs from a woman pregnant for the second time and with a prior abortion. No ideal comparison exists. Abortion does not protect against the increased risk of low birth weight in first births.[3]

As discussed previously, recall bias is a pervasive problem with studies looking back in time (case-control studies). Women are hesitant to disclose abortion information to researchers; only studies with contemporaneous documentation of abortion are free of this underreporting.[4,5]

Control of confounding (a mixing or blurring of effects) is often inadequate. As noted repeatedly in this book, women who obtain abortions differ in many ways from typical healthy women. Teasing out the potential impact of these various characteristics, e.g., race, poverty, and malnutrition, from any effect of abortion is difficult.

Risk factors for premature birth

Many personal characteristics and other factors have been linked with prematurity. In a Lancet review article on the epidemiology of preterm birth,[1] the strongest association was with race: African-American women consistently have higher risks than do women of other ethnicities. A history of preterm birth carries a substantial risk of recurrence. Multiple gestations, e.g., twins or triplets, are strongly linked with prematurity. Vaginal bleeding during pregnancy, a short interval between pregnancies, psychological stress, depression, and tobacco use are other reported risk factors. Subclinical (unnoticed) uterine infection is thought to account for a substantial proportion of premature births. Of note, legal abortion was not mentioned as a risk factor in the Lancet review.[1]

Other reviews of the epidemiology of prematurity have inventoried known risk factors. These include reviews done by scientists at the Centers for Disease Control and Prevention,[6] the Bill and Melinda Gates Foundation,[7] and universities.[8] In each of these recent reviews,[1,6-8] legal abortion is not mentioned. The absence of abortion indicates that dispassionate scientists and organizations do not consider abortion a risk factor for preterm birth.

Early studies

Decades ago, epidemiologists at the CDC in Atlanta reviewed the world's literature and concluded that abortion was not causally related to prematurity or other complications of pregnancy.[3,9] Since then, large population-based studies have consistently found no association between abortion and preterm birth or low birth weight.[10-12]

Recent meta-analyses

Two recent syntheses of the published studies of this question have been reassuring. A U.S. group reviewed studies from 1995 to 2007 and found an overall relative risk of preterm birth after one abortion of 1.3 (95% CI 1.0-1.5). As noted previously, weak associations of this size (<2.0) are generally noise, not signal.[13] Moreover, miscarriage was associated with similar risks of preterm birth as was abortion. As the authors noted, "Women who undergo elective abortions have more acute and chronic psychosocial stress, resulting from poverty, unsafe neighborhoods, lack of partner and social supports, domestic violence or racism…"[14]

"Bias and/or confounding variables, however, are possible explanations. Low socioeconomic status, higher gravidity [number of pregnancies], high-risk sexual behaviors, smoking and illicit drug use, prior adverse pregnancy outcomes, being single with poor social support, and black race are more common among women with prior induced abortions than controls."[14]

A British group did a similar review but with broader scope: all articles published from the inception of several electronic databases until August of 2008. With one abortion, the odds ratio of preterm birth adjusted for confounding was 1.3 (95% CI 1.1-1.4). However, tests for heterogeneity (ability to combine studies into a single estimate) found large variation in results between studies (I^2 test for heterogeneity = 78%). In this situation, aggregating studies is improper; the results are too dissimilar to lump together.[15,16] As in the other recent meta-analysis,[14] this weak association falls well below the discriminatory threshold for studies of this type.[17,18] Figure 15-1 shows that this weak association lies within the zone of probable bias, not the zone of potential interest.[17] The authors were circumspect in their inferences: "…it has been identified that women with a history of I-TOP [Induced Termination Of Pregnancy] were unmarried, young and from socio-economically disadvantaged group (sic)….Certain studies controlled for confounders [bias], whereas other studies failed to do so."[19] The authors further warned, "We caution interpretation being causal as confounding effects of socio-economic factors, which are important, were considered in very few studies only." Stated alternatively, bias could easily have accounted for the weak association observed.

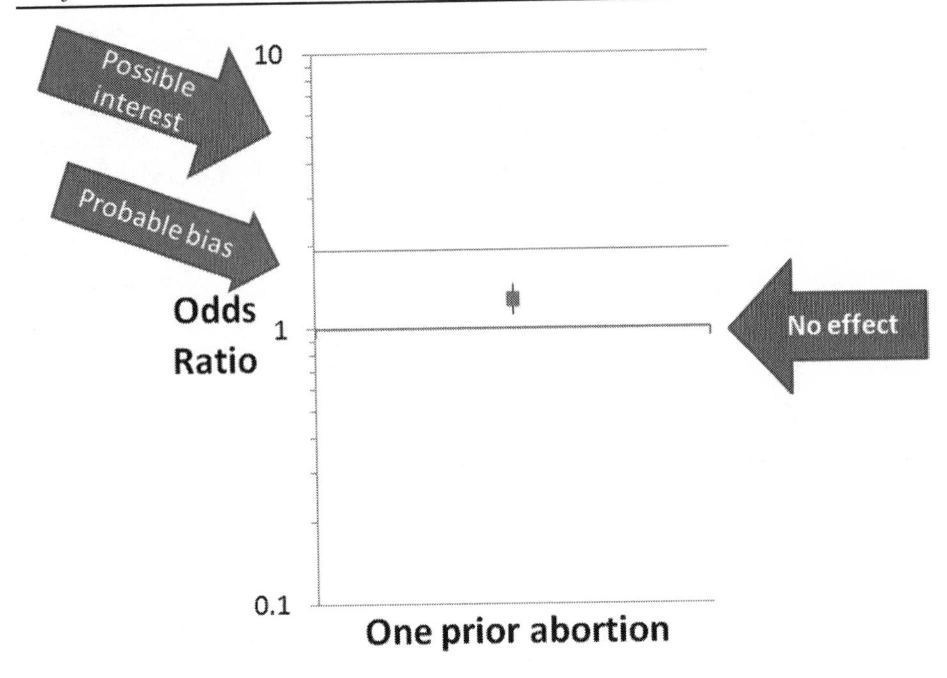

Figure 15-1
Odds ratio* of preterm births among women with one prior abortion in a recent meta-analysis[19]
*Summary odds ratios inappropriate because of high heterogeneity (variability) between studies[16]
Source: Shah[19]

Finnish findings

Abortion opponents commonly cite a Finnish database study[20] as evidence of the putative dangers of abortion. Hence, it deserves brief mention. As background, Finland has half the population of North Carolina. To extrapolate from Eastern North Carolina (or Finland) to women in the rest of the world would be inappropriate. Second, database studies of this sort are simply inadequate for studying causal associations.[21] This exercise is similar to trying to understand the causes of car wrecks using only the information gleaned from driver's licenses. Moreover, gross misclassification of events in the Finnish abortion database has been documented previously.[22]

Overall, the study found no relationship between prior abortion and preterm birth (<37 weeks). For women with one or more abortions compared to none, the odds ratio was 1.0, meaning no effect at all. No dose-response relationship was evident: for women with one prior abortion the adjusted odds ratio was 1.0, for two prior abortions 1.0, and for three or more 1.4. All of these are weak associations, well below the discriminatory limits of administrative database research.[17] The only statistically significant increase (40%) was seen among women with three or more prior abortions. Common sense indicates that women with three or more abortions before first delivering a child are atypical women (whether in Finland or in the U.S.), thus limiting extrapolation. The authors concluded by warning that, "...observational studies like ours, however large and well-controlled, will not prove causality."[20]

An "immutable risk factor"

Abortion opponents eagerly point out that a comprehensive monograph on preterm birth listed abortion among "immutable medical risk factors associated with preterm birth."[23] This 791-page review of the literature was written under the auspices of the Institute of Medicine, of which I am a member. In table B-5, "multiple 2nd trimester spontaneous abortion (sic)" and "prior first trimester induced abortion" appear on a list of "immutable" features such as "infant sex" and "short stature."

To suggest that the Institute of Medicine considers abortion causally related to preterm birth reveals that this monograph was not read.[23] The word "abortion" appears 11 times in 791 pages, excluding the references. It appears nine times as "spontaneous abortion," once in Table B-5, and once in "induced abortion" related to the Born-Alive Infants Protection Act of 2001. Nowhere in this tome, except Table B-5, is the suggestion made that abortion is related in any way to prematurity.

Conclusions of medical and public health organizations

Because of the potential importance of a link, if it existed, a number of organizations have examined the evidence. Importantly, none of these has any ax to grind regarding abortion; their missions are to improve health and well-being. As early as 1979, the World Health Organization had concluded, based on its own research, that suction curettage abortion was unrelated to prematurity.[24] That assessment remains true today as well. In 2012, the WHO updated its technical and policy guidance for health systems.[25] Nowhere in this 134-page monograph do the words "prematurity" or "preterm" appear.

The American College of Obstetricians and Gynecologists independently reached the same conclusion: abortion has no causal relationship to prematurity. This assessment is evident in multiple ACOG publications. The ACOG Frequently Asked Questions document Preterm Labor and Preterm Birth[26] identifies the following risk factors: prior preterm birth, a short cervix, pregnancies close together, operations on the cervix, pregnancy complications, and lifestyle factors. In its Frequently Asked Question publication on Induced Abortion, prematurity is not listed among risks of the procedure.[27] ACOG's patient education pamphlet by the same title also makes no mention of prematurity as a potential complication of abortion.[28] Likewise, ACOG's Practice Bulletin for clinicians entitled "Prediction and prevention of preterm birth" does not mention abortion.[29]

The U.K. counterpart of ACOG, the Royal College of Obstetricians and Gynaecologists has noted that, "Women should be informed that induced abortion is associated with a small increase in the risk of subsequent preterm birth, which increases with the number of abortions. However, there is insufficient evidence to imply causality."[30]

The American Academy of Pediatrics concurs with ACOG. The AAP notes that, "Extensive reviews conclude that there are no documented negative psychological or medical sequelae [complications] to elective, legal, first-trimester abortion among teenaged women."[31] The Academy went on to say that, "...legal abortion results in fewer deleterious sequelae for women compared with other possible outcomes of unwanted pregnancy."[31]

The Centers for Disease Control and Prevention finds no causal association between abortion and prematurity. The CDC's web site "Preterm birth" makes no mention of abortion.[32] The Center's "Infographic" entitled "Factors associated with preterm birth" identifies 11 factors; abortion is not on the list.[33]

The American Public Health Association concurs. The most recent APHA policy statement on abortion notes that abortion "...is one of the most common and safest gynecologic interventions in the United States."[34] Again, "prematurity" and "preterm" are not mentioned.

The March of Dimes, whose motto is "working together for stronger, healthier babies," also reports no link between prematurity and abortion. On its web page "Finding the causes of prematurity," abortion does not appear.[35] Instead, it discusses immune responses to inflammation.

Maybe, maybe not

The claim by abortion opponents that abortion is causally related to premature birth may depend on the audience. One opponent wrote in a Raleigh, NC, newspaper about "the indisputable evidence demonstrating that a prior abortion increases a woman's risk for future preterm birth."[36] The same physician co-authored an analysis of late preterm births in North Carolina, published in the medical literature the same year.[37] The analysis adjusted for 34 different personal and medical factors identified as potential "risk factors…based on published literature…" Abortion was not one of them.[37] Nor did the word appear anywhere in the published report. The scientific standards of newspapers and peer-reviewed medical journals seem to differ.

Obstetrical textbooks

Williams' Obstetrics has helped to educate generations of U.S. physicians.[38] Now in its twenty-third edition, it remains a widely used reference work. Chapter 36 focuses on preterm birth. In the chapter, the section on "Antecedents and Risk Factors" summarizes the evidence concerning the factors related to prematurity. Table 15-1 lists every factor mentioned in the section. Abortion is absent.

Gum disease

Medical research, like other human activities, has its fads. Over the past decade, the hypothesis that gum disease might contribute to premature birth has been a hot topic, with keen interest among odd collaborators: obstetricians and dentists. Inflammation stemming from infection is believed to account for a substantial proportion of all cases of premature birth.[39] Could inflammation in the mouth trigger preterm labor as well?

Periodontitis, an inflammation of the gum tissue surrounding the roots of teeth, cannot be diagnosed by inspection alone; instead, one must probe the space between tooth and gum (the gingival crevice) with a probe to identify detachment. The depth of the detachment increases with the severity of disease.[40] Caused initially by several types of bacteria, periodontitis ultimately results in loss of bone.[41] The inflammation and tissue destruction have been theorized to release inflammatory agents (e.g., endotoxins and cytokines); these, in turn, might nudge the pregnant uterus into premature labor.[42]

Table 15-1
"Antecedents and Contributing Factors"
for Premature Birth Cited in Williams' Obstetrics, 23d Edition[38]

Bacterial vaginosis (bacterial imbalance in vagina)
Birth defects
Cigarette smoking
Genetic predisposition
Gum (periodontal) disease
Illicit drug use
Inadequate weight gain
Intrauterine infection
Maternal age (young or old)
Occupational factors (strenuous work, long hours)
Physical abuse
Poverty
Prior preterm birth
Psychological factors (depression, anxiety, stress)
Race (African-American)
Short interval between pregnancies
Short stature
Threatened abortion (vaginal bleeding)
Vitamin C deficiency

The floodgates open

Researchers using observational studies (not randomized controlled trials free of bias) quickly found statistical associations between periodontitis and prematurity. Early case-control studies found strong associations, with odds ratios ranging from three to eight (a three- to eight-fold increase in risk among women with periodontitis). Cohort studies reported relative risks from two to 20, the latter a powerful association.[42] Although some reports found weak or no association, most studies pointed to a link. The fundamental question, however, remained unanswered: is gum disease *causally* related to prematurity or only a marker for other causes?[42]

Randomized trials: issue settled

The only way to tease out the answer to this important public health question was to test the hypothesis in a formal randomized controlled trial. Women with periodontitis in pregnancy would be allocated to treatment in early pregnancy or to no treatment (or sham treatment, such as teeth cleaning). If periodontitis were causally related to prematurity, reducing the extent of disease should reduce the risk of preterm birth.

Because of the worldwide publicity[43] surrounding the hypothesis, a number of trials were quickly done and published. Soon thereafter, researchers began to summarize these trials in formal meta-analyses;[44-49] these syntheses increase the power (ability) to find treatment benefits if they existed. As of this writing, 12 trials have been included in meta-analyses, and the answer is striking. In the five trials[40,50-53] that were well-done (with low risk of bias), no benefit was seen. In contrast, in lesser-quality trials, benefit was seen. Bias in the suboptimal trials accounted for the apparent benefit. Failure to follow the rules of conduct can introduce large amounts of bias.[54,55] Regrettably, many trials do not follow good research practices.[56]

The strong associations seen in some observational studies disappeared under the cold light of better science. These formal experiments refuted the notion that periodontitis can cause prematurity. Instead, it is merely a marker for other factors related to preterm birth.

Dirty finger nails

Dirty finger nails are illustrative. They reflect poor general hygiene, lower socioeconomic status, and other personal characteristics that jeopardize health. Dirty fingernails may be causally involved in the transmission of intestinal diseases through hand-to-mouth contamination. Studies have found that children with dirty fingernails have a five-fold increased risk of dehydration from diarrhea in Bangladesh.[57] Similarly, dirty finger nails were weakly associated with intestinal parasites in Ethiopian children.[58] Unhygienic finger nails have also been linked to hookworm infection and anemia in the same country.[59]

However, dirty finger nails have also been linked to carious teeth (cavities).[60] The dirtier the nails, the greater were the number of decayed, missing or filled teeth. The authors noted that this association was not causal; instead, dirty finger nails were a simple marker of hygiene, which predicts dental health. In the same vein, tattoos and piercings (other than the ear lobe) are markers of risk-taking behaviors in adolescents. These behaviors include the use of "gateway" or introductory drugs, use of hard drugs, unprotected sex, and suicide.[61] To conclude, however, that tattoos and piercings cause drug abuse would be inappropriate.

Association, not causation

Abortion does not cause prematurity. Periodontitis does not cause prematurity. Dirty finger nails do not cause dental cavities. Tattoos do not cause illicit drug use. Yellow fingertips (from holding cigarettes) do not cause heart attacks. All of these are potential markers, but not causes, of poor health outcomes.

No major medical or public health organization, including the Institute of Medicine, holds that abortion causes premature birth. Standard textbooks, such as **Williams' Obstetrics**,[38] concur. To suggest otherwise reflects a lack of understanding of the science or an intentional misrepresentation of it.[36,62,63]

Take-home messages

- *Many risk factors for premature birth have been identified; abortion is not one of them*
- *Women having abortions are different in important ways from women continuing to delivery*
- *Well-done studies consistently find no association between abortion and prematurity*
- *Major medical and public health organizations in the U.S. concur that abortion is not causally related to preterm birth*

Chapter 16
Fetal feelings?

> *"We can never know whether fetuses feel pain, because they can never tell us."*
>
> *Goodman*[1]

While dabbing her eyes with a tissue, my patient, in her late thirties, hesitated, then asked, "Just one more thing..... Will my fetus have any pain?" She and her husband had chosen a second-trimester abortion after her amniocentesis indicated a lethal chromosomal anomaly. Despite the grim prognosis—no medical intervention would save the life of her child if born—the decision to end the pregnancy had been difficult for them. The pregnancy had been desperately wanted. "No, that's not medically possible," was my initial response, and then I shared with the couple the explanation below. They were relieved by the news. I hope you will be as well.

Concern about the potential for fetal pain has both medical and political roots. The discovery that operations on newborn babies cause hormonal responses that are similar to those seen in adults raised attention.[1] Opponents of abortion then inappropriately extrapolated this experience to the fetus, their goal being further regulation or outright prohibition of abortion.[1-4]

As noted by Goodman[1], no direct answer about fetal pain will ever be possible. However, several compelling, indirect lines of evidence indicate that concern about possible perception of fetal pain is misplaced: no fetus can feel "pain" in the true sense during abortion. First, the neural pathways that carry sensations to the brain are not yet in place. Second, even if these pathways were connected, a fetus lacks the emotional context required for pain perception. Moreover, the question is irrelevant to abortions in the U.S.

What is pain?

The International Society for the Study of Pain defines it as, "An unpleasant sensory and emotional experience associated with actual or potential tissue damage, or described in terms of such damage."[5] The biological role of pain is to help avoid injury. As noted above, pain involves two distinct elements: sensation of the stimulus and emotional reaction to it. Both elements are required. One must have achieved a certain level of neural functioning, and one must have had prior experience with pain in order to understand it.[6-8] The areas of the higher brain that process these functions are anatomically distinct and located some distance from each other.

Neural pathways: laying down the cables

Nerves from the periphery of the body make their way to the spinal column early in embryonic development. As a result, an embryo can respond to being touched as early as 8 weeks. However, this response involves only the spinal cord, not the brain. If a clinician taps your patellar (knee) tendon with a rubber hammer, your knee jerks in response. This reaction is governed by your spinal cord, with no conscious thought on your part involved. An anesthetized person will respond to painful stimuli by moving away and mounting a hormonal response, yet will feel no pain. Similarly, when an amoeba (a one-celled animal) is given a mechanical shock, it stops moving in response—without any perception of pain.[9] The common jellyfish has a basic neural network but no head or central nervous system.[10] It swims directionally in response to sensory stimuli, but few would argue that it is conscious.

If, as some claim, the mere act of responding to an unpleasant stimulus constitutes pain, then house thermostats are suffering by the millions.[7] When the room temperature strays beyond a comfortable threshold, the thermostat on the wall responds to this noxious stimulus with an electrical signal prompting corrective action. However, even the most sentimental homeowner would not suggest that the thermostat is suffering. These examples from biology and physics indicate that nociception (noxious stimulation) is necessary but insufficient to be pain.

Connecting the periphery to the brain takes longer than connecting to the spinal cord (Table 16-1)[11]. The nerve fibers that carry impulses up to the brain develop late in pregnancy. Around 20 weeks, the connections between the spinal cord and thalamus (a switchboard or way-station in processing impulses) are complete, and these nerves are fully insulated (covered with a sheath of myelin) by week 29 (Figure 16-1).

Noxious impulses cannot reach the brain until week 24 to 26.[8,12] The thalamus begins its links to the brain (located above) around weeks 24-26. These functional connections to the pain-receiving areas of the brain are completed during weeks 26 to 34. Until these connections from the thalamus to the brain's sensory regions are in place, fetal perception of stimuli is not biologically possible. As shown in Figure 16-2, nearly all abortions in the U.S. take place at gestational ages before the possibility of fetal perception of noxious stimuli.

Table 16-1

Milestones in the development of the fetal nervous system

Event	Duration of pregnancy (weeks from last menses)
Pain receptors in the skin appear	7 to 20
Pain fibers from the periphery reach and synapse with the spinal cord	10 to 30
Connections between spinal cord and thalamus (switching center) develop	20
First electrical activity in cortex of brain	20
Pain pathways develop protective sheaths (myelin)	22
First nerves reach base of cortex (cortical plate)	20 to 22
Connections between thalamus and cortex complete	26 to 34
First symmetric and organized brain waves develop	26
Evoked electrical activity in response to stimulation	29
Electrical patterns of sleep and awake evident	30

Source: Vanhatalo[11]

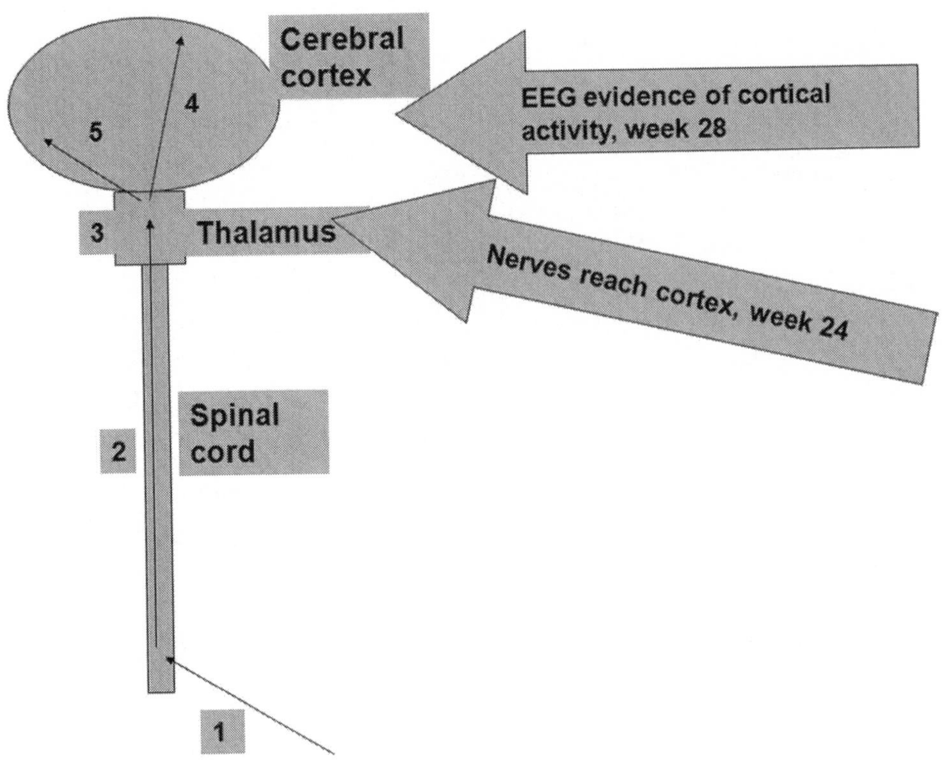

Figure 16-1

Nerve pathways to the brain

Nerve fiber (1) carries sensation to the spinal cord. There, it synapses (communicates with) a second nerve fiber (2) called the ascending neuron. This nerve passes up to a lower brain structure called the thalamus, where it synapses with the next neuron (3). From the thalamus, the sensation is transmitted to two separate areas of the brain cortex. One area (the somatosensory cortex) (4) perceives pain. Another (the limbic system) (5) processes the affective aspects of pain. Thus, the stimulus must reach the cortical (higher) regions of the brain for pain to be felt.

Sources: Vanhatalo,[11] Royal College of Obstetricians and Gynaecologists[8]

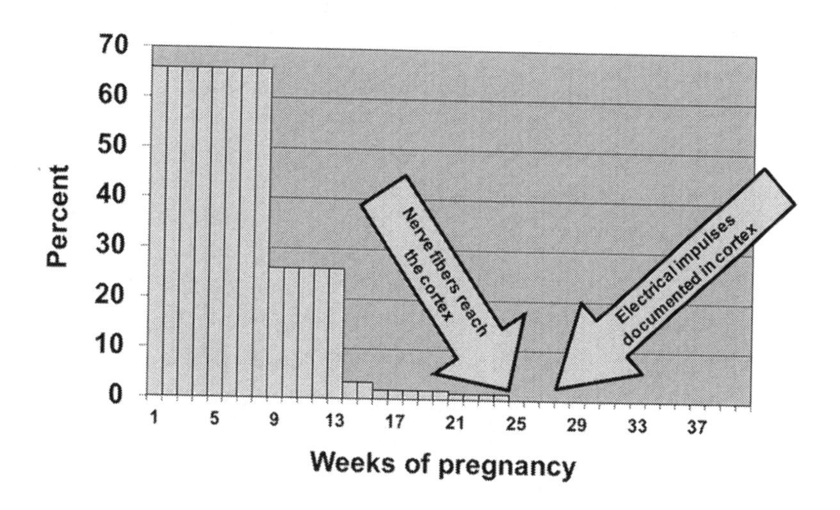

Figure 16-2
Distribution of abortions in the United States by duration of pregnancy and timing of pain fibers reaching the fetal brain
Sources: Pazol,[13] Vanhatalo[11]

Fetal sentience?

Although the wiring may be in place, is the fetus sentient?[8] Can it think or feel? Electroencephalograms (brain-wave studies) of fetuses provide indirect evidence concerning the earliest possible timing. These studies indicate that brain electrical activity begins around 20 weeks. However, it is disorganized. Only by week 26 does the pattern become synchronized. The most relevant information relates to the ability of the brain to register electrical activity in response to stimulation of the periphery. Only at weeks 28 to 29 can these evoked electrical potentials be seen. Hence, functional messaging from the periphery to the higher brain regions begins at weeks 28 to 29.[6,8,11]

The pain palette

Pain is uniquely human. It represents an amalgam of thought, sensation, and emotion. These three elements cannot be disentangled, and each is essential to the definition. As noted by Derbyshire,[14] one can image pain as a palette of unique colors. Red, blue, and yellow paint mixed together yield brown. In contrast, red and yellow mixed together produce orange. Changing the amount of blue paint added will alter the brown tint, but taking blue away altogether from the mix yields orange, not brown. By analogy, taking away altogether the emotional component of pain produces something that is no longer pain. It might be noxious, but it cannot be pain.

Children and adults learn about pain by living in the world. Daily life provides numerous examples of how emotion, thought, and experience modulate pain. For example, anxious patients feel much more pain in the dentist's chair than do relaxed patients having the same dental work. Women attended in childbirth by a spouse, doula, or other caregiver require less pain medicine than do those without support.[15] Noxious stimuli are processed and interpreted by our experiences and feelings. For example, placebo medicines often produce dramatic reductions in pain perception. The noxious stimuli continue to reach the brain, but the perception of pain is radically altered by the expectation of relief. Persons hypnotized may endure tissue injury without reporting any sensation of pain. Patients after amputation may perceive "phantom pain" in a limb that does not exist. Indeed, all pain occurs in the brain: the statement that "my hand hurts" really is not descriptive.[16]

Thus, pain is a subjective, qualitative experience. Each person's pain perception is unique, reflecting his or her history. Philosophers concur with physicians that in the absence of a perception of "self" and a legacy of experience, no pain can occur.[17] Without an awareness of oneself and one's body, no pain can be felt. Can a fetus sense noxious stimuli after 28 or 29 weeks? Probably. But can a fetus feel pain? Given the prevailing biological and philosophical definition, the answer must be no.[6,8,17-20]

Medical Research Council Report

In 1999, the U.K. Department of Health empanelled an expert committee to address fetal pain. The panel included several representatives of a variety of disciplines, including obstetrics, pediatrics, anesthesia, biology, anatomy, and pharmacology. The committee concluded that both neural pathways and consciousness (awareness) are required for perception of pain. They noted that reflex activity of a fetus should not be construed as "pain." Perception of pain "will mature over many pre-and post-natal months to produce complete pain awareness."[16]

A systematic review from the University of California, San Francisco

A multidisciplinary, systematic review of the world's literature on fetal pain concluded that perception of pain is unlikely before the third trimester of pregnancy.[12] One of the authors was an anesthesiologist who pioneered anesthesia techniques for fetal surgery. Published in the prestigious Journal of the American Medical Association, the manuscript underwent rigorous peer review before acceptance. The article attracted considerable media attention, since its conclusions directly contradicted legislation then pending in the U.S. Senate. The Brownback bill (S. 51, 109th Congress)[21] stated that "At least 20 weeks after fertilization, an unborn child has the physical structures necessary to experience pain." The article proved to be an embarrassment for those supporting legislation requiring doctors to advise women about fetal pain. Since the science was unassailable, critics instead attacked the authors and the editor of the journal.[22]

Medical organizations

A Working Party of the Royal College of Obstetricians and Gynaecologists was asked to review all the relevant evidence; its summary report was published in 2010.[8] The authors included developmental biologists, pediatricians, obstetricians, midwives, and lay persons. The report concluded that, "Connections from the periphery to the cortex [higher brain] are not intact before 24 weeks of gestation....The lack of cortical connections before 24 weeks, therefore, implies that pain is not possible until after 24 weeks."[8]

The report suggested responses to common questions from patients. In reply to the question, "Will the fetus/baby feel pain?" the suggested answer began with, "No, the fetus does not experience pain."

Among the conclusions of the report was that, "A further important feature is the suggestion, supported by increasing evidence, that the fetus never enters a state of wakefulness in utero and is bathed in a chemical environment that induces a sleep-like unconsciousness, suppressing higher cortical [brain]activation."[8]

The American College of Obstetricians and Gynecologists has also refuted the claims of fetal pain. In 2005, ACOG noted that "no legitimate scientific data" support the theory that a fetus can feel pain at 20 weeks of pregnancy. The College also warned of the dangers of government intrusion into medical matters: "Requiring a physician to read a government-mandated script that is not supported by scientific information violates the established doctrine of medical informed consent."[3]

In response to another wave of legislation at the federal and State level concerning the myth of fetal pain,[4] ACOG weighed in again. "Here are the scientific facts concerning fetal pain: A rigorous 2005 scientific review of evidence published in the Journal of the American Medical Association (JAMA) concluded that fetal perception of pain is unlikely before the third trimester." The statement went on to say that, "Personal decision-making by women and their doctors should not be replaced by political ideology."[20]

Third-trimester abortion: an oxymoron

First, abortion is defined as removal of the fetus from the uterus prior to viability.[23] Landmarks of 20 weeks or 500g weight are commonly used in dictionaries.[24] Since viability (often with severe life-long impairment) may begin around 23 weeks, any later termination is after viability and thus not an "abortion." Alternative terms should be used for procedures after viability, such as "premature birth"[23] or "termination of pregnancy."[25] These are obstetrical interventions but not abortions.

Second, claims that large numbers of "third-trimester abortions" were being done prompted the CDC to investigate. Its researchers carefully reviewed all reports of "third-trimester abortion" (defined as 25 weeks' gestation or later) in the State of Georgia over a two-year period.[26] Nearly all such reports were coding errors. Fetal deaths that occurred naturally accounted for a large proportion of such "abortions," and errors in estimating the date of last menstrual period accounted for most of the remainder. For example, if a woman had no menstrual period for four months after stopping oral contraceptives and conceived thereafter, an abortion requested 3 months later would appear to be seven months in duration (counting from the last menses). Only three cases were found to be at 25 weeks or later; two were due to fetal anencephaly (lacking a brain), and one case had insufficient documentation to determine the cause.

Third-trimester terminations

Third-trimester pregnancy termination, albeit rare, does exist. Advances in prenatal diagnosis in recent decades have led to increases in third-trimester pregnancy terminations for medical reasons. Chervenak has been the most articulate in defining acceptable fetal grounds for ending a pregnancy after potential viability.[27-29] Fetal conditions that are uniformly fatal during pregnancy or soon thereafter and those incompatible with cognitive function justify offering termination as an option. As pointed out by Chervenak, this course of action is consistent with the beneficence-based (altruistic) obligations of both the pregnant woman and physician. Since neither survival nor brain function can occur, neither the woman nor physician has an obligation to prolong fetal life. A few conditions (Table 16-2) meet these strict criteria. The advisability of considering fetal sensation during these rare procedures is unclear. Options include feticide by injection of drugs into the amniotic fluid or fetus[30] or administration of general anesthesia to the woman (which, in turn, may anesthetize the fetus).

Table 16-2

Examples of conditions for which third-trimester pregnancy termination may be appropriate

Condition	Explanation
Trisomy 13	An extra chromosome 13, which is uniformly lethal
Trisomy 18	An extra chromosome 18, which leads to death during pregnancy or soon after birth
Renal agenesis	Absent kidneys, which is uniformly lethal
Thanatophoric dysplasia	The most common lethal skeletal growth abnormality[31]
Alobar holoprosencephaly	Complete failure of the brain tissue to develop into left and right hemispheres, with fused thalamus and basal ganglia (lower brain centers)[32]
Anencephaly	Absent brain

Sources: Chervenak[27-29]

Noxious childbirth?

> *"...I wonder what sensation adults might feel if squeezed for 10 minutes or so through a passage just a little smaller than their heads. Or is someone about to advocate general anaesthesia for all babies as they are born?"*
>
> *Goodman*[1]

To recommend that fetuses having vaginal delivery receive analgesia or anesthesia establishes a dual standard of care that society is unlikely to embrace.[21] Clinicians traditionally avoid any consideration of fetal analgesia or anesthesia in labor and delivery, because of potentially depressive effects on the newborn.

Providing analgesia or anesthesia to a fetus poses a daunting challenge. Direct injection of narcotics into the blood vessels of the fetus is difficult, and injection into fetal muscle would take time to have effect. Injection into the amniotic fluid yields drug levels in the fetus that are too low to be therapeutic.[21] Administration of narcotics to the woman is not workable, since the drugs cross the placenta to the fetus slowly and in low concentration.[8] To reach adequate levels in the fetus, the woman would have to receive an overdose. Administering general anesthesia to pregnant women is more risky than to other women because of different anatomy and physiology during pregnancy. Hence, administering effective analgesia to a fetus without having harmful effects on the woman "is a considerable challenge."[33]

Ironically, the stressful process of labor and vaginal birth is beneficial for the newborn. Vaginal delivery leads to higher levels of catecholamines (e.g., adrenalin), cortisol (a stress hormone) and endorphins (pain relievers) than does elective cesarean delivery.[8] The stress of vaginal birth confers health benefits. Infants born after labor have a lower risk of breathing problems (e.g., transient tachypnea of the newborn) soon after birth and in later life than do infants born by cesarean delivery without labor.[8,34,35] The stress of labor and vaginal delivery helps the fetus.

Moreover, a new standard might have to apply to newborns as well. At present premature babies in hospital endure numerous noxious procedures as a part of daily care: heel sticks with a needle to draw blood for tests, insertion of catheters into veins, lumbar punctures, insertion of breathing tubes, etc.[16] At present, little concern is given to the potential pain of these tiny patients. Their discomfort would seem more worthy of concern than that of fetuses without a functioning central nervous system.

In summary, because of late development of neural pathways and the lack of self-awareness, fetal pain cannot occur during abortion.[8,20] The lack of concern about noxious experiences during vaginal birth (such as forceps delivery)[36] and in the nursery (such as blood-drawing[16] and circumcision[37,38]) suggests that the current legislative focus on the fetus may have motivations other than beneficence.[1,4,39]

Take-home messages

- *A fetus cannot perceive pain for two reasons:*
  - *The nerves carrying noxious stimuli do not reach the base of the brain until around 24 to 26 weeks of pregnancy*
  - *Perception of pain requires filtering of noxious stimuli through a lifetime of experiences, which the fetus does not have*
- *Major medical organizations on both sides of the Atlantic Ocean concur that fetal pain is not a concern during abortion*

Chapter 17
"Partial-birth abortion:" a distinct non-entity

> *"Moreover, what makes the term 'partial-birth abortion' politically powerful is its inaccurate conflation of two polar-opposite results of pregnancy, birth and abortion."*
>
> *Annas*[2]

The call came on a sunny Sunday afternoon in June. An obstetrician at a local hospital had a patient in trouble. The pregnant woman was desperately ill with HELLP syndrome, a severe form of preeclampsia (toxemia) in the middle of pregnancy.[1] Like the patient described earlier in this book, she had high blood pressure and kidney disease, the liver enzyme levels in her blood were skyrocketing, and her platelets (blood elements that help clotting) had nearly disappeared.

Toxemia is related to the placenta, and the definitive treatment is to empty the uterus…the only known way to save the woman's life. Despite several days of attempts to induce labor, delivery was not imminent. Her obstetrician asked if I would come and do a D&E abortion to save her life. He would arrange emergency surgical privileges for me at his hospital within the hour. I agreed to come immediately.

After speaking with the patient and her family, I took her to the operating room, anticipating performing a standard D&E. On examination under anesthesia, I found her cervix open. Given her grave condition, I reached into her uterus, turned the immature fetus and delivered it feet-first. The intact D&E took less than three minutes, and blood loss was negligible. She and her family were deeply appreciative, and she left the hospital a week later.

An operation ruled illegal by Congress in 2003 (intact D&E) proved to be the best resolution of her life-threatening illness. Labor induction had failed, a standard D&E would have taken longer and risked heavy bleeding with her clotting disorder, and a cesarean delivery would have been still more dangerous.

"Partial-birth abortion": a distinct non-entity

Coined by opponents of abortion, "partial-birth abortion" has no basis in science or medicine. Indeed, it does not appear in medical dictionaries[3] or medical coding manuals.

Background of the brouhaha

The "partial-birth abortion" strategy stemmed from the failure of abortion opponents in a key 1992 Supreme Court decision and a new threat in Congress. In *Planned Parenthood of Southeastern Pennsylvania v. Casey*, the U. S. Supreme Court resoundingly confirmed the core of *Roe v. Wade* by noting that States could not outlaw abortion before fetal viability and could do so after viability only if the life or health of the woman were not threatened by the pregnancy. Hence, the judicial prospects for overturning *Roe v. Wade* seemed bleak, and a new tactic was needed.[2]

In the spring of 1993, supporters of abortion rights decided to shift from the defense to the offense. They had introduced into the Senate the Freedom of Choice Act (S.25) which would prevent State and local governments from restricting women's access to abortion. In other words, the bill would codify the decision of the Supreme Court in *Roe v. Wade*, two decades earlier.[4] Dr. Martin Haskell of Ohio had described his approach to intact D&E at a medical meeting in 1992, and the National Right to Life committee obtained a copy of his paper, which was then shared widely. They decided to use this little-known procedure as the centerpiece for their attack on the Freedom of Choice Act by arguing that passage would "lead to an increase in the use of this grisly procedure."[4]

Framing the issue proved critical.[5] Initially, Right to Life used the medical term "dilation and extraction" in more than 4 million brochures and ads. However, this was viewed as medical, bland, and not sufficiently provocative. Other options were considered: "brain-suction abortion" and "partial-birth abortion." The former was deemed too extreme, "provoking a nervous, almost comical, response."[4,5] The final choice, "partial-birth abortion," became a resounding public-relations success: it framed the rhetoric in an altogether different way. Indeed, as noted by one legal scholar, the juxtaposition of abortion and birth in one phrase is intentionally deceptive and thus unnerving.[2] The outcomes of abortion and birth could not be more different.

The first "partial-birth abortion" ban (H.R. 1833) was introduced into the House of Representatives in June of 1995. It passed both chambers of Congress but was vetoed by President Clinton in April of the following year. The House mustered sufficient votes to override the Presidential veto, but the Senate did not, so the bill died in September of 1996.

The graphical images introduced into Congressional testimony and carried nationwide by television were powerful public-relations tools.

Indeed, the normal response of lay persons to the details of extirpative operations is repulsion, whether the tissue being removed is a fetus, an ovarian tumor, or a gangrenous foot. Graphic descriptions and images of any surgery—by their very nature—are generally not for the faint of heart.[4] In these Congressional discussions of surgical technique, the preoccupation with the fetus led to the woman being forgotten. Moreover, the discussion subtly shifted from the outcome to the process of abortion.

In 1997, a similar ban was introduced once again (H.R. 1122) and, again, reached a dead end. As before, it quickly passed in both chambers of Congress and was vetoed by President Clinton. It achieved a vote to override the veto in the House but not in the Senate. With the end of the Congressional session, the bill died.

The next major development in "partial-birth abortion" legislation came from the U.S. Supreme Court in 2000. Nebraska, like many States, had passed abortion bans modeled after the federal bills. As of September, 1998, 28 States had passed copycat legislation; of the 20 bans challenged in court, 19 had been enjoined or severely limited.[4] Leroy Carhart, M.D., a physician in Nebraska, challenged the State law, and the case (*Stenberg v. Carhart*) reached the U.S. Supreme Court. On June 28, 2000, by a vote of 5 to 4, the Court ruled the Nebraska law (and all other State laws banning "partial-birth abortion") unconstitutional.

Two fundamental defects led to the ruling. First, the Nebraska law had no exception to preserve the health of the woman, as required by *Roe v. Wade*. Second, the statute imposed an "undue burden" on a woman's liberty to end her pregnancy before fetal viability. Because the language of the statute was so inarticulate and vague, the bill would threaten access to dilation and evacuation (D&E), the most common means of second-trimester abortion.[6] As noted by Justice Ginsburg, the law would not save any fetus or protect women's health; hence, Nebraska had no reasonable interest in enacting such legislation.[7]

Undeterred by this clear direction from the Supreme Court and with a new President in the Oval Office, abortion opponents renewed the attack. In June of 2002, H.R. 4965 was introduced into Congress. It passed in the House but was not considered in the Senate.

Success, however, was just around the corner. In February of 2003, the Partial-Birth Abortion Ban Act of 2003 was introduced, passed in both chambers, and then signed into law by a jubilant President Bush. With Roman Catholic Cardinal Egan at his side, Bush (who has no medical training) announced on November 5, 2003, that the now-outlawed procedure was "...harmful to the mother, and a violation of medical ethics."[8] Within 48 hours, federal judges had enjoined the new (illegal) law to protect the rights of women and physicians while legal challenges took place in three States: California, New York, and Nebraska.

A judicial "hat trick"[9]

Because these fundamental defects had not been addressed, three separate Federal District Courts ruled unanimously that the 2003 federal ban was unconstitutional. All three judges cited the lack of a health exception as grounds for invalidation. Judge Hamilton in San Francisco and Judge Kopf in Nebraska also found the law unconstitutional because of the "undue burden" it posed for women. (Judge Casey in New York did not address "undue burden" since he had found it unconstitutional for other reasons.)

The judicial criticism of Congressional scholarship was harsh. As noted by Judge Hamilton, "...oral testimony before Congress was not only unbalanced, but intentionally polemic....This Court heard more evidence during its trial than Congress heard over the span of eight years....Even the Government's own experts [in this case] disagreed with almost all of Congress's factual findings."[9]

The federal government appealed the decisions, which reached the U.S. Supreme Court in 2006. On April 18 of the following year, the Court held (by a 5-4 decision) in *Gonzales v. Carhart* that the law did *not* violate the Constitution.[10] The replacement of Justice Sandra Day O'Connor by Samuel Alito in early 2006 tipped the balance.

Frequency of intact D&E

Trying to determine the frequency of these procedures poses a challenge. Claims that intact D&E procedures are infrequent are difficult to substantiate, since the Centers for Disease Control and Prevention does not gather information this detailed. Justice Breyer, in his ruling in *Stenberg v. Carhart*, estimated from 640 to 5000 cases nationwide each year.[7] The Guttmacher Institute provided the best national estimate, based on its surveys of providers. In 1996, intact D&E was rarely used, accounting for 0.03 to 0.05% of all abortions. This translated into an estimated 650 such operations.[4]

Safety evidence, at last

Despite claims to the contrary, intact D&E has comparable safety as does traditional D&E. Chasen and colleagues at Weill Cornell Medical College in New York City reported the outcomes of 120 women having intact D&E procedures compared with a contemporaneous comparison group of 263 having traditional D&E at the same facility.[11] Women in the former group had pregnancies two weeks further advanced, on average (23 weeks vs. 21 weeks). The large majority of women had fetal indications for the abortion, including abnormal fetal chromosomes, fetal structural anomaly, and fetal death. Early complications occurred with similar frequency – 5% – with both techniques. Median procedure times were comparable (22 min.), as were median estimated blood losses (100 mL, less than half a cup). Data on long-term consequences were more limited: 62 women had later pregnancies and deliveries at the same medical center. No second-trimester miscarriages occurred in either group, and the risk of premature birth was not significantly different. A follow-up from the same institution found no differences in later obstetrical outcomes between intact and standard D&E.[12] Because of the infrequency of both the operation and complications thereafter, large trials are unlikely to be done.

These reports provided long-needed data to address widely cited concerns about putative risks of intact D&E. These claims included cervical incompetence, uterine injury, abnormal placental attachment, and amniotic fluid embolism.[13,14] Although the power of this cohort study was limited to address rare complications, no large discrepancies were evident between the two surgical approaches. The authors concluded that, "Attempts to regulate intact D&X [intact D&E] on the basis of concern for maternal well-being cannot be supported by available evidence."[11]

An accompanying editorial[15] noted that former Senate majority leader William Frist (R-TN) claimed that enactment of the "partial-birth abortion" ban would "save lives."[16] A former cardiothoracic surgeon, Sen. Frist presumably had no relevant gynecologic experience. No evidence was available to support his assertion then; this remains true today.[11]

The American College of Obstetricians and Gynecologists

The professional organization representing board-certified obstetrician/gynecologists (ACOG) has been resolute in its defense of

> *"The intervention of legislative bodies into medical decision making is inappropriate, ill advised, and dangerous."*
>
> *American College of Obstetricians and Gynecologists*[17]

women and their doctors. A 1997 ACOG Statement of Policy on Intact Dilatation and Extraction said unambiguously that "The physician, in consultation with the patient, must choose the most appropriate [abortion] method based on the patient's individual circumstances."[17] The College went on to note that, "The potential exists that legislation prohibiting specific medical practices, such as intact D&X, may outlaw techniques that are critical to the lives and health of American women." The College cited examples of how the vagueness of "partial-birth abortion" legislation might compromise other pregnancy care. For example, draining excessive fluid from the head of a hydrocephalic fetus during term delivery that resulted in fetal death might fall under the proposed law. Alternatively, assisting with a miscarriage during which the fetus dies might be prosecuted under the legislation.

ACOG has consistently argued that these difficult decisions are the province of patients and doctors, not legislators. The College filed an *amicus curiae* brief in the successful challenge to Nebraska's abortion ban, and it has consistently opposed State and federal legislation of this type. The College has complained that, "legislators try to circumvent the Court's requirements by issuing their own opinion to the nation's physicians and patients that such a procedure is never needed to protect a woman's health—notwithstanding opposing opinions from the medical community. The medical misinformation currently circulating in political discussions of abortion procedures only reinforces ACOG's position: in the individual circumstance of each particular medical case, the patient and physician—not legislators—are the appropriate parties to determine the best method of treatment."[18]

Congress, doctors, and patients

According to the Constitution, protecting the health and safety of the public is the domain of individual States. Physicians hold State, not federal, medical licenses. The Congress does not have the authority to regulate the practice of medicine, unless one views the practice of medicine as a form of interstate commerce, over which the Congress does have jurisdiction.[2] Most of us consider our health care to be fundamentally different from interstate trucking.

Maternal-fetal medicine experts at Massachusetts General Hospital offered several clinical scenarios that might warrant an intact D&E.[19] Eisenmenger syndrome (a congenital heart defect leading to heart failure) has a 50% maternal mortality risk if a pregnancy continues to delivery. A woman with chronic renal failure has a one-in-twenty chance of losing her remaining kidney function if she carries a pregnancy to term. A diabetic woman with progressive retinal disease risks blindness if pregnancy is continued. In these very ill women, who is best qualified to decide on treatment options?

Sexism in medicine

Predictably, Congressional intrusions into medicine are sexist: they disproportionately relate to women.[7] Examples include Congressional legislation regarding duration of hospital stay after childbirth and a resolution promoting mammography for women in their forties, despite limited and conflicting medical evidence.[20] As warned by a former editor of the New England Journal of Medicine, Congress "is not the appropriate forum for making complex medical decisions."[20]

The government's intrusion into decision-making between women and their physicians, however, is consistent with the discrimination women typically encounter in other private medical matters. In one study, courts honored the right-to-die requests of 75% of men but only 14% of women.[21] Compared with men, women's end-of-life wishes[22] or cardiac symptoms[23] are consistently more likely to be discounted or ignored. The State-mandated care of a brain-dead pregnant woman in Texas,[24] in violation of her and her family's wishes, was reminiscent of the government intrusion into the care of Terry Schiavo.[25]

Medical fallout: harms to women

The Partial-Birth Abortion Ban Act of 2003 has hurt women both directly and indirectly.[26] A direct consequence was restricted access to care. In Massachusetts some providers began doing unnecessary medical procedures, one stopped providing D&E abortions without specific indications, and another reduced its services. Costs at three facilities rose as a consequence.[27,28]

> *"Congress should stay out of the examining rooms, the maternity ward, and the operating rooms."*
>
> *Kassirer*[20]

An indirect consequence was preventable suffering and complications. In response to the vagueness of the Act, some physicians and abortion providers adopted a defensive approach. If the fetus is not alive at the time of an abortion, then the provisions of the Act do not apply. Feticidal injections of potassium chloride or digoxin into the amniotic fluid or into the fetus to stop fetal heart activity began to be more widely used before second-trimester abortion.[27,29] This involved an unnecessary procedure for the woman, time and expense on the part of the physician, added costs of drugs, and needless pain. Some providers thought this to be a reasonable trade-off to avoid running afoul of the new law.

Soon, harms of this approach began to emerge. Women who had these feticidal injections were more likely to have vomiting, infection, and miscarriage outside the medical facility.[30,31] The putative medical benefits of feticidal injections did not materialize: a well-done trial found no improvements in the difficulty and duration of the operation.[31] Given clear evidence of harm and no known benefit (other than protecting the physician), feticidal injection in this context was deemed unethical.[32]

Regardless of the initial political motivation behind this aberrant legislation, its net effect has been needless suffering of women, which is unethical.[33] When legislators play doctor,[20] women suffer.

Take-home messages

- *Partial-birth abortion is a political, not a medical term*
- *The safety of intact D&E is comparable to that of standard D&E*
- *In some instances, intact D&E is the preferable abortion method*
- *Banning intact D&E led to needless and harmful procedures*
- *Basing laws on aesthetics rather than evidence is wrong*

Chapter 18
State legislatures: practicing medicine without a license

> *"...we can expect to be [sic] a dramatic decrease in the numbers of abortions nationwide, since this is when the flood of pro-life legislation as a result of conservative gains in the 2010 midterm elections goes into effect....tools we can use to close clinics..."*
>
> *Cheryl Sullenger*
> *Senior Policy Advisor*
> *Operation Rescue*[1]

Ms. Sullenger (see box on right) knows about shutting down abortion clinics. A convicted felon, she was sentenced to three years in federal prison after pleading guilty to trying to firebomb a San Diego clinic in 1987.[2,3]

Over the past four decades, civil disobedience, arson and bombings,[4] and murders of health care providers[5] have not ended safe, legal abortion in the U.S. A recent strategy, oppressive State legislation, is proving more effective. The strategy is to make medical care so expensive and burdensome that access becomes severely restricted, especially for those in greatest need.

This is a perversion of public health practice. In public health, one identifies a problem, determines the cause, implements corrective action[6] such as new legislation and then monitors its effect. In recent years, States have been briskly responding to a non-problem. Under the specious banner of patient safety, State legislatures, unencumbered by medical evidence and often in direct contradiction of such evidence, have been piling further restrictions on the safest outcome of pregnancy.[7-11] The real goal is more sinister: returning women to the back alley.

Misogynistic Motives

Few politicians involved have the honesty to admit their intent publicly. For example, Republican State Senator Warren Daniels, sponsor of a Targeted Regulation of Abortion Providers (TRAP) law in North Carolina, reported innocently that, "I guess I don't know exactly how it will affect clinics across the State."[12] I suspect he does. TRAP laws are associated with a reduction in providers.[13] Pastor Bobby Eubanks of South Carolina, who helped a legislator craft a bill requiring abortion providers to have hospital admitting privileges nearby, was more forthcoming; he predicted that "...not a hospital in South Carolina" would grant the requisite privileges.[14]

> *"The strategy that the anti-choice movement is focusing on now is to pass laws that are purportedly to protect women's health, but in actuality are there for the purpose of shutting down abortion clinics and driving abortion back onto the black market. The reason to frame anti-abortion legislation as being for 'women's health' is an obvious one: the misogyny of the anti-choice movement is bad for its reputation, and they'd very much like the public to believe the hatred of women isn't driving this."*
>
> *Marcotte*[15]

Public health impact

Since 2010, when the GOP took control of many State legislatures, 205 anti-abortion laws have been enacted. This is more laws of this ilk than were passed in the prior decade.[16] Since 2010 more than 50 abortion providers in 27 States have ended service, though not all were due to legislation.[17] Women in rural areas now face greater challenges in reaching competent care; in Mississippi and both Dakotas, only one clinic provides abortion to the entire State.

Needless regulations hurt public health. For example, mandatory waiting periods for abortion are associated with harm to young black women. Using data from the federal Pregnancy Risk Assessment Monitoring System, researchers at the University of Rochester studied the impact of State mandatory waiting periods.[18] These State-imposed delays were related to higher rates of unplanned births among teens, especially those of color. Adolescent childbearing carries added risks to the mothers and babies, and the cost to society is large. Overall, minors who lived in States with mandatory waiting periods had twice the risk of unintended births as did teens in other States. The inability to receive same-day service, as with most other medical and dental procedures, proved critical.

TRAP laws

TRAP laws constitute governmental intrusion into the practice of medicine. Singling out abortion providers, these laws dictate medically unnecessary standards, building features, and personnel requirements. Often, the laws direct State health departments to revise existing licensing requirements and to hold clinics to the irrelevant standards of general hospitals or ambulatory surgery centers. The particulars may include wider hallways, awnings over entrances, showers for physicians, closets for janitors, and pest control in the front yard. No woman in the U.S. has ever died from inadequate door width or lack of an awning at an abortion clinic.[19-21]

Twenty-seven States regulate abortion providers in medically unnecessary ways. Of these, 24 require that abortion clinics meet the standards of ambulatory surgery centers, which provide surgical care from ophthalmology to orthopedics. Among the unnecessary regulations are procedure room size and corridor width (13 States). Thirteen require providers to have an affiliation with a local hospital. None of these laws is reasonably related to improving patient safety.[9,23]

Legislative harassment extends beyond TRAP laws. The Guttmacher Institute tallies these legislative burdens on a regular basis.[24] At the time of this writing, 39 States require that abortions (including medical abortion with pills only) be provided by a physician. This is irrational, since non-physicians have been shown to be competent providers in settings as remote as Nepal.[25] In the U.S., pharmacists administer millions of immunizations each year in drug stores.[26] Forty-one States specify a gestational-age limit. Eight States limit private insurance plan coverage of abortion. Seventeen require medically incorrect counseling before abortion, including disinformation about breast cancer, fetal pain, and mental woes. Contrary to the recommendation of the American Academy of Pediatrics,[27] 39 States require parental involvement.

Like the actor Tom Cruise (who bought an ultrasonography machine for his home),[28] politicians have developed a keen interest in sonar. Twenty-three States dictate medical practice concerning ultrasound; few of these legislators, if asked, could find the "on" switch for the machine. When Texas ordered vaginal penetration with an ultrasound probe against the will of women and their physicians, comparisons were made to other forms of sexual violence.[29] In North Carolina[30] and Oklahoma,[31] ultrasound provisions have been permanently enjoined as unconstitutional.

Although physicians have competently counseled millions of abortion patients since *Roe v. Wade*, untrained politicians decided doctors needed their help. As a result 27 States have directed their health departments to write educational materials. Twelve States add erroneous details about fetal pain. Five advise women that "personhood" begins at conception (an error explored earlier in this book). Twenty-six require unnecessary delays in receiving care, which increase risk and expense.

Public health problems in North Carolina

Two patients died in North Carolina dental offices in 2013.[32] Linked to improper use of sedatives during dental work, the deaths were unprecedented. Although the N. C. Dental Board investigated, the N.C. General Assembly did not further regulate dental offices despite this epidemic.

In the winter of 2013-14, more than 70 North Carolinians died from influenza.[33] Few of those who died had received the recommended immunization against the disease.[34] The General Assembly enacted no new laws to promote or require immunization to save scores of lives.

In 2012, 23 persons died in North Carolina boating accidents.[35] According to the U.S. Coast Guard, the boating fatality ratio for N.C. that year was 5.9 deaths per 100,000 registered boats. (In contrast, the corresponding figures for abortion deaths and abortion fatality ratios in the State were zero and zero in 2012, as they have been for decades.) In the 2013-2014 session, the General Assembly considered Senate Bill 636 (Wildlife Resources Commission Penalty Changes), but it was referred back to a committee and was not enacted.[36] Thus, no new laws addressed this ongoing boating safety problem.

In contrast, the General Assembly has been busy enacting laws to address *no* abortion deaths, a chronic problem in North Carolina. Most recently, the legislature, working late at night, attached to a motorcycle safety law (!) further regulations for the already well-regulated abortion clinics in the State.[37] North Carolina faces many problems (including its legislature); [38,39] abortion safety is not one of them.

The flurry of anti-abortion legislation in my home State is tragic, especially given North Carolina's shameful legacy of social engineering. Through its State Eugenics Board, established by the General Assembly in 1933, more than 7,000 North Carolinians were sterilized by 1977.[40,41] Most were women, poor, and African-American.[42] The State belatedly acknowledged this gross violation of civil rights and apologized in 2002.

The camel's nose under the tent

When laws are based on ideology and not evidence, and when public health is based on revelation rather than reality, medical care suffers. Both the privacy and sanctity of the doctor-patient relationship are compromised. As noted by U.S. medical specialty societies, intrusion of uninformed lay persons (viz., State legislators) into the practice of medicine is dangerous and unwarranted. Consider guns. Although firearms are an important cause of accidental deaths of children, Florida enacted the Firearm Owners' Privacy Act in 2011. It would have limited physicians' ability to discuss gun safety with parents; the Act was permanently enjoined the following year, since it violated First Amendment free-speech protection.[43]

How would the American Bar Association respond if a State legislature prohibited lawyers from discussing wills and trusts with a client? What would the Bar Association do if a State scripted the content of the private discussions between lawyer and client? How would the nation's lawyers respond if a State made depositions longer than 20 pages illegal?

> *"Government exists to protect us from each other. Where government has gone beyond its limits is in deciding to protect us from ourselves."*
>
> *Ronald Reagan*[45]

Research has identified predictors of TRAP law passage. Public attitudes about abortion, the proportion of Roman Catholics, and ideology do not predict the passage of TRAP laws in States. Instead, Republican control of both houses of the State legislature plus the Governor's office significantly predicts the passage of such laws.[13] TRAP laws are partisan politics, not public health. Given the well-established health benefits of abortion,[9,23] State laws that deter and delay abortion care violate the ethical principles of beneficence, autonomy, and justice.[44]

A brave new world?

Emboldened by recent legislative incursions into gynecology, a hypothetical religious group opposed to blood transfusion seizes the moment. It introduces in State legislatures bills to regulate the practice of transfusion medicine. The true agenda is to limit—and ultimately end—safe, legal transfusion. This is not because blood transfusion is dangerous (indeed, it is often lifesaving), but because this medical procedure conflicts with their personal values.

Numerous "A Woman's Right to Know about Blood" bills are passed. Women (but not men, who are less flighty and capricious) must receive counseling 24 hours in advance of transfusion. A State-scripted pamphlet (ghost-written by non-medical enthusiasts) describes the dangers involved. The pamphlet implies that receiving a blood transfusion is riskier than continuing to hemorrhage, which is a natural process. Hemorrhage is part of a predetermined plan, and physicians should not tinker with it. As part of the informed consent, women (but not men) must view photographs showing the development of a red blood cell from its stem cell precursor to a reticulocyte, the red cell about to be launched from the bone marrow into the bloodstream. The transfusing physician has to narrate the morphologic changes step-by-step to ensure that the woman fully comprehends red cell development. She must be told that each red cell is a living entity, separate and unique from its host (or hostess).

To ensure blood safety, the blood bank must now be housed in a building separate from the germy hospital; it has wide hallways, an awning, a janitor's closet, and no ants in the yard. To be cautious, the shelf life of donated blood is restricted to just 24 hours. Transfusion of blood after 24 hr of donation is made illegal.

Blood bank medical directors must have admitting privileges at their hospital (or another within 30 miles), despite the fact that they do not admit or treat patients. Numerous States rule that insurance, both private and governmental, cannot cover blood products. Opponents of transfusions mount a web site with a list of known offending physicians and the addresses of their hospitals.

As planned, the costs of medical care rise, and access to life-saving blood transfusions deteriorates. Affluent women anticipating the need for blood transfusions travel to States with less draconian restrictions. Poor women of color, unable to travel or pay inflated costs,[17] die instead.[46]

Take-home messages

- *Opponents of abortion are using State laws to make abortion care inaccessible*
- *Given four decades of safe service provision, the recent flurry of abortion regulations is unwarranted*
- *Claims that TRAP laws are designed to improve patient safety are patently dishonest*
- *Reducing access to abortion violates the ethical principles of beneficence, autonomy, and justice*

Section V

Looking back, looking ahead

J.M.S.

In 2009, a 17-year-old woman found herself pregnant by an older man. He is reportedly now facing charges for using her for pornography. She lived in a remote part of Utah, a long distance from an abortion clinic in Salt Lake City. Her housing was wretched. She lived in a building with no electricity or running water. To have an abortion in Salt Lake City, she would need money for the procedure, transportation, a responsible adult to drive her and accompany her home, as well as money for an overnight stay (because of Utah's 24-hour waiting period).

Since legal abortion was beyond her reach, she reportedly paid a man to beat her in the hopes that it would cause an abortion. She and the fetus survived the beating, but the State charged her with solicitation to commit murder. The charge was later dropped since her actions did not fall under the existing abortion laws of the State. In response, the State legislature changed the law so that women like her would become criminals. The case has been appealed to the State's Supreme Court.[1]

1. Goldberg M. The return of back-alley abortions. http://www.thedailybeast.com/articles/2011/06/03/abortions-return-to-back-alleys-amid-restrictive-new-state-laws.html, accessed January 18, 2014.

Chapter 19
Turning back the clock in Romania

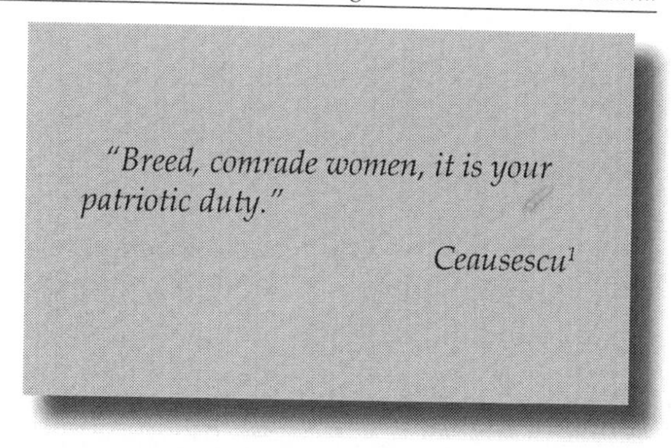

One of the poorest nations in Europe, Romania spawned two of history's worst tyrants. Although they lived half a millennium apart, Vlad the Impaler (Figure 19-1) and Nicolae Ceausescu (Figure 19-2) left similar legacies of suffering and death. Vlad, also known as Vlad III, Dracula, and Tepes ("spike" in Romanian), was born in Transylvania (part of Romania) in 1431.[2] His ruthlessness was legend, and his exploits prompted the classic novel **Dracula** by Bram Stoker. In a widely cited incident, he purged his kingdom of undesirables by inviting the hungry to a feast in his castle. After the repast, he locked the dining hall doors and set the room on fire, burning his guests alive. He disemboweled his mistress for lying about being pregnant. When Turkish ambassadors refused to remove their Phrygian caps in his presence, he had the hats nailed to their heads.

Figure 19-1
Vlad the Impaler (1431-1476)

Figure 19-2
Nicolae Ceausescu (1918-1989)

Fototeca online a comunismului românesc, photo #BA231, 1/1978 (accessed 23:37, 16 September 2010 (UTC))
This image is available from the Romanian Communism Online Photo Collection ***under the digital ID*** *44670X5163X5732*

Although death by impalement was not his invention, Vlad elevated this torture to an art form.[3] A sharpened stake would be inserted into the victim's rectum or vagina, after which the victim was suspended on the stake (vertical impalement). The weight of the body would gradually thrust the stake deeper into the body until death ensued; this took from hours to several days. Alternatively, the victim could be impaled horizontally (Figure 19-3). He enjoyed arranging the stakes in decorative patterns, such as circles; mass impalement was a specialty. On a single St. Bartholomew's Day, he reportedly impaled 30,000 merchants for violating trade laws, leaving the dead to decay on stakes outside the city walls to deter further disobedience. In another well-chronicled example, he impaled 20,000 Turkish prisoners ("the Forest of the Impaled") outside the city of Tirgoviste to thwart an attack by advancing Turks. It worked. During his six-year reign, the best estimate of the numbers killed this way is about 100,000.

Figure 19-3
Vlad the Impaler dining among his victims
Source: Woodcut from title page of 1499 pamphlet by Markus Ayer in Nuremburg, Germany

Like Vlad, modern-day Ceausescu left a trail of misery and death. His rule over Romania was exceptional not only for its carnage but also for the unique natural experiment it provided: Ceausescu turned back the clock on abortion access. As described in Chapters 2 and 3, when abortion is made legal and accessible, maternal health promptly improves.[4-7] Ceausescu showed the world that the opposite is true as well.

Abortion and the pregnancy police

Legal abortion began in Romania in 1957 (Figure 19-4).[8,9] The government authorized abortion upon request within the first trimester of pregnancy; the procedure had to be performed in a hospital, and the woman was required to pay a small fee. The government's rationale had two goals: self-determination regarding reproduction and improvement of women's health. In response to poor economic conditions and liberal access to abortion, Romanian abortions increased,[10] and the birth rate plummeted. The birth rate fell from 22 births per 1,000 population in 1957 to only 14 in 1966.[9]

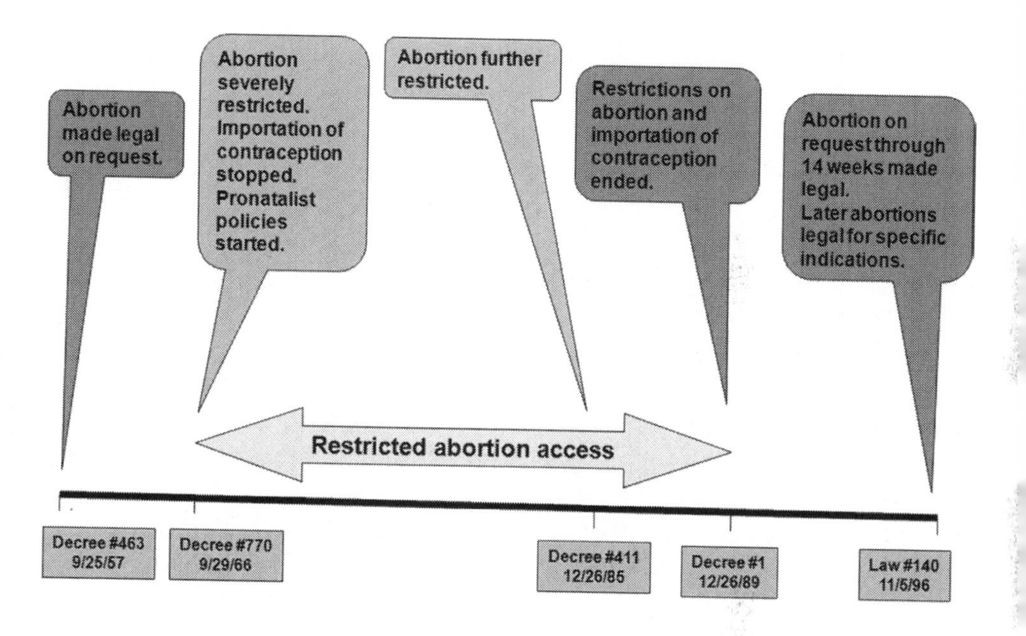

Figure 19-4
Timeline of abortion legislation, Romania, 1957 - 1996
Sources: United Nations Secretariat,[8] Hord[9]

Soon after becoming President in 1965, Ceausescu implemented pronatalist policies designed to reverse this trend. Without warning, in 1966 he severely restricted access to abortion. As specified in Council of State Decree 770, only women meeting the following criteria could have an abortion: age 45 years or older, or four or more children to support; the woman or husband had a severe disease that could cause birth defects or that could be transmitted to the child; severe disability that would prevent proper care of a newborn; life endangerment; and rape or incest.

(The age limit was lowered to 40 years in 1972 and raised to 42 years in 1984.)[8] Importation of contraceptives was outlawed, and women and providers both could be prosecuted for illegal abortions.

The program provided other incentives for marriage and childbearing. Family allowances were increased, and the income tax burden fell by about one-third for families with three or more children. Couples with a child under age 16 years found divorces more difficult to obtain.

The intrusions of the State into the private lives of women were grotesque. Women of reproductive age were forced to submit to monthly gynecological examinations to detect early pregnancies; those who declined lost their governmental benefits, such as health care and social security. Factories were given birth quotas, and factory-employed physicians received their full pay only if women at the factory reached the specified birth quota each month. To promote marriage, unmarried Romanians older than 25 years were taxed an extra 10% of earnings. If a couple had not produced a child within two years of marriage and if no medical explanation for infertility were documented, both the man and woman had to pay more taxes.

The State went so far as to establish "pregnancy police." The Romanian State Security Police set up a special unit to monitor illegal abortions, and officers were stationed in maternity wards and women's clinics to investigate possible illegal abortions. Physicians had to report all pregnant women, and their pregnancies were then monitored by the police until delivery.[8] Miscarriages were investigated to rule out intentional abortions. Women who obtained illegal abortions could be fined or imprisoned, and doctors could go to jail for a dozen years and lose their licenses to practice if found guilty. Given the juxtaposition of police enforcement on reproductive health care, women faced unparalleled dangers. Not only did they have to resort to illegal abortions, but they also feared seeking medical care when complications developed, because of possible prosecution. Women were driven deeply underground.[9]

"The Romanian experience demonstrates the futility and folly of attempts to control reproductive behavior through legislation. A law that forbids abortion does not stop women from aborting unwanted pregnancies."

Stephenson[14]

In 1985, after these policies had failed to drive up the birth rate as hoped, Ceausescu tightened the vise further. Decree 411 raised the age limit back to 45 years, and the number of children under a woman's care from four to five, all under 18 years of age. All contraception was banned.[9] In contrast, abortion rates were declining in Czechoslovakia, East Germany, and Hungary because of the availability of modern contraceptives, some locally made.[11]

Ceausescu was overthrown in 1989, and after a perfunctory trial, he and his wife were executed by firing squad on Christmas Day. The very next day, the new government took steps to mitigate the damage of 23 years of tyranny: it repealed the abortion restrictions. Contraception and sterilization restrictions were soon dropped as well. The flood gates opened. However, laws concerning abortion were not enacted until 1996. Law number 140 specified that abortions could be performed on request during the first 14 weeks of pregnancy when provided by a physician in an approved medical facility. Later abortions were allowed for specific indications.[8]

The harm to women, children, and Romanian society of the draconian pronatalism was prompt and profound. By every indicator, Ceausescu's policies proved disastrous. The impact was felt in illegal abortions, stillbirths, neonatal deaths, abandoned children, and orphans….but, ironically, not in the birth rate.[12]

Births in pronatalist Romania

In 1967, the first year after abortion was severely restricted, the birth rate nearly doubled. The crude birth rate (number of births per 1,000 population) jumped from 14 in 1966 to 27 in 1967.[13] Thereafter, it progressively declined to levels similar to those before the abortion and contraception restrictions were imposed. In the late 1980s, the crude fertility rate reached a plateau at around 16…. all in the absence of easy access to contraception and abortion.

The back alley again

Lack of family planning services took a heavy toll on the sex lives of Romanians. Coitus became dangerous for both married and unmarried couples with limited access to contraception or abortion. Many had intercourse only infrequently and under great emotional stress.[9,13] A follow-up survey done in 1991-1992 found that among 2000 women queried in five Romanian hospitals, from 65% to 86% of respondents reported that the restrictive abortion law had affected their sexual relationship.[10] While a reduction in sexual intercourse may have accounted for some of the fertility decline after 1967, illegal abortion accounted for most. Women returned to clandestine abortion once again, as they had after World War II. The cost of an illegal abortion was exorbitant...the equivalent of two months' salary.[9] Most of these providers were clinically incompetent; most providers imprisoned for performing illegal abortions were women with no medical background.

The carnage began promptly. As portrayed in Figure 19-5, maternal mortality began rising soon after the restriction of abortion. As increasing numbers of women resorted to unsafe abortion to control their fertility, the number of abortion-related deaths per 100,000 live births climbed from 23 in 1965 to a peak of 151 in 1982…an astounding six-fold increase in less than two decades.

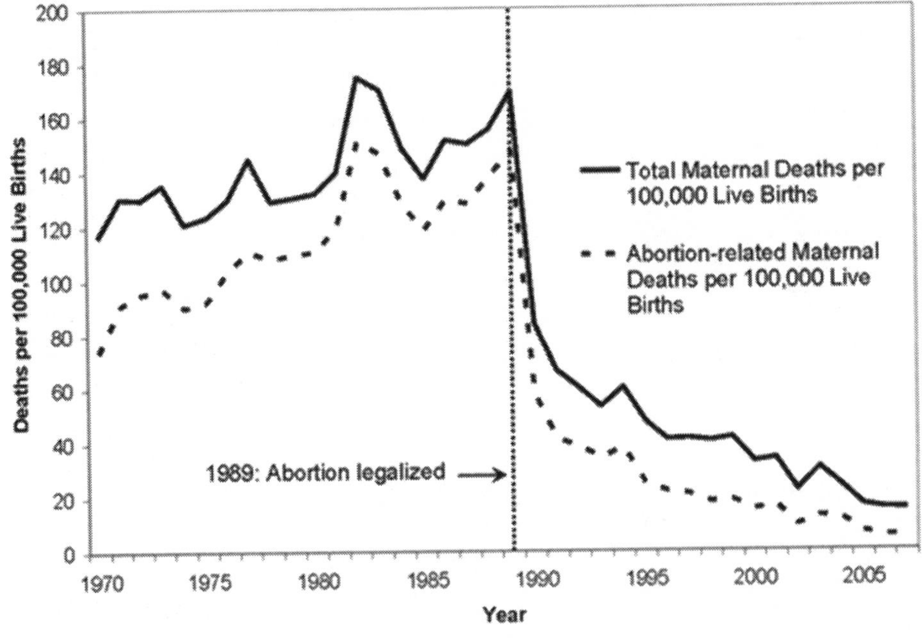

Figure 19-5
Total vs. abortion-related maternal deaths, Romania, 1970-2005
Source: Benson[15]

Moreover, the proportion of all maternal deaths that was caused by abortion (as opposed to obstetrical complications, such as preeclampsia) increased as well. The proportion of maternal deaths attributed to abortion grew from 27% in 1965 to a plateau in the 1980s in the 80% range. Illegal abortion became the dominant cause of maternal deaths in Ceausescu's Romania.[13] In the pre-Ceausescu years of 1960-1965, an average of 76 women per year died from septic abortion. Between 1966 and 1989, an average of 302 women died per year from this cause…due to governmental policies.[13] Romanian women quickly were able to control their fertility by substituting dangerous, back-alley abortions for safe abortions, which were no longer available due to government edict. The results were exactly as one would expect.

During these dark years, Romania became a dangerous place for women. The overall maternal mortality ratio rose from 85 deaths per 100,000 live births in 1965 to a peak of 175 in 1985, the highest in all of Europe.[13] The ratio was ten times that of any other European nation.[16] Again, this reflected primarily deaths from unsafe, illegal abortions procured by desperate women.

As with unsafe abortion worldwide, for every death, a much larger number of women were injured, often permanently. While Romanian mortality data are considered accurate, morbidity data are difficult to obtain.[14,16] Abortion became the leading cause of kidney failure in women nationwide. Among women with kidney failure after illegal abortion, hysterectomy was required in 44%, death occurred in 15%, and 8% had permanent kidney damage.[17] According to United Nations estimates, as many as 20% of the 4.9 million Romanian women of reproductive age had some type of permanent impairment stemming from illegal abortion.

Abortion access and maternal deaths are inversely related (Figure 19-5). When abortion is restricted, women die in large numbers. When abortion is made legal and accessible, women's lives are saved.

In 1989, when Ceausescu was deposed, abortion access improved and the abortion mortality risk plummeted. The maternal mortality ratio dropped from 170 deaths per 100,000 live births in 1989 to 83 in the very next year.[13] This freefall drop was entirely due to safe abortion. The corresponding decrease in abortion mortality ratio was also dramatic: from 150 in 1989 to 57 in 1990.[13] With wider access to modern contraception, the abortion rate fell from 160 per 1,000 women in 1990 to 10 per 1,000 women in 2010.[18] During this interval the risk of maternal death from abortion decreased from about 150 to 5 deaths per 100,000 live births.[18]

Stillbirths

Stillbirths increased when abortion was restricted. In the first year of the restrictions, the late fetal death ratio (per 1,000 live births) increased 23%.[19] This reversed a trend underway since the previous decade. The ratio increased from 15 fetal deaths per 1,000 live births in 1966 to 19 the following year. It then fell progressively to 13 by 1970.[19]

Infant deaths

Infants paid a price for the pronatalist policies as well. The risk of death in the first year of life in Romania had been declining since 1950, and by 1965, it had fallen to 44 deaths per 1,000 live births.[19] The figure remained stable at 47 in 1966 and 1967, after which it rose 27% to 60 in 1968. Like fetal deaths, this risk then declined to 40 by 1972. Further analyses of these trends indicated that the excess in neonatal deaths was concentrated in the first day of life.

This setback, not surprisingly, embarrassed the Ceausescu regime. To save face, the government stopped publishing official birth rates in 1983. Moreover, starting in 1986 a 30-day delay was imposed on birth registration, to allow neonatal deaths to escape detection in national statistics. Only after Ceausescu was deposed did the truth emerge.[13]

The prompt but transient effect on fetal and infant deaths may have several explanations. The sudden increase in births after abortion was restricted may have overwhelmed obstetrical facilities. The increased economic incentives for births may have improved the reporting of live births followed by early death. Yet another, more sinister, explanation is that these fetal deaths and early neonatal deaths may have resulted from attempts at illegal abortion late in pregnancy.[19,20]

Orphans and abandoned children

Pronatalist policies led to wholesale abandonment of unwanted children. "Due to its severe economic problems, Romania's pro-natalist program in place for more than two decades had anti-natalist consequences."[21] Some families, burdened with children they did not want and could not afford, simply turned them over to the State for what has been termed "warehousing."[14] Others put them out in the street. There, abandoned children begged, stole, or prostituted themselves to survive.[21] For years in Bucharest, hundreds of children from all over the country lived on the streets and supported themselves through begging or crime.... another shameful legacy of Ceausescu.[22]

> *"In Bucharest, the capital city, hundreds of young children, drawn from all over the country, wander the streets begging, stealing, and/or prostituting themselves. They are a legacy of the Ceausescu fertility policy."*
>
> *Keil*[21]

Unknown thousands of children were placed temporarily or permanently in orphanages by impoverished parents.[23] According to estimates from the United Nations Children's Fund, about 150,000 to 200,000 children were institutionalized when Ceausescu was deposed.[14] Grim conditions in these orphanages led to high rates of developmental disability, infectious diseases (including HIV infection), and death. In one impoverished orphanage for "irrecuperable" children, the mortality rate approached 40% in the winter, due to lack of heating. The bathing of children was done once a week in a group tub and without soap.[23] While some children have been returned to their families and others adopted, many mentally and physically handicapped children will remain in such institutions for life, a lasting legacy of pronatalism.[14]

The longer a child stayed in an institution, the worse the prognosis. Among children adopted by American families, the longer a child had been in an orphanage in Romania, the greater the risk of acquiring an infectious disease, such as hepatitis, and the likelihood of growth retardation and failure to thrive.[23] While prospects for institutionalized children were bleak, as noted by Doctors Without Borders in 1996, "their chances of surviving outside the institutions are null."[23]

Orphanages became fertile breeding grounds for AIDS. Shortly after the fall of Ceausescu, the U.S. Centers for Disease Control and Prevention established surveillance of HIV infection in partnership with the Romanian Ministry of Health. By the end of 1990, 1168 cases of AIDS had been identified, of which 94% were in children less than 13 years of age. Most (62%) were living in institutions. By 1995, most children who died of AIDS had come from orphanages. Romania in the mid-1990s had the highest number of pediatric AIDS cases in Europe.

This tragedy stemmed from the common practice of giving children intramuscular injections of drugs and small amounts of blood. For example, fractions of a single unit of donated blood might be given to as many as 20 children. Should the unit be HIV infected, transmission was efficient.[23] Routine intramuscular administration of vitamins to infants at three to six months of age likely contributed to the HIV epidemic as well.

Painful lessons

Ceausescu's ruthless policies were designed to produce more workers for the proletariat. They backfired horribly. After a transient rise in the birth rate, fertility returned to traditional levels as women returned to the back alley. Over 10,000 women died as a direct consequence, and the indirect deaths from premature birth, HIV, and economic deprivation were large as well.

Poland: following suit?

While Romania provides the most egregious example of the harm of restricting abortion access, Poland provides corroborating evidence. After decades of liberal access to abortion, Poland severely restricted access to abortion in 1993.[24] Approved by a margin of one vote in the upper house of the Polish parliament, this law overturned the liberal 1956 legislation. As happened in Romania, damage appeared quickly. Within a year of enactment of anti-abortion legislation, women were seeking clandestine abortion either in Poland at high prices or across the borders. "Gynecological tourism" is still booming as a result. As the number of abortions in hospitals declined from about 31,000 in 1991 to 777 in 1993, "miscarriages" (many illegally induced) rose quickly.[24] The number of women who abandoned their babies after delivery rose to 153 nationwide, and the 49 abandoned in Warsaw in 1993 was twice the number in the preceding year.

The global movement to safe abortion

In recent years, many countries have liberalized abortion laws. Seventeen countries liberalized their abortion laws between 1997 and 2008: Benin, Bhutan, Cambodia, Chad, Colombia, Ethiopia, Guinea, Iran, Mali, Nepal, Niger, Portugal, St. Lucia, Swaziland, Switzerland, Thailand, and Togo.[25] Mexico City[26] is an outpost of legal abortion in a country with otherwise illegal and unsafe abortion. These changes are prerequisites to improved health for women. For example, in 1995 Guyana became the first country in South America to liberalize its abortion laws, with prompt and predictable benefit. Despite inaction on the part of the government, hospital admissions for septic and incomplete abortions decreased by 41% within six months of the law's enactment. Similarly, the volume of blood transfused to treat abortion complications dropped by 35%.[7]

> *"Against this public health backdrop, the U.S. has been moving backward in recent years. Those in the U.S. striving to turn back the clock on women apparently consider women's deaths and suffering acceptable 'collateral damage.'"*
>
> *"...one Romania is enough."*
>
> *"Everyone who has had a happy sexual life or children they love should try and understand the world of... ...Ceausescu's Romania. Maybe then they will comprehend what happens when safe abortion is not available."*
>
> The Lancet[27]

South Africa's experience is illustrative. An estimated quarter million illegal, unsafe abortions took place annually. In 1996, the Choice on Termination of Pregnancy Act was enacted. Abortion— free of charge in the public sector— was provided upon request for broad indications. Because of slow uptake, the government aggressively expanded access starting in 1998, and midwives joined the cadre of trained providers.

The impact of the liberalized South Africa law was as would be expected, and it was rapidly apparent (Figure 19-6). Between 1994 and 2007, the risk of death from abortion fell more than 50-fold. Although use of modern contraception expanded during this interval as well, the temporal link to legalization of abortion provides strong indirect evidence of the health benefits of safe, legal abortion.

These natural experiments provide compelling evidence of the health benefits for both women and children of liberal abortion laws. When abortion becomes safe and accessible, health improves, as in the United States,[4-6] Guyana,[7] and South Africa.[15] When it is restricted, women and children suffer, as in Romania[27] and Poland.[24] The evidence is clear and incontrovertible.

> *"As subtle as an invisible army, Ceausescu's regime put women's bodies under siege and captured their children."*
>
> Morrison[23]

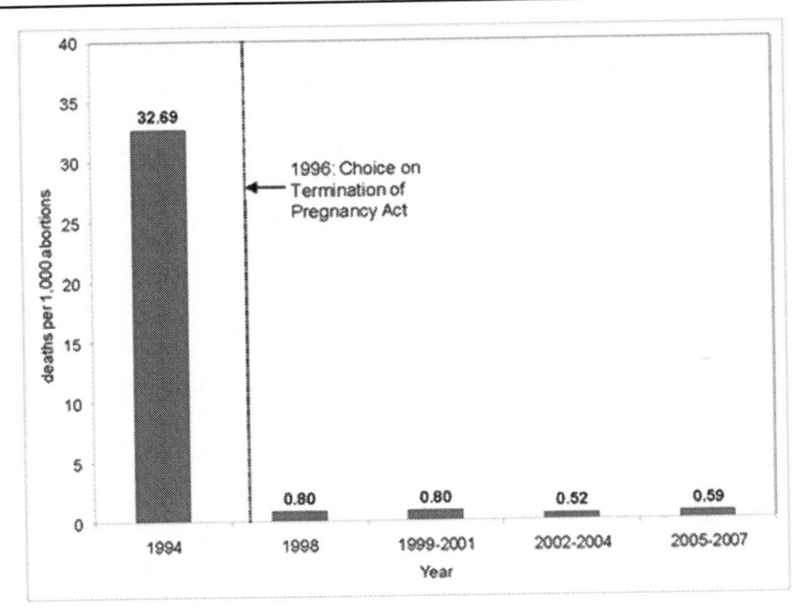

Figure 19-6
Abortion-related maternal deaths in South Africa, 1994-2007
Source: Benson[15]

In the Twentieth Century, a misguided despot brought misery and death to once-prosperous Romania. Intentional suffering and deprivation on this scale had been unknown there since the Fifteenth Century, when another twisted tyrant,[2] Vlad the Impaler, wreaked havoc in the Carpathians. Backward laws, rather than a stake, became the instrument of choice.

Take-home messages

- *When Romania made abortion unobtainable, women resorted once again to the back alley*
- *As a result, the Romanian maternal mortality ratio rose to become the highest in all of Europe*
- *When Ceausescu was deposed and safe abortion re-introduced, women's health improved*
- *In country after country, safe abortion improves the lives of women and children; when access is restricted, suffering and death result*

Chapter 20
On the road again?

Should Roe *go.....*

During the "sandwich years" of regional abortion availability, a 17-year-old single Arkansas woman discovered she was pregnant. Two weeks shy of her 18th birthday, she would have had to have parental consent for an abortion. Instead of asking her parents or waiting for two weeks, she went to New York City. She was found to be 16 weeks' pregnant and underwent a labor-induction abortion. After the abortion, she developed a fever and was started on intramuscular and oral antibiotics. Her last recorded temperature before leaving the hospital was 102 degrees.

Still sick after returning home, she consulted her physician. She did not disclose the recent abortion. Her physician injected her with penicillin, and she returned daily for repeat injections. By her third visit, she was having shaking chills, and she finally told her physician of the abortion. Hospitalization was advised, but for unknown reasons she was not admitted until the next day, by which time her fever was 106 degrees. Despite aggressive antibiotic therapy, she died of overwhelming sepsis with *Staphylococcus aureus*.[1] Had she received her abortion close to home, she would likely be alive today and about 60 years old.

This preventable abortion-related death from the early 1970s illustrates the medical risks of long-distance care. Should *Roe v. Wade* be overturned, the mass medical tourism of those years would likely recur.[2] One estimate projected an extra 157 miles to obtain an abortion in a State where it was legal.[3,4] The ability to obtain health care, including office visits, hospitalization, and pharmacy supplies, near one's home has undeniable advantages. Conversely, having to travel long distances, often to other States, delays care, compromises medical follow-up, and drives up costs. This burden falls disproportionately on America's poor.

Many in Congress and in State legislatures are intent on overturning *Roe v. Wade*. Given the composition of the U.S. Supreme Court, *Roe* may well be one vote away from being overturned, despite the disincentives to change a law on which many other rulings have relied (*stare decisis*). How might reversal of *Roe* affect grassroots America? The past may indeed be prologue.[5]

Were *Roe v. Wade* overturned, abortion would not become illegal. Instead, the United States would return to a patchwork quilt of State laws, like the "sandwich years" of 1970-1972. Many women would be on the road once again in an odyssey to get decent medical care. Without federal assurance of a Constitutional right to choose, States would be open, indeed, invited, to enact legislation. States enacted 380 different abortion restrictions between 1995 and 2004, while *Roe v. Wade* was the law of the land.[6] An avalanche followed: more anti-abortion laws were enacted from 2011-2013 than in the prior decade.[7] This brisk legislative response to a public-health non-issue indicates that ideology, rather than public welfare, is the driving force. If *Roe v. Wade* were overturned, bedlam would ensue in State legislatures.

While some States have moved proactively to ensure reproductive rights, others have worked to undermine them.[8] Seven States have enacted legislation that guarantees women access to safe abortion, independent of *Roe v. Wade*. These include the early leaders in the history of women's rights: California, Connecticut, Hawaii, Maine, Maryland, Nevada, and Washington. These States would remain abortion centers even without national availability.

In contrast, several States have existing laws that would make abortion illegal again in the absence of *Roe:* Louisiana, Mississippi, North Dakota, and South Dakota. In other States, laws banning abortion that antedate *Roe v. Wade* remain on the books; these might be invoked to drive women across State lines[9] or into the back alley once again. These include Alabama, Arizona, Arkansas, Delaware, Massachusetts, Michigan, Mississippi, New Mexico, Oklahoma, Vermont, West Virginia, and Wisconsin. Since some of these States are generally supportive of abortion rights, the likelihood of pre-*Roe* laws being used to undermine women's rights might be low.[7] In addition, eight States have laws with the explicit purpose of limiting abortion to the extent possible.[8]

Maltreatment of women

Antipathy to women is not limited to reproductive rights. However, a clear correlation exists between State views on abortion and the health of women living there. The national women's Health Care Report Card is a joint product of the Oregon Health and Sciences University and the National Women's Law Center.[10] Led by Dr. Michelle Berlin of the University, a panel of experts reviewed the extent to which each State and the nation met 26 health status benchmarks and 68 policy goals. These diverse measures ranged from proportion of women having Pap smears to cholesterol screening to diabetes-related services. The report card provides an objective snapshot of the health status of women.

Thirteen States are generally supportive of abortion rights, with zero or one major legal restrictions in place (Table 20-1).[7] Ten have two or three major restrictions, which can be viewed as intermediate. However, 27 States can be considered hostile to abortion rights, manifested by four to ten restrictions. In 2010, the most recent year for which data were available, 11 States and the District of Columbia got a failing grade for women's health. Not surprisingly, ten of the 11 failing States are those most antithetical to abortion rights (Table 20-1). One State (8%) supportive of abortion received a failing grade for women (West Virginia). In contrast, 10 States (37%) hostile to abortion failed. A disparity this large is unlikely to be due to chance. Maltreatment, disdain, and neglect of women come in many forms, some subtle but nonetheless deadly.[11-13] Opposition to abortion is clearly one of them.

Travel distances matter

In the early years of legal abortion, the Centers for Disease Control and Prevention documented the inverse relationship between travel distances and abortion rates.[14] The further a woman had to travel in Georgia to obtain an abortion, the less likely she was to get one. This held true both for rural and urban women, although the effect was most dramatic among African-American teenagers. Similarly, lack of transportation led to many poor, elderly, or sick African-Americans being trapped in New Orleans during Hurricane Katrina in 2005, some with fatal consequences.

Recent analyses of abortions provided in New York State from 1971-1972 have confirmed the impact of travel distance. The farther women had to travel to get care, the lower was their abortion rate.[9]

Travel distances influence women's reproductive decisions. Where a woman lives has a powerful effect on her pregnancy outcomes.[15] Between 1978 and 1988, as distance to an abortion provider (either in- or out-of-State) increased, abortion rates fell significantly.[16] In contrast, these measures did not affect birth rates. A similar study of Texas teenagers found that abortion rates of those aged 13 to 17 years were sensitive to variations in travel costs.[17] As seen earlier in Georgia,[14] counties with higher travel costs had lower abortion rates for teenage women. This held true whether measured per woman or per pregnancy.

Decades after the pioneering CDC study,[14] the transportation problem persisted for many American women. In 1993, about one-quarter of all U.S. women who obtained a non-hospital abortion had to travel more than 50 miles each way to a provider.[18] Of those, 16% had to travel more than 50 miles each way, while another 8% had to travel more than 100 miles each way to reach care. The travel distances were greatest in the East South Central census division, which includes Mississippi and Alabama.

Another survey of more than 8,000 abortion patients in 2008 found persistent travel challenges.[3] The mean distance traveled for care was 30 miles, with a median of 15 miles. Of note, 6% of women had to travel over 100 miles to receive care. Among rural women, 31% had to travel this distance.

Mandatory delays

Should a woman need a hysterectomy and should the surgical staff and operating room be available, the operation can be done the same day. In contrast, prompt abortion care depends on where a woman lives. Unable to prohibit abortion, legislators in many States have erected regulatory roadblocks to deter women. These laws require various numbers of hours or days after receiving State-scripted information before the procedure can be done. The most daunting of these (Mississippi) requires the prospective patient to make two separate trips to the provider for in-person counseling; in contrast, some States allow this information to be communicated in more convenient ways.

Mississippi's mandatory delay was enacted early (August 8, 1992) and has been enforced continuously.[19] The law requires the following information to be given to the woman at least 24 hours before her procedure: name of the physician, medical risks, probable gestational age of the pregnancy, and the medical risks of continuing the pregnancy. This information must be given in person. In addition, the physician or a designee must advise the woman that medical assistance benefits may be available for pregnancy care, the father is liable for child support, and pregnancy prevention services are available. She must also be told that she has the right to read State-produced materials that include social service agencies and brochures that describe "the unborn child [sic]" at two-week intervals. The woman must sign a document saying that she has been given the required information.[20]

The mandatory delay law in Mississippi has hurt women's health, as documented by several reports. In 1994, investigators at the Guttmacher Institute found that the average monthly number of abortions performed in Mississippi dropped from 717 in January through July, 1992, to 507 in August through December, a 30% decline. Taking into account seasonal fluctuations in the numbers of abortions, the revised estimate was a 22% decrease after enactment of the law.[20]

This figure is deceptive, however, since it obscures the effect of travel. In response to the new law, women took to the road once again. Seventeen percent more Mississippi women had abortions out-of-State (Alabama and Tennessee) in the latter part of 1992, implying a true decline of 13%. If women fled to Louisiana to the same extent as they traveled to Alabama and Tennessee, the figure would revise downward to 11%. Moreover, 30% fewer residents of other States came to Mississippi to have abortions.

Did the law have the desired effect on pregnancy outcomes? If the law had prompted women to continue their pregnancies, then a transient rise in the numbers of live births should have been evident nine months later. None occurred. Thus, the net effect on the birth rate was negligible; the law simply forced women across State lines[9] for care.

A second adverse outcome was immediately evident: the 24-hour State-imposed delays translated into medically harmful delays in getting care. The proportion of abortions in Mississippi performed after 12 weeks' gestation rose by 17% in the early months after enactment of the law.[20] The duration of pregnancy (gestational age) at the time of abortion is one of the two most powerful predictors of the risk of complications and death (see Table 5-1).[21] The risk increases progressively with time; delays, whether due to regulations, transportation, or financial concerns, have the same net effect of increasing risk.

Women in Mississippi fared less well than did their contemporaries in other southern States. A second analysis of Mississippi's law compared women's experience there with women in Georgia and South Carolina between 1989 and 1994.[19] Neither of the latter two States had a mandatory delay, although both began parental notification rules during those years. In the year after enactment of Mississippi's mandatory delay, abortions among women living in Mississippi declined more than in either comparison State. Again, the proportion of abortions done after 12 weeks' gestation increased in comparison with Georgia and South Carolina. As had been documented before,[20] the proportion of Mississippi women who left the State to get care increased dramatically when compared with

> *"....some women in all states have difficulty gaining access to abortion services, and a law that created a barrier for a substantial proportion of Mississippi women would undoubtedly have a similar effect on many women in other states."*
>
> *Joyce*[19]

> *"...the fact that so many women left the state for their abortion once the law took effect suggests that many found the two-visit requirement to be too costly or perhaps considered the information provided to be unnecessary."*
>
> *Joyce*[22]

States without this regulatory barrier.

A third analysis confirmed the adverse effect on abortion timing. Joyce and Kaestner analyzed Mississippi abortion patients from 1989 to 1995 in two categories: those whose closest provider was in Mississippi, and those whose nearest provider was across State lines.[22] Women who relied on Mississippi physicians were forced into later, more hazardous procedures. The proportion of abortions after 12 weeks' gestation was at least 2.6% higher among women whose nearest provider was in-State. The net effect of the law on the timing of abortion was an overall delay of 0.61 weeks. This, in turn, increased the number and severity of complications that women experienced, and increased the net costs of the procedures as well.

Medicaid funding restrictions

In the early days of the Hyde Amendment, which cut off federal financing of abortions for poor women, women living along the Mexican border were more likely to suffer complications of illegal abortions than were other U.S. women.[23] According to surveillance of complications in 1977 and 1978 conducted by the Centers for Disease Control and Prevention, the funding restriction did not lead to a detectable nationwide increase in complications of illegal abortion; instead, complications clustered along Mexican border States. This suggested that poor women along the border resorted to illegal abortions in Mexico because of traditional patterns of care.[24]

As anticipated, Medicaid restrictions were associated with a delay in getting care. Medicaid-eligible women with complications after legal abortion in non-funded States had a two-week later mean gestational age than did their counterparts with legal abortion complications in funded States.[23] Since the risk of complications increases by about 20% and of death by about 40% for every week of delay, the immediate impact was to shift women to later, more dangerous procedures.[21,25]

Some States later withdrew their support of abortion for poor women, and investigators tracked the impact on women. In 1985,

Colorado, North Carolina, and Pennsylvania restricted the use of State funds in this way. In the first year thereafter, birth rates, which had been steadily declining for years, increased in each of these States.[26] This suggested that some unwanted pregnancies, which would have otherwise been aborted, yielded live births predominantly to poor women.

More recent studies have been divided on the impact of Medicaid funding restrictions. By 1997, 34 States had restricted funding. Between 1978 and 1992, these States had abortion rates that were 2% lower than in States that maintained State funding.[27] After correction for the effect of the supply of providers and demographic characteristics of women in those States, the difference in abortion rates was not statistically significant, indicating that this might be due to chance. No effect on birthrates was seen. Another study concluded that State abortion funding restrictions are associated with a reduction in abortions and either no change or a decline in births.[28] This suggested a net drop in the number of pregnancies occurring in those States. As anticipated, the effect was concentrated among the poor.

Other studies found more dramatic effects. From 1974 to 1988, Blank and others identified a 19-25% drop in abortions among poor women after funding restriction.[29] These investigators also documented lower abortion rates in-State and higher rates in nearby States, indicating that women were taking to the road again. A detailed examination of North Carolina, which had on-again, off-again funding between 1980 and 1994, found that one-third of pregnancies that would have been aborted were carried to delivery in the absence of funding.[30] The adverse impact of the funding restriction was felt most acutely by African-American women aged 18 to 29 years. The authors were surprised to learn that the few hundred dollars needed for an abortion often proved decisive in the decision; many of these women had severe financial hardship. Thus, studies of the impact of Medicaid funding have had mixed results: more delays in obtaining abortions, no change or a decrease in abortion rates, and no change or an increase in births. The impact of these restrictions appears limited to poor women and those living in Mexican border States.

> *"It is rather remarkable that the necessity of paying a couple-of-hundred-dollar fee for an abortion is sufficient to persuade (or compel) some women to incur the much larger financial and personal costs of bearing an unwanted child."*
>
> Cook[30]

Parental notification

When a 17-year-old becomes pregnant, she needs no one's permission to become a mother while still a child. Should she prefer *not* to become a teenage mother, however, the decision is not hers alone to make in most of America.

Parental involvement laws increase travel distances and lead to later, more risky and expensive procedures. Researchers have examined these laws' impact in individual States, several States, and the nation. The first such report examined the experience of minors from Massachusetts after the April, 1981, implementation of a parental consent law.[31] In the ensuing 20 months, only half as many Massachusetts minors obtained abortions in-State. Instead, more than 1800 young women traveled to five surrounding States for care. Only a small number of minors had children rather than having abortions because of the law. The Massachusetts law apparently had little impact on women's decisions about their pregnancies, but it drove large numbers out-of-State.

A similar study of the parental notification law in Minnesota found an increase in the proportion of late abortions, especially among women aged 15 to 17 years, after implementation.[32] The Mississippi law was associated with greater delays in care among minors (an average of three days) and a higher proportion of second-trimester abortions.[33] A study of laws in Minnesota, Missouri, and Indiana found no effect on births but a decrease in in-State abortions and greater travel.[34] Another study in South Carolina, Tennessee, and Virginia found little effect on the pregnancy resolution of minors.[35]

The Texas parental notification law had harmful effects on health.[36] Its enactment was associated with the anticipated drop in abortion rates among those aged 15 to 17 years. Correspondingly, the teenage birth rate increased. According to the federal government, teenage motherhood is associated with an array of medical and social harms for both mother and child.[37] Three-quarters of births to women aged 15-19 years are unintended, and the societal costs exceed *$10 billion* per year in the U.S.[37]

Studies involving a larger number of States have generally confirmed these observations. A national assessment in 1994 found that parental involvement laws reduced minors' abortion rates relative to older women.[38] This study estimated that these laws increased adolescent fertility by about 10%. Noting that most adolescent pregnancies are unintended, the authors inferred that the net effect was an increase in unintended childbearing among adolescents. Other authors have determined that parental involvement laws lead to interstate travel for care,[39] fewer abortions among adolescents,[40] and an increased proportion of later, more hazardous procedures.[41] A recent economic analysis suggested fewer abortions resulting from fewer pregnancies among minors, and no impact on births.[42] These studies, however, are handicapped by inability to account adequately for interstate travel for care.

Single-State studies of parental involvement show that minors take to the road. While in-State abortion rates among minors promptly decline, the decrease is spurious, since many abortions shift to a neighboring State.[31,33,35] Studies that have focused on aggregate behavior as opposed to minors' behavior are not well equipped to measure the effect on adolescents.[16,29,41] In studies focusing on minors, results have been mixed.[38,40,43] Nonetheless, no study has reported a benefit: less travel, lower costs, or less delay in care.

What has been the public health impact of these piecemeal attempts to limit women's access to abortion? While evidence is divided on the impact on fertility, these laws clearly force women across State lines and delay their abortions to later, more risky, procedures.[9] Thus, these laws compromise women's health and drive up the cost of health care. Net harm is the result; thus, they violate the ethical principle of non-maleficence.[44] The same can be expected to occur in a patchwork quilt of State laws should *Roe v. Wade* disappear. The harms were pointed out decades ago by the American Medical Association:

> *"Because poor and low-income women are most likely to have difficulty with financial arrangements for travel and the costs of the procedure, they are more likely to delay the procedure and are therefore at greater risk of abortion-related complications or death....As access to safer, earlier legal abortion becomes increasingly restricted, there is likely to be a small but measurable increase in mortality and morbidity among women in the United States."*
>
> *American Medical Association*[45]

Table 20-1

Number of abortion restrictions*and women's health report card grade** by State

– 0-1 Restrictions –		– 2-3 Restrictions –		– 4-10 Restrictions –	
State	**Grade**	**State**	**Grade**	**State**	**Grade**
California	Unsatisfactory	Alaska	Unsatisfactory	Alabama	*Fail*
Connecticut	Unsatisfactory	Colorado	Unsatisfactory	Arizona	Unsatisfactory
Hawaii	Unsatisfactory	Delaware	Unsatisfactory	Arkansas	*Fail*
Maine	Unsatisfactory	Iowa	Unsatisfactory	Florida	Unsatisfactory
Maryland	Unsatisfactory	Illinois	Unsatisfactory	Georgia	Unsatisfactory
New Hampshire	Unsatisfactory	Massachusetts	Satisfactory minus	Idaho	Unsatisfactory
New Jersey	Unsatisfactory	Minnesota	Unsatisfactory	Indiana	Unsatisfactory
New Mexico	Unsatisfactory	Montana	Unsatisfactory	Kansas	Unsatisfactory
New York	Unsatisfactory	Nevada	Unsatisfactory	Kentucky	*Fail*
Oregon	Unsatisfactory	Wyoming	Unsatisfactory	Louisiana	*Fail*
Vermont	Satisfactory minus			Michigan	Unsatisfactory
Washington	Unsatisfactory			Mississippi	*Fail*
West Virginia	Fail			Missouri	*Fail*
				Nebraska	Unsatisfactory
				North Carolina	Unsatisfactory
				North Dakota	Unsatisfactory
				Ohio	Unsatisfactory
				Oklahoma	*Fail*
				Pennsylvania	Unsatisfactory
				Rhode Island	Unsatisfactory
				South Carolina	*Fail*
				South Dakota	Unsatisfactory
				Tennessee	*Fail*
				Texas	*Fail*
				Utah	Unsatisfactory
				Virginia	Unsatisfactory
				Wisconsin	Unsatisfactory

*For 2013
**For 2010

Sources: Guttmacher Institute[7]
National Women's Law Center[10]

Take-home messages

- *Receiving medical care close to home has many benefits*
- *State laws are already forcing women into interstate travel for abortion care*
- *Should* Roe v. Wade *be reversed, women would once again travel long distances to get abortion care*
- *States with the poorest general health of women tend to be those most inimical to abortion rights, which is unlikely to be a coincidence*

Chapter 21
Abortion denied

> *"...unwantedness in early pregnancy has a not negligible effect on later development, influencing quality of life and casting a shadow on the next generation."*
>
> *David*[1]

What are the consequences when a woman's request for abortion is not granted? Theoretically, babies that are wanted and loved should have better life prospects than those who are not. Some empirical indirect evidence supports this hypothesis. For example, a study in Washington State found that males born to unmarried teenage mothers had an 11-fold higher risk than others of becoming chronic juvenile delinquents.[2] One explanation proposed is an indirect association with poverty and low maternal educational attainment. Unwanted children are at increased risk of delinquency,[3] and most births to teenagers, to unmarried women, and to those living in poverty are unintended.[4] Alternatively, some women who reluctantly become mothers have difficulty nurturing and supervising their infants. Others have found that the combination of a complicated birth and early maternal rejection is strongly associated with juvenile delinquency.[5]

However, longitudinal follow-up studies with proper control groups are necessary to assess the impact of denied abortion on later outcomes for mother and child. Several landmark studies from Sweden and Czechoslovakia (Table 21-1) have provided critical insights.

Göteborg, Sweden

Denied abortion was associated with developmental problems in children born thereafter. The effect was more pronounced between birth and 21 years, moderating in adulthood. This was the conclusion of the first study to track children systematically until they were 35 years old. Conducted in Göteborg, Sweden, the study by a psychiatrist and his social worker was a landmark contribution.[6,7] It guided similar research in other countries.

Sweden liberalized its abortion laws in 1939 to allow the procedure for selected indications. These included maternal and fetal health indications as well as crimes, such as rape and incest. Overall, 197 women had their abortion requests denied by the local hospital. Sixty-eight of these pregnancies ended in abortion, many illegal. A total of 120 children, 66 boys and 54 girls, were available for follow-up. Control children were the next-born child of the same sex at the hospital. The controls were born to other women—mothers who had not sought abortion and been denied.

Table 21-1

Cohort studies of children born after denied abortion

Location	Birth years	Numbers studied	Findings
Göteborg, Sweden[1,6,7]	1939-42	120 children; 120 controls with same sex born at same time	Unwanted children had uniformly worse social and psychiatric outcomes; these differences diminished in adulthood
Stockholm and Uppsala, Sweden[1]	1948	204 children; 88 residing in Stockholm at age 18 yr. had 88 paired controls, matched by gender and birth date	Unwanted children had higher rates of crime and economic assistance
Sweden[1,8]	1960	90 children; 90 control children matched by mother's age, number of mother's births, and sex	Unwanted children had poorer school performance, more psychiatric referrals, and more antisocial behavior
Prague, Czechoslovakia[1,9-12]	1961-63	220 children; 220 controls pair-matched for age, sex, birth order, number of siblings, and school	Unwanted children followed to age 35 yr. were consistently disadvantaged; when siblings were included as controls, unwanted children had significantly more psychiatric treatment as adults

Sources: David[1], Forssman[6,7], Dytrych[9], David[10], Kubicka[11], Matajcek[12]

Use of public records, rather than personal interviews, reduced bias in determining outcomes. Of note, the only matching was by gender and birth date. The two groups differed in several ways, including age of the mother and social class; these may have biased the comparisons, as noted by some.[13] However, when the authors evaluated subgroups of children with concordant social class, the findings persisted, so differences in socioeconomic status are unlikely to have accounted for the results.[7,14]

As shown in Figure 21-1, children born after denied abortion had uniformly worse outcomes in early life than did other children. Psychiatric consultation and hospitalization were twice as common among children born after denied abortion than among their peers. Registration with children's aid bureaus for delinquency was two times more common. Registration for crime in the Central Penal Register was three times more common, while registration for drunken misconduct was 50% higher. The likelihood of receiving public assistance between the ages of 16 and 21 years was six times higher, and the likelihood of having "subnormal educability" was two times higher. Conversely, 58 of the unwanted children had no adverse outcome, in contrast to 82 of the control children.[6]

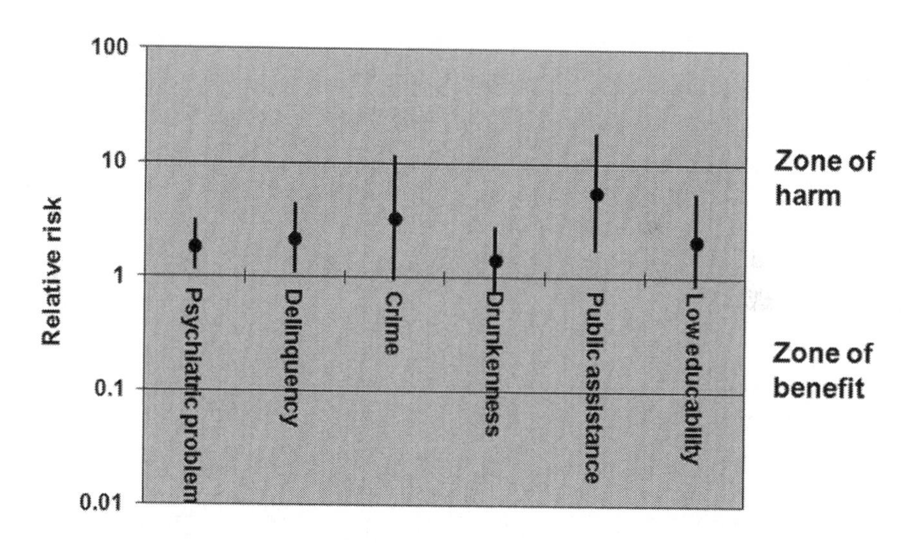

Figure 21-1
Relative risk and 95% confidence intervals for selected outcomes among children born after denied abortion, Göteborg, Sweden
Source: Forssman[6]

> *"...children born after their mothers have been refused permission for legal abortion are born into a worse situation than other children....the very fact a woman applies for legal abortion means that the prospective child runs a risk of having to surmount greater social and mental handicaps than its peers, even when the grounds for the application are so slight that it is refused."*
>
> *Forssman*[6]

These adverse outcomes diminished with time. When a follow-up was done of the index children at age 35 years, those born after denied abortion had persistently higher rates of social and psychiatric problems, although the size of the differences had decreased. After studying these children for over a third of a century, the authors concluded that those born after denied abortion have an increased risk of psychosocial problems in early life, although these attenuate in adulthood.[1]

Stockholm and Uppsala, Sweden

Children born after denied abortion in other parts of Sweden fared similarly. In a similar follow-up study from Stockholm and Uppsala counties, Hook[15] tracked 249 women whose applications for abortion had been denied by the Swedish National Board of Health in 1948. Hook followed to age 18 years those unwanted children who were still living in Stockholm; she then chose controls for each: the same-sex classmate with the nearest birthday. More unwanted children (20%) had been registered for criminality than controls (12%), and significantly more had received public assistance.

Sweden nationwide

A later national study from Sweden also found an association between denied abortion and adverse outcomes in children. Several authors examined the records of women refused abortion in 1960. Blomberg[8] examined the development over 15 years of 90 pairs of same-sex children, both wanted and unwanted. Their mothers were of similar age, childbearing experience, and social class. Overall, unwanted children fared less well. They had significantly poorer school performance, more psychiatric referrals, and more registration with child welfare authorities for social problems such as alcohol abuse or theft.

Prague, Czechoslovakia

> *"The unwanted children grew up in a more insecure environment, performed worse in school and more often needed treatment for nervous and psychosomatic disorders. There was also a tendency toward worse social adjustment."*
>
> *Blomberg*[8]

A long-term follow-up study of denied abortion in Czechoslovakia confirmed the earlier findings from Sweden.[1,9] Abortion was legalized in 1957, and local District Abortion Commissions had to approve requested procedures. Only 2% of applications were ultimately rejected after initial request and after appeal; this operationally defined a cohort of women whose pregnancies were unwanted. To avoid bias, controls were chosen for index children after matching for age, sex, birth order, number of siblings, and school class. Ultimately, 220 children, equally divided between sexes, were available for follow-up, with an equal number of controls. A unique strength of this prospective cohort study is that researchers performing all subsequent assessments of children in both groups have remained blinded as to wantedness status. Over the ensuing decades, teams of researchers have administered the same battery of intelligence and psychological tests, plus personal evaluation of each child by a pediatrician and psychologist.

Age 9 years

Early on, unwanted children had worse outcomes.[9] By third grade more unwanted children disliked school, and their diligence was rated significantly poorer by their teachers. Unwanted children performed less well on the Wechsler Intelligence Scale for Children, although the differences were not statistically significant. Unwanted children, especially boys, had significantly higher maladaptation scores than did the control children.

"Psychological subdeprivation" has been used to describe the underlying mechanism: the basic psychological needs of the unwanted child were met less well than those of wanted peers.[1] The authors postulated that unconditional acceptance of the child is the cornerstone of a trusting, loving relationship. In these children, the relationship was not that of outright rejection but, rather, a qualified acceptance of the child

> *"The child of a woman denied abortion appears to be born into a potentially handicapping situation."*
> *David*[1]

after denied abortion. This would, in turn, manifest in incomplete, ineffective, or ambivalent relationships between mother and child. Overall, the adverse impact was not profound, but the direction was consistent.

Ages 14-16 years

A second round of evaluations began when the children were 14 years old.[1,16] By that time, 43 unwanted children had been referred to child psychiatric and school counseling centers, in contrast to 30 control children. The unwanted children were seen significantly more often because of serious behavior disorders.

By ages 14 to 16 years, school performance had continued to deteriorate among the unwanted children, so that the differences now became statistically significant (unlikely to be due to chance). The discrepancy was larger for boys than for girls. According to teachers' ratings, unwanted children were less conscientious, more excitable, and less obedient than the control children.

Ages 21-23 years

Civic registers confirmed more problems in the families of the unwanted children. Fifty-seven families of unwanted children had been registered because of parental divorce, delinquency of the child, or other reasons, in contrast to 37 control families. Similarly, ten unwanted children were in Alcoholism Treatment Centers in Prague, in contrast to four control children.

Criminal activity was more frequent among unwanted children as well. As of 1985, 41 sentences had been imposed on 23 unwanted children, in contrast to 19 sentences on 11 control children. The authors noted that the discrepancy between outcomes for unwanted and wanted children widened with time, extending into adult life and thus potentially affecting subsequent generations.

Age 30 years

Over the years of the study, the researchers began to appreciate that despite their matching of controls to index children, residual unmeasured confounding (bias) was likely distorting the results.[10,11] For example, the families of unwanted and wanted children may differ in important ways other than pregnancy intention. Hence, the research team broadened the control group to include siblings of the index unwanted children and wanted children as well. This was done for the fourth and fifth follow-up waves, when the participants were 30 and 35 years old. In this way, the researchers hoped to control for family influences (e.g., parental strife or divorce) and examine more clearly the potential impact of an unwanted birth. While an improvement, this technique has problems as well: the birth of an unwanted child may affect the family dynamics and financial status in ways that spill over to other siblings, thus blunting the effect of the unwanted birth.

In this more rigorous test, denied abortion was significantly associated with psychiatric treatment in adulthood.[11] As anticipated, however, siblings of unwanted children also had higher proportions of poorly socialized individuals than did the wanted children and their brothers and sisters. A third finding was a lack of association with alcohol abuse, heavy smoking, and criminal behavior.

Age 35 years

A final follow-up confirmed the higher likelihood of psychiatric treatment as an adult.[10,11] Again, no association was noted with alcohol, tobacco, and crime. Over the five waves of data-gathering, unwanted children fared less well than their peers. Rather than manifesting a large number of highly negative indicators, these children were underrepresented in indicators of excellence. Simply put, they clustered toward the lower end of the bell-shaped curve. In the more rigorous final analyses using siblings as controls, the higher risk of psychiatric treatment, especially inpatient care, remained statistically significant.

> *"Emerging from subdeprivation, the relationships of the young [unwanted pregnancy] adults with their families of origin, friends, coworkers, supervisors, and especially with their sexual or marital partners are dogged by serious difficulties."*
>
> *David*[1]

Obstetrical outcomes of unwanted pregnancies

> *"It seems that most women who really have made up their minds will obtain an abortion elsewhere even after denial."*
>
> *Hunton*[19]

Among women who do not seek illegal abortion or experience miscarriage, pregnancy outcomes do not appear to carry an increased risk of complications. This question has been addressed in several countries. Investigators at the Centers for Disease Control and Prevention followed the obstetrical courses of 316 women who were refused a second-trimester abortion at Grady Memorial Hospital, Atlanta, Georgia, in 1978-79.[17] Of the 258 women with known pregnancy outcomes, 20% got a legal abortion elsewhere. Of the 206 who continued their pregnancies, 156 had live births at term, 32 delivered premature infants, eight had miscarriages, one had a stillbirth, and nine had missing information. The obstetrical complication rates were typical for women delivering at the hospital.

A Croatian study of pregnancy outcomes after denied abortion documented striking pregnancy attrition.[18] The obstetrical records of women refused abortion from 1970-71 in three regions of Croatia were reviewed to determine outcomes. Most women denied abortion did not subsequently deliver the baby. In Zagreb, of 269 denials, only 30 women gave birth. In Varazdin, the corresponding numbers were 42 and 6, while in Zadar, they were 67 and 11. Since miscarriage is uncommon after the first trimester (the most common reason for denial here), most of these pregnancies were presumably terminated through illegal abortion. Among the small number of deliveries, only one low-birth weight baby was recorded.

Similar experience was documented in New Zealand.[19] Among 200 Auckland patients denied abortion at a private abortion service, only 69 (35%) carried the pregnancy to delivery. A substantial number traveled to Australia to get services. Some acknowledged to the research staff that they had induced abortion themselves.

Emotional outcomes of unwanted pregnancies

Several international studies addressed women's emotional responses to continuing a pregnancy after denied abortion. When interviewed seven years after the denied abortion, 249 Swedish women reported a range of responses.[15] About a quarter had coped well with the pregnancy, about half had serious emotional strain for some time after the birth (now overcome), and another quarter still showed signs of poor adjustment and mental illness.

Another study from South Africa found that those denied abortion reported more anxiety and guilt than the women who had received the requested procedure.[20,21] Perhaps most worrisome are data from 73 women interviewed from one to three years after being denied abortion.[22] Most women (59%) reported that they had accepted the child and were glad they had not aborted. However, a sizable minority (34%) reported that they resented the child, which posed a continuing burden. Thus, women may harbor hostility, easily elicited by interviewers, for years after the birth of the child.

The Turnaway Study

The Turnaway Study is an ongoing U.S. investigation of the socioeconomic impact of being denied an abortion because of gestational age limits.[23,24] Two comparison groups are included: women just below the gestational-age limit who obtained an abortion and women who had a first-trimester abortion. Preliminary results showed harmful effects of being denied an abortion. One year later, women turned away were more likely to be receiving public assistance than those who had an abortion (76% vs. 44%). They were also less likely to be working full time and more likely to be living in poverty.[23]

Summing up

Refusal of abortion does not eliminate the problem underlying the request. Many women will, at considerable expense, travel to get care. Others abort themselves or turn to the back alley. Of those children born, few are placed for adoption,[23,24] and most are reared by their biological mother. In studies to date, around 30% of women reported negative feelings toward the child.[25]

In long-term follow-up studies of varying rigor from Sweden and Czechoslovakia, children born after denied abortion have consistently fared less well than wanted children. The uniformity of this observation suggests that the effect is real. The association between refused abortion and suboptimal child development is consistent and persistent . . . lasting at least into adult life.

Take-home messages

- *Studies in Europe indicate that children born after denied abortion fare less well than their wanted peers*
- *These adverse effects persist at least into young adulthood*
- *Substantial numbers of women denied abortion harbor hostility toward the child*
- *In the U.S. denial of abortion is associated with later poverty and receipt of public assistance*

Chapter 22
Bad old days redux: self-induced abortion

Although abortion has been legal nationwide for four decades, opponents have made access increasingly difficult. Because of financial constraints, State-imposed barriers to prompt care, and geographic distances, some desperate women have taken matters into their own hands.[1] Although the methods available to women today are superior to those before *Roe v. Wade*, abortions unsupervised by clinicians carry needless risks. This chapter describes the prevalence of self-abortion, reviews the methods being used, describes some dangers to women, and examines some of the legal consequences for women.

Prevalence

Many women are aware of methods to induce abortion, and a small minority has used them in the era of legal abortion. A survey of 610 women in a Latina population in New York City in 2000 found that about a third (37%) were aware of misoprostol as an abortifacient.[2] Twenty-nine women (5%) acknowledged having used the drug to induce an abortion. Women who were foreign born, those who had had a prior abortion, and those who knew someone who had used misoprostol were more likely to report self-abortion. Ironically, although three-quarter of the respondents were Medicaid recipients, only half knew that Medicaid pays for abortions in New York State.

A nationwide survey of abortion patients in 2008-2009 found little use of misoprostol for this purpose.[3] The Guttmacher Institute study included over 9,000 women receiving care in 95 different clinics. Overall, only 101 women (1%) admitted to using misoprostol for self-induced abortion. Older women (30-34 years), those with two or more prior abortions, and those foreign born were more likely to have used this drug compared with other women.

Methods

With the exception of misoprostol (Table 22-1), the array of ineffective and potentially dangerous abortifacients has changed little since the era of illegal abortion (Chapter 1). Black-market misoprostol has dramatically improved the safety of self-induced abortion. For example, in South American countries, where abortion is generally illegal, use of black-market misoprostol has been temporally linked to steep declines in abortion complications.[4,5] When given in appropriate doses, misoprostol can produce a medical abortion in most women who are in early pregnancy. This is far safer than more invasive approaches, such as inserting foreign bodies into the uterus.

Table 22-1

Examples of self-induced abortion methods in the era of legal abortion

Category	Example
Medications	Aspirin
	Hormonal contraception
	Humphrey's 11 (black cohosh, windflower, and sepia)
	Laxatives
	Oral contraceptives
	Quinine
	Vitamin C
Uterine stimulants	Misoprostol
Poisons	Bleach
Food, beverages	Coffee with lemon
	Malta (non-alcoholic malt beverage)
	Vinegar and oral contraceptives
	Warm Coca-Cola with baking soda
Herbs	Black and blue cohosh
	Black walnut
	Comfrey
	Motherwort
	Oak bark
	Pennyroyal oil
	Rue and sage tea
	St. John's wort
Trauma	Abdominal blunt trauma
	Gunshot to abdomen
	Strenuous exercise
	Vaginal, including coat hangers

Sources: Grossman[1], Netland[6], Coles[7], Ensign[8], Friedman[9], Buchsbaum[10], Anderson[11]

Dangers of self-induced abortion

Gunshot wounds

Desperate women sometimes take desperate actions. In the 1980s, two young women were brought to the emergency department at Parkland Hospital in Dallas after having shot themselves in the abdomen to induce abortion.[10] A 15-year-old woman shot herself with a 22-caliber pistol, which killed her fetus at 25-26 weeks' gestation. At operation, she was found to have about one pint of blood in her abdomen from the bullet wound in her uterus.

A 22-year-old woman at 15 weeks' gestation used a 22-caliber pistol to end her pregnancy as well. The bullet passed through her uterus and lodged in her large bowel. She required an exploratory operation, repair of the bowel injury, and a sigmoid colostomy (bag on her side for feces) during her recuperation.

In 1994, a 19-year-old in Clearwater, Florida, fired a 22-caliber pistol into her uterus at six months to end her pregnancy.[12] She reportedly had been turned away from a clinic because of lack of money. An emergency cesarean delivery was done, and the baby died the next month. The woman had a three-year-old child, no means of support, and was living with her grandmother, who would not approve of another child. The boyfriend offered financial help, but none was forthcoming.

Poisoning

Women can overdose—and sometimes kill themselves—during attempts at self-abortion. One of the most dangerous herbal preparations used for this purpose is pennyroyal oil. As little as two teaspoons can cause toxicity.[11] CDC investigators identified three hazards from using this agent.[13] First, if unsuccessful, the woman might continue to increase the amount until toxic dosages are reached. Second, if the abortion failed and the pregnancy continued, the fetus would be exposed to the agent. Finally, if the abortion failed, she would be delayed getting a legal abortion. Delays drive up the risk and expense.[14]

In 1978, two women in Denver took pennyroyal oil to induce abortion.[15] An 18-year-old swallowed two one-half ounce bottles and rapidly became sick. She was hospitalized and received prompt and aggressive medical treatment. Despite the best of medical care, she died one week later. On autopsy, her liver was destroyed, and she had four quarts of blood in her abdominal cavity. Ironically, the woman had a negative pregnancy test on admission. She had not been pregnant.

A 22-year-old took two teaspoons of pennyroyal oil for abortion. Feeling ill, she went to the emergency department at Denver General Hospital. Her pregnancy test was positive. She was observed in the hospital for two days then discharged to home. No complications occurred, and she was lost to follow-up.

A fatality occurred more recently on the West Coast.[11] A 24-year-old woman took pennyroyal oil and black cohosh root for two weeks to cause abortion. When this was unsuccessful, she took a larger dose and had a seizure. When the paramedics arrived, she was in cardiopulmonary arrest. She was resuscitated and received aggressive treatment in the hospital, including an exploratory abdominal operation. The surgeons discovered an ectopic (tubal) pregnancy with evidence of infection. She died 46 hours after taking the last dose of pennyroyal oil; autopsy confirmed that her liver had been destroyed by the pennyroyal oil. Two other adults took pennyroyal oil as an abortifacient and survived.[11]

Other concoctions

A 33-year-old woman in New Orleans took quinine and Humphrey's 11, an herbal formulation. Soon thereafter, she developed nausea, vomiting, and severe ringing in her ears. She was seen at a hospital and had her stomach washed out through a tube. She then received charcoal by mouth to help absorb any residual abortifacient.[6]

A 21-year-old at 11 weeks' gestation took 20 Humphrey's 11 tablets and came to an emergency department because of pain, nausea, and vomiting. She required only observation.[6]

A 32-year-old woman took 80 aspirin tablets and two glasses of vodka to induce an abortion. Aspirin is a toxic drug. Lethal quantities can be purchased over the counter. At the hospital, she had her stomach washed out and received charcoal as well. She had elevated levels of aspirin in her blood, and she reported deafening ringing in her ears (a symptom of high blood levels of aspirin). She was discharged the next day and lost to follow-up.[6]

Another 17-year-old at nine weeks' gestation in New Orleans took three oral contraceptive tablets. She washed the pills down with six ounces of vinegar. Ironically, the green pills she took were the placebos in the package. She was released from the emergency department and lost to follow-up.[6]

Coat hangers

The symbol of illegal abortion has not disappeared. Women continue to insert a coat hanger into the uterus through the cervix to cause abortion. A 16-year-old took a home pregnancy test and interpreted it as positive.[16] To end the pregnancy, she inserted a coat hanger through her cervix. The next day, she became seriously ill with pelvic infection and sought help at an emergency department. Her pregnancy test was negative, indicating that she had not been pregnant. She was treated with antibiotics for the pelvic infection caused by the wire.

A 24-year-old mother of two developed severe abdominal pain after inserting a coat hanger into her cervix and rupturing her membranes (bag of waters).[17] She was pregnant with twins at 21 weeks. Because her vital signs were unstable, she underwent an exploratory operation. A quart of blood was found in her abdomen, and a hole in the top of her uterus, caused by the coat hanger, was evident. She had a stormy course in the hospital. The fetuses died, and because of overwhelming infection, the woman had to have her uterus, tubes, and ovaries removed to save her life.

Quinine

In Lafayette, Indiana, a 19-year-old woman became ill after taking 5 quinine tablets in an attempt to abort herself.[18] Suffering from acute quinine toxicity, she was seen in a local emergency department. After three days, she was discharged without complications and subsequently had a legal abortion.

Legal consequences

Amber Abreu

Women who abort themselves outside of the health care system can wind up in court. Amber Abreu, an 18-year-old in Massachusetts, attempted to abort herself in 2007 by taking three misoprostol tablets.[19] She was apparently 25-26 weeks pregnant, at which time some newborns can survive. She succeeded in inducing labor, and her daughter Ashley was born at Lawrence General Hospital weighing a little over one pound. The severely premature infant died four days later at a large university hospital in Boston.

Police intervened when they heard conflicting accounts about the misoprostol dose. She was charged with improperly procuring a miscarriage and was held on $15,000 bail. She was sentenced to probation and ordered to receive therapy.[20]

Katrina Pierce

A victim of domestic violence, this 24-year-old reportedly took 30 Tylenol® (acetaminophen) and five 800 mg tablets of Motrin® (the equivalent of 20 drug-store strength tablets) to abort her 13-week fetus.[21] Sheriff's deputies were called to her home in West Monroe, NY, in response to a 911 call. Three children were found at home. She was taken to a hospital in Syracuse.

She was subsequently charged with "second-degree self-abortion,"[21] because her statements fit the description of the State law:

> *"Section 125:50 – A female is guilty of self-abortion in the second degree when, being pregnant, she commits or submits to an abortional [sic] act upon herself, unless such abortional [sic] act is justifiable pursuant to subdivision three of section 125.05."*[21]

Gabriela Flores

Ms. Flores, a mother of three, was a Mexican migrant worker in South Carolina in August of 2004.[9] Two of her children were living in Mexico, and one was living with her. Too poor to support another child, she obtained misoprostol.[22] She aborted the four-[20] or five-month[9] fetus in her home. A neighbor alerted the local sheriff, and Ms. Flores was taken to jail. She spent four months in jail while awaiting a decision from the State as to charges. She was ultimately charged with "unlawful abortion." Although the penalty can include up to two years in prison and a $1,000 fine,[9] she pleaded guilty and was sentenced to 90 days in jail.[20]

Kawana Ashley

Ms. Ashley, the third gunshot victim above, was criminally charged when her immature infant died at two weeks of age from complications of prematurity.[23] The medical examiner's office called the death "homicide." Ms. Ashley was charged with third-degree murder and manslaughter; each crime carried a sentence of up to 15 years. After legal wrangling for three years, the case finally made its way to the Florida Supreme Court. The Court ruled that she could not be charged with these crimes for self-induced abortion.[24]

Jennie McCormack

In 2011, this 32-year-old mother of three reportedly took pills to induce an abortion at five to six months' gestation.[22,25] Ms. McCormick resorted to self-abortion because of a lack of money; her sister ordered the pills over the Internet. Police eventually were called and found the aborted fetus. Since self-induced abortion is illegal in Idaho, she was charged with a crime. The punishment could include up to five years in prison and a fine of $5,000.

Ms. McCormack's attorney, Richard Hearn, sought to enjoin the State from prosecuting her. After a preliminary injunction, the Attorney General appealed to the Ninth Circuit Court of Appeals. The Court upheld the injunction. This ruling may be an important precedent: women should not be prosecuted for inducing abortion outside the medical setting.[25]

Summing up

When confronted by an unwanted or threatening pregnancy, some women are desperate. They will poison, physically abuse, and shoot themselves. Some die. Without medical supervision some of these women tried to induce abortion although they were not pregnant. Some developed life-threatening obstetrical complications, such as an ectopic pregnancy, that required medical help.[26] The more difficult access to and funding of abortion becomes, the greater the likelihood that women will revisit the bad old days of Chapter 1. Women deserve better.

Take-home messages

- *Without access to affordable abortion care, some women try to abort themselves*
- *Misoprostol offers a safer option than was available in the illegal abortion era*
- *Shooting oneself in the pregnant abdomen indicates the desperation involved*
- *Medically unsupervised attempts at self abortion with toxic drugs like pennyroyal oil can prove fatal*

Chapter 23
Children and barbarians

Many Americans are familiar with Harvard philosopher George Santayana's warning: those who cannot remember the past are doomed to repeat it. Few, however, know the rest of the quotation (right). Ignoring the past, he observed, is the hallmark of children and barbarians. What can contemporary American society learn from the natural experiment with legal abortion in the past four decades?

> *"Progress, far from consisting in change, depends on retentiveness... when experience is not retained, as among savages, infancy is perpetual. Those who cannot remember the past are condemned to repeat it....This is the condition of children and barbarians, in whom instinct has learned nothing from experience."*
>
> *Santayana*[1]

Three abortion eras

Abortion has always been available in the U.S. Rich women could generally find safe venues to terminate their pregnancies; the poor could not.[2] Countless American women suffered as a result. In the era before *Roe v. Wade*, estimates of the number of illegal abortions in the U.S. ranged from 200,000 to 1.2 million per year.[3,4] This wide range reflects the scarcity of reliable figures about clandestine procedures. What is clear is that more than 1,000 U.S. women died annually—more than three a day—from illegal abortion as recently as the 1940s.[5] Today, similarly brutal conditions prevail for women in many developing countries where archaic laws and uncaring medical providers condemn women to back-alley procedures.[6]

When State laws began to liberalize abortion access in the U.S., an unprecedented medical tourism began. Soon, tens of thousands of women were traveling to abortion centers in New York City, Washington, D.C., and California.[7] Impoverished women sacrificed disproportionately to pay for these lengthy trips. Some pawned personal belongings; others turned to theft, drug sales, or prostitution to pay for safe care. In States where abortion became legal—and, hence, safe—an astounding transformation occurred in municipal hospitals: septic abortion wards emptied, then closed.[8,9] The tight temporal association with changing laws made the conclusion unavoidable: legal abortion improved health.

When *Roe v. Wade* brought safe abortion to the nation, the benefits were prompt. Medical technology improved rapidly, surgical skill grew, women obtained abortions earlier in pregnancy, and the safety of abortion rose as a result.[10] By the early 1970s, induced abortion was demonstrated to be safer than childbirth. This "relative safety" concept was cited by the U.S. Supreme Court in its *Roe v. Wade* decisions.[11] The Court concluded that the State should not force a woman to bear a child (against her wishes) if that outcome were more likely to kill her than would a desired abortion. That relative safety concept holds today.[12,13]

The role of pregnancy loss

Early pregnancy loss is important for our species. As many as 70% of human conceptions may never develop, and a third of *recognized* pregnancies terminate spontaneously through miscarriage. This extraordinary reproductive inefficiency has a compelling purpose: increasing the likelihood that fetuses surviving the natural winnowing of pregnancy are normal and healthy.[14] This miscarriage screening process works well,[15,16] and nearly all newborns are healthy as a result. From a biological perspective, induced abortion is an extension of miscarriage—a continued winnowing designed to ensure than children are well born.... healthy, wanted, and loved.

Most abortions today are performed by suction curettage under local anesthesia in freestanding clinics.[17] The safety record amassed over the past four decades is unequivocal.[18,19]

Because of miscarriage and abortion, only a fraction of pregnancies in America continues to delivery. Although commonly viewed as "natural" (and thus "normal"), pregnancy poses both discomforts and risks. While the risk of death is low, complications are startlingly common. More than half of all pregnant women suffer one or more complications from childbearing.[20,21] From 3% to 9% have a prolonged stay after delivery or are re-admitted to the hospital because of complications.[22] One woman in 10,000 dies from pregnancy and childbirth each year, which is 14 times the risk with abortion.[13]

Abortion in grassroots America

Every third woman in America will have had an abortion during her life.[23] Rather than being something that "other women do," abortion is an everyday, if little discussed, event in the lives of women and their families. This reflects the fallibility of contraception, fallibility in its use, coercive sex, and failures of early embryonic development. While abortion is common throughout grassroots America, women disadvantaged by poverty, illness, and domestic violence disproportionately rely on abortion to control their fertility.[24] These are the same women whose access to effective contraception is similarly handicapped.

> *"I would have gotten my gas turned on. Right now, I'm using electric heaters and keeping my babies huddled up."*
>
> *"...if they [my children staying with their grandparents during my abortion] come back early, there won't be any food in the house."*
>
> *Medicaid-eligible abortion patients*[25]

One cruel barrier to abortion is cost. The Hyde Amendment cut off federal funding for poor women's abortions in the late 1970s. Rather than discouraging poor women from having abortions, the law further impoverished women: they continued to obtain abortions with their own meager resources.

Claims that public tax dollars should not pay for private abortions led many States to discontinue State support for abortion. In reality, every dollar spent on abortion averts $2 to $10 in future costs to society.[26,27] Thus, from a hardnosed fiscal point of view, the failure to pay for poor women's abortions is imprudent. Compared with other proven public health programs, such as childhood immunization, abortion is a sound public investment.

Legal abortion has improved family formation in America. Numbers of "shotgun" marriages, out-of-wedlock births, and unwanted babies placed for adoption all declined after the legalization of abortion.[28,29] Premature births and early infant deaths plummeted as well. Simply stated, wanted babies fare better in life.[30] Another boon to families has been prenatal diagnosis, which is predicated upon the availability of safe, legal abortion in the event a serious defect is discovered. Most pregnant women at risk of having a fetus with birth defects want prenatal testing.[31] Most women who learn of serious defects choose abortion. This process is yet another extension of the biological winnowing of miscarriage. Because of prenatal testing with abortion as a back-up, deaths of newborns and children with profound handicaps have decreased dramatically.[32]

Abortion alarmism

Because abortion is safe, opponents have been forced to invent bogus concerns. The Internet is a rich source of sites filled with half-truths and outright lies about abortion and its aftereffects. The common feature of these sites is a religious, rather than medical, sponsor. Saving souls appears more important than saving women's lives. These sponsors have questionable medical expertise but an unquestionable political agenda. Authors with diploma-mill doctorates[33] using discredited, misleading methods[34] are prolific contributors to this alternative literature. Egregious errors are rampant;[35] whether these stem from poor methods or intentional misrepresentation, the net result is the same: women seeking reliable information to cope with a stressful situation are confused.

Intentionally misleading frightened women is improper and unethical.[36] Moreover, it is cruel. The putative abortion and breast cancer link is a notable example of Internet disinformation. Flawed studies led some authors without medical or epidemiological training to conclude that abortion increases the risk of subsequent breast cancer; large, national studies free of recall bias refute any such link.[37,38] Despite the conclusion of objective medical organizations, such as the World Health Organization and the National Cancer Institute, the disinformation campaign is unrelenting.

Other specious concerns include "partial-birth abortion." "Partial-birth abortion" is a non-entity. It has, however, become a *cause celebre* of abortion critics. In reality, most abortions in America occur so early that the embryo cannot be seen with the unaided eye. Despite claims to the contrary, intact D&E is safe.[39,40]

Fetal pain is another specious issue. Abortion is an early event in pregnancy; development of a functioning fetal nervous system is a late event in pregnancy. The world's medical literature indicates that fetal pain is biologically impossible during abortion.[41]

The "abortion trauma syndrome" is not a recognized psychiatric diagnosis. Few women note lasting emotional problems after abortion, and those who do tend to be women whose psychological health was compromised before. Thorough reviews of the literature by the American Psychological Association[42] and Surgeon General Koop's report agree that adverse emotional consequences after abortion are "miniscule" from a public health perspective.[43]

Natural experiments—in both directions

> *"The lessons of Prohibition remain important today. They apply..... to such issues as censorship...abortion, and gambling."*
>
> *Thornton*[45]

Natural experiments concerning access to safe abortion are consistent. When abortion became legal in Sweden, the United States, South Africa and elsewhere, women's health quickly improved. When abortion was made illegal in Romania and Poland, women's health promptly suffered. The process works in both directions. When Ceausescu deprived women of safe abortion, the maternal mortality ratio in Romania soared to the highest in Europe. Driven into the back alley once again, Romanian women controlled their fertility—but paid for it with their lives. When Ceausescu was deposed and abortion was made legal again, maternal mortality plummeted.[44]

Legislating personal behavior

> *"The lesson...is that it is counter-productive to try to legislate morality."*
>
> *Digital History*[46]

Another social experiment in America, the prohibition of alcohol (1920-1933), proved disastrous. The Eighteenth Amendment to the Constitution drove underground the production, distribution, and consumption of alcohol. Organized crime took over sales and distribution, with Al Capone one of many beneficiaries of the "ignoble experiment." "Speakeasies" soon outnumbered the legitimate bars of earlier times, and thousands of Americans were blinded or brain-damaged by "bathtub gin" made with dangerous ingredients. Alcohol remained widely available in America; it just became more dangerous, and criminals were the beneficiaries.

Santayana's warning

Anti-abortion groups, working feverishly in State legislatures, are bent on further restricting or overturning *Roe v. Wade*. What might be the consequences? Carnage on the scale of Romania[44] or the Third World[47] is unlikely. More plausible is a return to the "sandwich years" of abortion in America (1970-1972), when women traveled long distances to get safe care.[48] National rates of abortion and birth would likely change little if women were forced on the road again.[49] However, costs would rise; abortions would occur at later, more risky stages of pregnancy; postoperative care would deteriorate. A black market for misoprostol and similar abortion-inducing drugs would flourish, and their use would be medically unsupervised.[50-52] As always, these changes would hurt poor and sick women disproportionately.[53]

And what of women who are denied abortion…or are unable to pay for it? Mandatory motherhood benefits neither the mother nor child. Unwanted children fare less well in life than do wanted children.[54,55] These disadvantages persist into adulthood.

The ethical debates swirling around abortion will continue. Many opponents claim to be taking the moral high ground. However, by depriving them of their civil rights, opposition to abortion hurts women and is thus unethical. It condemns women to mandatory motherhood. This attitude is not new. The systematic maltreatment of women has been institutionalized by governments and religions for several millennia.[56-58]

The clarity and cogency of the argument against abortion should be sufficient to sway public opinion. However, over the past four decades, this has not been the case. Opponents of abortion have resorted to eight murders,[59,60] arson, firebombing,[61] intimidation of women and clinicians,[62] governmental intrusion into the physician-patient relationship,[63] imposition of obstacles that deter and delay abortion, and increased costs.[64,65] A broad campaign of deception and chicanery, including crisis pregnancy centers and disinformation sites on the Internet,[66] has influenced decisions about abortion and its safety. Without the smokescreen about abortion safety, the ongoing attack on women and health care providers might be recognized for what it is: misogyny directed against our wives, sisters, and daughters. Ironically, the same political conservatives who oppose "big government" and its interference in our daily lives are sponsoring anti-abortion legislation mandating *more* intrusion of government into the private lives—and bodies—of American women.

While the ethical dimensions of abortion will continue to be debated, the medical science is incontrovertible: legal abortion has been a resounding public-health success.[18,19] The development of antibiotics, immunization, modern contraception, and legalized abortion all stand out as landmark public-health achievements of the Twentieth Century.

> *"When medical historians of the future look back on this century, the increasing availability of safe, legal abortion will stand out as a public health triumph."*
>
> Cates[67]

Abortion cannot be legislated out of existence. It will not go away. The tragic experience with Prohibition underscores this reality.[45] The question confronting our nation today is simply what women will be forced to pay for their abortions....in terms of dollars, disease, degradation, and—for some—death. As Santayana observed, society must learn from its experiences; if not, we remain savages.[1] If we lose our collective memory of the "bad old days" and turn back the clock on women in America—as happened in Romania[44]—we do so at great peril.

Glossary and Appendices

Glossary

Term	Definition
Abortion	Expulsion of a fetus or embryo prior to the stage of viability (20 weeks' gestation or fetal weight <500 g)
Abortion mortality rate	Number of abortion-related deaths per 100,000 procedures
Abortion rate	Number of abortions per 1,000 women.
Abortion ratio	Number of abortions per 1,000 live births
Abortus	A fetus or embryo removed or expelled before viability. Other descriptors include weight less than 500 g or measuring less than 25 cm.
Amniocentesis	Aspirating fluid from the pregnancy sac by inserting a thin needle through the abdominal wall
Amniotic fluid embolism	Obstruction and constriction of the blood vessels of the lung by amniotic fluid entering the woman's circulation, causing obstetrical shock
Analgesia	A moderation of painful stimuli (e.g., by giving narcotics) so that, while still felt, the stimuli are no longer painful
Anencephaly	Absence of the cranium (skull) and missing or rudimentary brain
Aneuploidy	An abnormal number of chromosomes, not a multiple of a haploid set
Atony	Relaxation or lack of tone
Autosomal dominant	A single abnormal gene on one of the 22 regular, non-sex chromosomes from either parent can cause the disease to occur in a child
Before-after study	One in which outcomes are compared before and after an intervention
Beneficence	The obligation to promote the well-being of the patient; a fundamental precept of bioethics
Bias	A systematic (as opposed to random) distortion of the truth in research
Body mass index	A measure of obesity: weight/height^2 Normal is less than 25
Bougie	A flexible cylindrical tube used for measuring or dilating
Cannula	A hollow plastic tube used to evacuate the uterus with suction
Case-control study	One in which persons with and without disease are tracked back in time to compare exposures that might be related to the outcome

Catheter	A hollow tube, commonly used to drain urine from the bladder
Cervical incompetence	Opening of the cervix, often in mid-pregnancy, without labor
Chemical pregnancy	A pregnancy so early that its presence can only be detected by sensitive pregnancy tests
Chimera	Individuals whose cells are derived from two or more zygotes
Choroideremia	An inherited progressive deterioration of part of the eye, leading to blindness
Clinical pregnancy	A pregnancy known to the woman by signs and symptoms, such as a missed period, fatigue, frequent urination, nausea, etc.
Cohort study	One in which persons exposed and not exposed to an intervention are followed forward in time to compare outcomes
Confidence interval	A measure of the precision of study results
Confounding	A distortion of the truth in a study caused by a mixing or blurring of effects. For example, a study compared the risk of heart attack in women using birth control pills to women not using birth control pills. The higher risk of heart attack in pill users was found to be due to the fact that more pill users were also cigarette smokers (the confounding factor)
Crude birth rate	Number of births per 1,000 total population (both sexes)
Cystic fibrosis	An inherited disorder of metabolism causing thick mucus, which obstructs the pancreas, intestines and breathing tubes of the lungs
Dilation and curettage (D&C)	Opening the cervix to a small diameter and scraping the uterine contents with a spoon-like metal instrument, the curette
Dilation and evacuation (D&E)	A curettage technique for second-trimester abortion; cervical dilation is often accomplished by use of osmotic dilators
Diploid	A cell having two chromosome sets (46), one from each parent; twice the haploid number (23)
Doula	A woman experienced in birth who provides emotional, physical, and informational support to another during labor, delivery, and the post-partum period

Down syndrome	A trisomy featuring an extra chromosome #21; associated with mental retardation and numerous other features, including heart disease
Eclampsia	Convulsions in a patient with preeclampsia
Ectopic pregnancy	A pregnancy implanted other than in the cavity of the uterus
Eisenmenger syndrome	A congenital heart defect, leading to insufficient oxygen in the blood (blue baby), high blood pressure in the lungs, and heart failure
Electroencephalogram	Recording of the electrical potentials of the brain, taken by attaching electrodes to the scalp
Embryo	The stage of human development beginning with the primitive streak (about 14 days after fertilization) until the end of the eighth week after fertilization, when all major structures are formed
Epidemiologist	A scientist who studies the causes, distribution, and prevention of diseases
Extremely low birthweight	A newborn weighing less than 1,000 g (about 2.2 pounds)
Gamete	A mature male or female germ cell, containing a haploid set of chromosomes. By combining with a gamete of the opposite sex, a diploid zygote is formed
General anesthesia	Pain relief in which the patient is unconscious (asleep), with no pain or memory of the event
Gestational age	Duration of the pregnancy, counting from the first day of the last menstrual period
Haploid	The chromosome complement of a normal ovum or sperm (23)
HELLP syndrome	A type of severe preeclampsia featuring Hemolysis (breakdown of blood cells), Elevated Liver enzymes (indicating liver disease), and Low Platelets (formed elements in the blood that assist in clotting)
Hematometra	An accumulation of blood in the uterine cavity
Hemophilia A	An inherited disorder of blood coagulation in males, leading to many spontaneous and traumatic hemorrhages

Hermaphroditism	An animal or plant having both male and female reproductive organs, from Greek mythology: Hermaphroditus was a son of Hermes and Aphrodite who became joined in one body with a nymph while bathing
Huntington disease	An inherited disorder leading to dementia and death, usually in the third or fourth decade of life
Hydatidiform mole	A placental disorder caused by fertilization of an egg with either no nucleus or an inactivated nucleus (classic mole); partial moles stem from fertilization of an egg by two sperm. The molar tissue resembles grape-like mush.
Hyde Amendment	Congressional bill that ended federal Medicaid payment for abortions for poor women
Hysterotomy	A mini-cesarean delivery, now rarely used for abortion
Identical twin	A twin resulting from a single fertilized egg that gives rise to two separate cell masses, each of which develops into an individual with the same genetic complement; a synonym for monozygotic
Illegal abortion	An abortion induced or performed by someone other than a licensed practitioner
Immature birth	An infant born earlier than a premature birth, for example, weighing less than 1,000 g
In vitro fertilization	An assisted reproductive technology used for infertile couples. An egg is harvested from the woman's ovary and is fertilized in a laboratory dish, after which it is placed into the uterus
Incomplete abortion	One in which part but not all the contents of the uterus have been expelled or removed
Infant mortality rate	The number of infant deaths per 1,000 live births
Information bias	A distortion of the truth in a study caused by collecting information differently for the groups being compared
Klinefelter syndrome	A 47 XXY chromosomal complement; patients are male but have breast development and appearance of a eunuch
Laminaria	Sticks of dried stalk of seaweed (about the size of a wooden matchstick); these are placed in the cervix for periods ranging from hours to days to open the cervix for an operation

Laparotomy	Making a surgical incision in the abdomen
Low birthweight	A newborn weighing less than 2500 g (about 5.5 pounds)
Marfan syndrome	An inherited connective-tissue disorder characterized by long limbs, heart defects, and eye problems
Maternal mortality ratio	Number of maternal deaths per 100,000 live births
Menses	A menstrual period
Methotrexate	A cancer chemotherapy drug used also to treat ectopic pregnancy and for early medical abortion
Mifepristone	An antiprogestin drug, also known as RU486 and Mifeprex® (Danco Laboratories, New York, NY) in the U.S. used for medical abortion. By blocking the action of progesterone, it induces a medical miscarriage
Miscarriage	Synonym for spontaneous abortion before the middle of the second trimester
Misoprostol	A prostaglandin pill, licensed for the prevention of ulcers in persons taking non-steroidal anti-inflammatory drugs; it also causes the cervix to soften and open and the uterus to contract
Monosomy	Absence of one chromosome of a pair (a deleted chromosome)
Morbid obesity	A body mass index greater than 40
Morbidity	Complications
Mortality	Death
Mosaicism	Two or more lines of cells containing different numbers of chromosomes
Neonatal	The first 28 days after birth
Neonatal mortality rate	The number of neonatal deaths per 1,000 live births
Neural tube defect	Failure of the early spinal cord to close in the early weeks of embryonic development; results range from anencephaly to slight vertebral defects (occult spina bifida)
Obesity	A body mass index (weight in kg divided by height in meters squared) greater than or equal to 30
Occult pregnancy	Hidden pregnancy; a very early pregnancy not evident to the woman; its presence can only be determined by laboratory tests

Odds ratio	A measure of association used in case-control studies; for uncommon events, it can be used interchangeably with relative risk
Oocyte	An immature female reproductive cell, not yet matured into an ovum
Oogonia	The precursor of eggs
Perinatal	An interval beginning at 22 completed weeks of pregnancy and ending 28 days after birth
Perinatal mortality rate	The number of stillbirths plus neonatal deaths per 1,000 total births
Peritonitis	Infection of the lining of the abdomen and its organs
Phenylketonuria	An inborn error of metabolism leading to increased levels of phenylalanine in the blood, which can cause brain damage
Polyploidy	The presence of three or more haploid sets of chromosomes
Post hoc ergo propter hoc	[After the thing, therefore on account of the thing.] A common error of logic: if B follows A in time, then B must have been caused by A
Preeclampsia	Development of high blood pressure and passage of protein in the urine, usually occurring after 20 weeks' gestation
Preembryo	The developing cells caused by the division of the zygote until the formation of the embryo (appearance of the primitive streak) at about 14 days after fertilization
Pregnancy	The state of a female from implantation of the embryo until birth
Preterm infant	An infant born before 37 completed weeks of pregnancy (often used synonymously with "premature" infant)
Prostaglandin $F_{2\alpha}$	A prostaglandin analogue used in the 1970s to induce abortion
Psychosis	A severe emotional and behavioral disorder, including schizophrenia and bipolar disorders
Quality-adjusted life years	An economic measure of cost-effectiveness

Recall bias	Distortion of the truth in research because of imperfect recollection of past events or exposures; a weakness of case-control studies relying on memory
Relative risk	A measure of association in research; the rate in the exposed group divided by the rate in the unexposed group
Retinitis pigmentosa	Progressive degeneration of the retina of the eye; some types of the disease are inherited
Rubella	Also known as German measles, this viral infection can cause birth defects if contracted during pregnancy
Saline	A salt solution placed in the uterus to induce abortion; a common concentration is 20% sodium chloride (table salt)
Schizophrenia	A common type of psychosis, featuring distorted perceptions or thoughts, affecting 2 million Americans
Selection bias	A distortion of the truth in a study caused by important baseline differences in the groups being compared.
Septic abortion	Abortion complicated by fever and infection
Sickle cell anemia	An inherited disorder of the blood cells, leading to severe anemia, pain crises, leg ulcers, bone death, atrophy of the spleen, and susceptibility to infection
Spina bifida	A gap, usually in the lower spine, through which a meningeal sac can protrude
Stillbirth rate	The number of stillborn infants per 1,000 infants born, including live births and stillbirths
Suction curettage (vacuum aspiration)	Opening the cervix to a small diameter and evacuating the uterine contents with a hollow tube attached to a suction source
Systematic review	A transparent, reproducible, and thorough search for all relevant literature on a given topic; more likely to be free of bias than narrative reviews with weaker methods
Tay-Sachs disease	An inherited disease leading to blindness and seizures in the first year of life; death occurs within a few years of birth
Teleology	The use of design or purpose to explain natural phenomena
Tenaculum	A surgical clamp used to stabilize the cervix

Term infant	An infant born between 37 and 42 complete weeks of pregnancy
Tetanus	A disease, also called lockjaw, featuring painful, persistent muscle contractions caused by a toxin of the bacterium, *Clostridium tetani*
Thalamus	A brain relay station located below the cerebral cortex
Thalassemia	An inherited disorder of the blood cells, leading to variable degrees of anemia
Triple-marker screen	A blood test (alphafetoprotein, human chorionic gonadotropin, and estriol) done in pregnant women to screen for abnormal fetuses
Triploidy	The presence of three haploid sets of chromosomes in all cells
Trisomy	The presence of an extra chromosome, that is, 47 instead of 46 (a common trisomy is Down syndrome, also called Trisomy 21)
Turner syndrome	A chromosome complement of 45 X; features include dwarfism, webbed neck, infantile sexual development and infertility
Ultrasound	High-frequency sonar (ranging from 1.6 to 10 MHz) used for diagnostic purposes; it can visualize internal organs such as the pregnant uterus
Urea	A nitrogen breakdown product, which can be placed in the uterus to induce abortion
Very low birthweight	A newborn weighing less than 1500 g (about 3.3 pounds)
Zygote	The single cell formed by fertilization of an ovum

References

American College of Obstetricians and Gynecologists. Ethics in obstetrics and gynecology. Second ed. Washington, D.C.: American College of Obstetricians and Gynecologists; 2004.

Cunningham FG, Leveno KJ, Bloom SL, Hauth JC, Rouse DJ, Spong CY. Williams' Obstetrics, 23d ed. New York: McGraw-Hill Companies, Inc.; 2010.

Anonymous. Stedman's Medical Dictionary. 27th ed. Philadelphia: Lippincott Williams & Wilkins; 2000.

Appendix A: Suggested websites for abortion information

Abortion clinics on-line	http://gynpages.com/
American Civil Liberties Union	http://www.aclu.org
American Congress (College) of Obstetricians and Gynecologists	http://www.acog.org/
American Medical Association	http://www.ama-assn.org/ama
American Medical Women's Association	http://www.amwa-doc.org
American Society for Reproductive Medicine	http://www.asrm.org/
Association of Reproductive Health Professionals	http://www.arhp.org/
Center for Reproductive Rights	http://www.reproductiverights.org
Centers for Disease Control and Prevention	http://www.cdc.gov/
Feminist Majority Foundation	http://www.feminist.org
Guttmacher Institute	http://www.guttmacher.org/
International Women's Health Coalition	http://iwhc.org/
Ipas	http://www.ipas.org/
NARAL Pro-Choice America	http://www.naral.org
National Abortion Federation	http://www.prochoice.org/
National Cancer Institute	http://www.cancer.gov
National Center for Health Statistics	http://www.cdc.gov/nchs/
National Library of Medicine (PubMed)	http://www.ncbi.nlm.nih.gov/entrez/query.fcgi?DB=pubmed
National Organization for Women	http://now.org
National Women's Health Network	http://nwhn.org/
People for the American Way	http://www.pfaw.org
Physicians for Reproductive Health	http://prh.org
Planned Parenthood Federation of America	http://www.plannedparenthood.org

POPLINE	http://www.popline.org
Reproductive Health Technologies Project	http://www.rhtp.org/
Royal College of Obstetricians and Gynaecologists	http://www.rcog.org.uk/
The Cochrane Library	http://www.thecochranelibrary.com
The Henry J. Kaiser Family Foundation	http://www.kff.org/
The Population Council	http://www.popcouncil.org/
Union of Concerned Scientists	http://www.ucsusa.org/
World Health Organization	http://www.who.int/en/

Appendix B: Conclusions of medical and public health organizations

World Health Organization

Recommendations related to regulatory, policy and human rights considerations[1]

- Laws and policies on abortion should protect women's health and their human rights.
- Regulatory, policy and programmatic barriers that hinder access to and timely provision of safe abortion care should be removed.
- An enabling regulatory and policy environment is needed to ensure that every woman who is legally eligible has ready access to safe abortion care. Policies should be geared to respecting, protecting and fulfilling the human rights of women, to achieving positive health outcomes for women, to providing good-quality contraceptive information and services, and to meeting the particular needs of poor women, adolescents, rape survivors and women living with HIV.

American College of Obstetricians and Gynecologists

Abortion policy[2]

- The College continues to affirm the legal right of a woman to obtain an abortion prior to fetal viability.
- The intervention of legislative bodies into medical decision making is inappropriate, ill advised, and dangerous.

American Academy of Pediatrics

The adolescent's right to confidential care when considering abortion[3]

- Extensive reviews conclude that there are no documented negative psychological or medical sequelae to elective, legal, first-trimester abortion among teenaged women.
- The scientific evidence indicates that legal abortion results in fewer deleterious sequelae for women compared with other possible outcomes of unwanted pregnancy.
- The most damaging impact of mandatory parental notification laws is that they can delay and obstruct the access of pregnant adolescents to timely professional advice and medical care.

- Compared with peers who terminate their pregnancies, adolescents who bear children are at significantly higher risk of educational deficits, economic disadvantage, and marital instability.

American Public Health Association

Support for sexual and reproductive health and rights in the United States and abroad[4]

- The local availability of accurate, comprehensive sexual and reproductive-health education and services, including safe abortion, improves health care decision making and reduces death and injury.

Provision of abortion care by advanced practice nurses and physician assistants[5]

- Abortion, recognized as one choice for women experiencing unintended pregnancy, is one of the most common and safest gynecologic interventions in the United States.

American Psychiatric Association

Abortion and women's reproductive health care rights[6]

- The American Psychiatric Association opposes all constitutional amendments, legislation, and regulations curtailing family planning and abortion services to any segment of the population.
- The American Psychiatric Association affirms that the freedom to act to interrupt pregnancy must be considered a mental health imperative with major social and mental health implications.

American Psychological Association

Mental health and abortion[7]

- In general, however, the prevalence of mental health problems observed among women in the United States who had a single, legal, first-trimester abortion for nontherapeutic reasons was consistent with normative rates of comparable mental health problems in the general population of women in the United States.
- Many of these same factors also predict negative psychological reactions to other types of stressful life events, including

childbirth, and, hence, are not uniquely predictive of psychological responses following abortion.

American Medical Association

Induced termination of pregnancy before and after Roe v Wade[8]

- Increasingly restrictive abortion laws are also likely to disproportionately affect young, poor, and minority women who may lose some of the access they currently have to the improved and sophisticated medical technology of current abortion procedures.
- As access to safer, earlier legal abortion becomes increasingly restricted, there is likely to be a small but measurable increase in mortality and morbidity among women in the United States.

American Medical Women's Association

Position paper on principles of abortion[9]

- The American Medical Women's Association will oppose efforts to overturn or weaken *Roe v. Wade*. We will oppose laws and court rulings that interfere with the doctor-patient relationship, either in requiring or proscribing specific medical advice to pregnant women. We will oppose measures that limit access to medical care for pregnant women, particularly for poor or underserved groups.

Position paper on principles of abortion and access to comprehensive reproductive health services [10]

- Since the advent of legal abortion in the United States, there has been a dramatic decrease in all pregnancy-related deaths and in pregnancy and abortion-related complications.
- AMWA understands that full access to abortion and family planning services, unrestricted by financial constraints, improves women's health and women's lives.

References

1. World Health Organization. Safe abortion: technical and policy guidance for health systems, second edition. http://www.who.int/reproductivehealth/publications/unsafe_abortion/9789241548434/en/, accessed December 28, 2013.
2. American College of Obstetricians and Gynecologists. Abortion policy. http://www.acog.org/-/media/Statements-of-Policy/Public/sop069.pdf?dmc=1&ts=20141130T1919471000, accessed January 6, 2014.
3. American Academy of Pediatrics Committee on Adolescence. The adolescent's right to confidential care when considering abortion. Pediatrics 1996;97:746-51.
4. American Public Health Association. Support for sexual and reproductive health and rights in the United States and abroad. http://www.apha.org/policies-and-advocacy/public-health-policy-statements/policy-database/2014/07/24/15/12/support-for-sexual-and-reproductive-health-and-rights-in-the-united-states-and-abroad, accessed January 6, 2014.
5. American Public Health Association. Provision of abortion care by advanced practice nurses and physician assistants. http://www.apha.org/policies-and-advocacy/public-health-policy-statements/policy-database/2014/07/28/16/00/provision-of-abortion-care-by-advanced-practice-nurses-and-physician-assistants, accessed January 1, 2014.
6. America Psychiatric Association. Abortion and women's reproductive health care rights. http://www.psychiatry.org/, accessed December 30, 2013.
7. American Psychological Association. Mental health and abortion. http://www.apa.org/pi/women/programs/abortion/executive-summary.pdf, accessed December 30, 2013.
8. Council on Scientific Affairs. Induced termination of pregnancy before and after Roe v Wade. Trends in the mortality and morbidity of women. Council on Scientific Affairs, American Medical Association. JAMA 1992;268:3231-9.
9. American Medical Women's Association. Position paper on principles of abortion. http://www.amwa-doc.org/wp-content/uploads/2013/12/Abortion1.pdf, accessed January 6, 2014.
10. American Medical Women's Association. Position paper on principles of abortion & access to comprehensive reproductive health services. http://www.amwa-doc.org/wp-content/uploads/2013/12/Abortion___Access_to_Comprehensive_Reprod_Health_Services1.pdf, accessed January 6, 2014.

Appendix C: Difficult choices

Alma

When an abdominal mass was discovered during a routine physical examination, Alma underwent her first abdominal operation. During an exploratory laparotomy, the surgeons found a rare germ-cell tumor of her ovary. A year later, a follow-up imaging study indicated an abnormal growth on her liver. Fearing metastatic spread of the ovarian tumor to her liver, the surgeons opened her abdomen a second time and removed part of the liver. On microscopic examination, the removed section of liver contained no cancer.

For a few years, Alma's life returned to that of a healthy middle-aged woman. However, severe abdominal pain and a worrisome X-ray of her bowels led to a third opening of her abdomen, this time for partial blockage of her small bowel. The surgeons removed scar tissue and attempted to free her bowel.

Within a few years, the partial small bowel obstruction recurred, requiring yet another difficult operation to remove the extensive adhesions between bowel and other structures. During the long and technically difficult operation, bleeding was excessive, and Alma received multiple blood transfusions.

After a lapse of several years, Alma experienced her third and most dangerous bowel blockage. Again, the operation was very difficult and bloody; once again, multiple blood transfusions were needed to save her life. She had a lengthy stay in the intensive care unit.

Her surgeons advised her never to attempt a pregnancy, since if she required a cesarean delivery, both she and her fetus might die from intractable bleeding. Despite this stern warning, Alma conceived while in a relationship in her late thirties. She reported that she had been using condoms when she became pregnant. She reluctantly requested an abortion to avoid the possibility of dying in childbirth.

Catherine

We were asked to see Catherine in consultation at the infectious disease hospital. She had recently been admitted with a severe type of tuberculosis. She was also HIV positive. Her infection was resistant to most of the usual drugs used to treat this life-threatening strain of tuberculosis. Given her HIV status and multiple-drug resistant tuberculosis, her pulmonologists (lung specialists) advised ending her pregnancy, now at 15 weeks' gestation. However, because she posed such a threat to other patients, we could not bring her to our hospital for the operation. After multiple consultations between teams of physicians, the decision was made to perform the abortion in a treatment room in the infectious disease hospital. On the chosen day, we loaded a suction machine and the requisite surgical equipment in the trunk of a sheriff's car, which took us to the infectious disease hospital, where she was quarantined. The operation went smoothly, and Catherine embarked on a long course of tuberculosis and HIV medications.

Diane

The call came at home on July Fourth. The cardiothoracic surgeons had just admitted a woman with previously undiagnosed Marfan syndrome. This inherited connective tissue disorder—according to some reports, President Lincoln was a sufferer— is associated with long, spindly fingers and weak blood vessel walls, making them prone to aneurysm (ballooning). Diane had developed chest pain, which led to a chest X-ray. On examination of the x-ray, doctors were startled to find an aortic aneurysm 9x15 cm (about 3x6 inches), about to burst. Bursting of the aneurysm would be fatal within minutes. Diane's situation was even more precarious: she was found to be 25 weeks' pregnant. Since the extra demands placed on the heart and blood vessels by the pregnancy posed an immediate threat to her life, she and the cardiothoracic surgeons requested an emergency abortion. We took our equipment to the cardiothoracic operating room and performed the procedure as the chest surgeons stood ready to open the Diane's chest should leakage or rupture occur. The operation went smoothly, and she was scheduled to get a new synthetic aorta within the coming week.

Felicia

Felicia was an Ensign in the U.S. Navy, stationed at an east-coast base. During her first prenatal visit, the routine blood tests were strikingly abnormal. Further tests the next week and a bone marrow biopsy confirmed the worst: Felicia had leukemia as well as an early pregnancy.

The oncologists caring for her were unwilling to start chemotherapy for leukemia until her pregnancy had been terminated, because of fear of the chemotherapy agents causing birth defects. Felicia agreed with this course of action but was shocked to learn that her naval hospital was unable to provide the needed abortion. She found our clinic on the Internet, and with her family traveled by car to get the care needed. Although her chemotherapy was covered by her naval insurance, her abortion was paid for by her parents, since her Ensign's salary was insufficient.

Hannah

Obese since childhood, Hannah's weight ballooned to dangerous values by her early twenties. Her Body Mass Index (weight corrected for height) climbed to 51 (normal being below 25). After what was believed to be a viral illness, she became short of breath and unable to climb stairs. She was hospitalized for evaluation, and she was found to have unexplained cardiomyopathy (heart failure). An echocardiogram indicated that her heart was only pumping 15-25% of the usual volume of blood. Fortified by a regimen of blood pressure medicine, heart medicines, and diuretics, she returned to her usual activities.

Not using contraception, Hannah became pregnant by her boyfriend. Although unplanned, she and her partner were happy with the prospect of having a baby. However, with the added demands on her heart from the growing pregnancy, her heart function began to fail once again. Moreover, because of her dangerous obesity, she developed diabetes of pregnancy as well.

After many discussions with maternal-fetal medicine specialists and cardiologists, she was advised that continuing the pregnancy carried a substantial risk of a major heart event, such as heart failure or death. The probability was estimated as 5 to 25%. Hannah was referred to our hospital for a second opinion and a discussion of the abortion option. After lengthy discussions, she and her partner were willing to assume the risks of pregnancy and decided the continue it. She was told that abortion remained an option should her health rapidly deteriorate.

Irene

Twelve-year-old Irene arrived in the clinic accompanied by her distraught mother, the family's minister, a social worker, and a detective from her local police department. Only when Irene's bulging abdomen had become evident a few weeks ago to a teacher at school had the pregnancy been discovered. Irene's mother's boyfriend Ray was reportedly the father of the pregnancy. Reluctant to talk to her mother, Irene admitted to her social worker that Ray had forced himself on her on many occasions while Irene's mother was at work. He threatened to kill Irene if she told anyone.

The family was pressing charges against him for statutory rape, and the detective had come along to request fetal tissue for DNA analysis. The abortion went smoothly, and the chain of evidence was started for DNA analysis. We wrote a generic excuse note for Irene's sixth-grade teacher to account for her absence.

Two years later, I received a subpoena from the district attorney in Irene's county requiring me to testify at trial. I arranged to take a day off from work to travel to this distant county. A week before the trial, however, the district attorney called to report that Ray had fled the State. He was never found.

Jackie

Born healthy, Jackie became brain-damaged because of head trauma sustained in a car wreck. Her intellectual development thereafter was profoundly compromised, and she was incapable of living independently as an adult. In her 30's, she was living in a group home 150 miles from our hospital.

Jackie was impregnated by an unidentified man at the group home. Recognizing that she was not able to rear a child, Jackie and her parents came to our clinic requesting an abortion. Preparations were begun, but the night before her operation, her transportation vanished. She had no family who could bring her to the hospital.

I suggested to her parents that she could get care in a local hospital. If they could provide me the name of the hospital, I would call the staff and assist with her management. When they arrived at the emergency department of a large hospital in her home town, she was turned away despite the fact that the abortion was in process. Rather than negotiating with the reluctant hospital, I had Jackie brought by ambulance 150 miles to finish her care with us. After the abortion, with her consent we placed an intrauterine device to prevent a repetition of this tragedy.

Kristin

In November, 2009, a mother of four became critically ill from pulmonary hypertension when 11 weeks pregnant. The blood pressure in the large vessels in her lungs was dangerously high. Pregnancy aggravates this condition, and the risk of the woman dying is commonly cited as 50%. In the estimation of the doctors caring for Kristin, her likelihood of dying if the pregnancy continued was "close to 100%."[1]

The ethics committee of St. Joseph's Hospital and Medical Center in Phoenix, owned by Catholic Healthcare West (now Dignity Health), was consulted. The committee concurred that an abortion was needed to save Kristin's life, and the procedure was done apparently without problems. Kristin survived and is caring for her children today.[2, 3]

The problem arose when the Bishop of Phoenix learned of the abortion afterwards. The Church severed ties with the hospital. Sister Margaret McBride, a hospital administrator and member of the ethics committee, was excommunicated. By December of 2011, McBride was no longer excommunicated.[1]

The hospital's Chief Executive Officer thoughtfully noted that, "Our first priority is to save both patients. If that is not possible, we always save the life we can save."[4] In his letter to the hospital, Bishop Olmsted had different priorities: "For the sake of salvation of souls and in the interest of justice.....adhere to my requests."[2]

1. Anonymous. Excommuncation of Margaret McBride. http://en.wikipedia.org/wiki/Excommunication_of_Margaret_McBride, accessed January 5, 2014.
2. Clancy M. Phoenix bishop warns St. Joseph's Hospital on health-care. http://www.azcentral.com/community/phoenix/articles/2010/12/15/20101215phoenix-bishop-st-josephs-hospital.html?nclick_check=1, accessed January 5, 2014.
3. Clancy M. Nun at St. Joseph's Hospital rebuked over abortion to save woman. http://www.azcentral.com/arizonarepublic/news/articles/2010/05/15/20100515phoenix-catholic-nun-abortion.html, accessed January 5, 2014.
4. Harris D. Bishop strips hospital of Catholic status after abortion. http://abcnews.go.com/Health/abortion-debate-hospital-stripped-catholic-status/story?id=12455295, accessed January 5, 2014.

Ruth

An only child, Ruth was born prematurely at 25 weeks to a teenage mother. Because of her extreme prematurity and related breathing difficulties, Ruth suffered permanent brain damage from lack of oxygen. She was left with severe intellectual and emotional impairment. Now 16 years old and a tenth grader, she attended special education classes designed to help her gain some independent functioning as an adult. Traumatized by hospitalization for an appendectomy at age 10, she had an overpowering fear of medical providers and needles.

When Ruth turned 15 years of age, her mother agreed that she could start to date. Ruth soon fell in love with a boy her own age. When a school nurse noticed that Ruth had missed a menstrual period, she did a pregnancy test which alerted Ruth's mother to the pregnancy. By this time, Ruth's boyfriend had moved out of State with his family. Ruth's mother said she was unable to cope with the prospect of rearing Ruth's baby while still caring for Ruth.

They traveled 150 miles to an abortion clinic when Ruth was 12 weeks' pregnant. Ruth began to shriek uncontrollably, and no care could be provided. She was referred to our hospital at 14 weeks in the hopes that general anesthesia in an operating room might enable the procedure to be done. Plans to prepare Ruth's cervix with laminaria (seaweed stick dilators) had to be abandoned when she became uncooperative. She would not allow an intravenous line to be started, which could have been used to give her tranquilizers and narcotics.

In consultation with Ruth's mother and the anesthesia department, oral misoprostol was used instead of laminaria to prepare the cervix. In the preoperative holding area, Ruth was given a large dose of an oral tranquilizer. When heavily sedated, she was given anesthetic gases by mask until she fell asleep. Only then could an intravenous line be started and routine blood work done. The abortion was uneventful thereafter, and, as requested by Ruth and her mother, an intrauterine device was placed.

Savita

Savita Halappanavar, a 31-year-old Indian dentist suffering a miscarriage at 17 weeks' gestation, sought care at a university hospital in Galway, Ireland on October 21, 2012. Savita was sent home from the hospital with advice for "physiotherapy."[physical therapy][1] She returned to the hospital a few hours later because of persistent discomfort.

On October 22, her membranes ruptured. At this point, the pregnancy was doomed. Once membranes rupture, bacteria can enter the normally sterile uterus and cause life-threatening infection. Since the fetus was many weeks from being able to survive outside the uterus, prompt abortion was medically necessary.

On October 23, Savita asked her obstetrician for an abortion. She was reportedly told that because the fetus was still alive, an abortion was illegal in Ireland. A midwife further explained that "it was the law, that this is a Catholic country."[1-3] However, Savita was a Hindu.

The fetus died the following day, and it was surgically removed from Savita's uterus. The delay proved fatal. Sepsis (generalized bloodstream infection) developed on October 25, and Savita died in the early morning of October 28 despite antibiotic therapy.

A formal inquiry into the death, published in June, 2013, identified numerous deficiencies in medical care. In the wake of this preventable death, the Protection of Life During Pregnancy Act of 2013 was signed into Irish law in July of that year.[2] The Act describes the circumstances under which abortion can be legally performed in Ireland.

1. McDonald H. Savita Halappanavar: the last week of her life. http://www.theguardian.com/world/2013/apr/19/savita-abortion-life-hospital-timeline, accessed January 5, 2014.

2. Anonymous. Death of Savita Halappanavar. http://en.wikipedia.org/wiki/Death_of_Savita_Halappanavar, accessed January 5, 2014.

3. Dalby D. Religious remark confirmed in Irish abortion case; http://www.nytimes.com/2013/04/11/world/europe/religious-remark-confirmed-in-irish-abortion-case.html?_r=0, accessed January 5, 2014.

References

Chapter 1 - The Bad Old Days

1. NARAL Pro-choice America Foundation. The safety of legal abortion and the hazards of illegal abortion. Washington, D.C.: NARAL Pro-choice America Foundation; 2005.
2. Lane SD, Jok JM, El-Mouelhy MT. Buying safety: the economics of reproductive risk and abortion in Egypt. Soc Sci Med 1998;47:1089-99.
3. Vlassoff M, Mugisha F, Sundaram A, et al. The health system cost of post-abortion care in Uganda. Health Policy Plan 2014; 29: 56-66.
4. Basinga P, Moore AM, Singh SD, Carlin EE, Birungi F, Ngabo F. Abortion incidence and postabortion care in Rwanda. Stud Fam Plann 2012;43:11-20.
5. Vlassoff M, Fetters T, Kumbi S, Singh S. The health system cost of postabortion care in Ethiopia. Int J Gynaecol Obstet 2012;118 Suppl 2:S127-33.
6. Kinsley M. Republican platform's abortion plank has a long history. http://articles.latimes.com/2012/aug/31/opinion/la-oe-0831-kinsley-abortion-gop-platform-20120831, accessed March 29,2013.
7. Cates WJ. Legal abortion: the public health record. Science 1982;215:1586-90.
8. Reagan LJ. "About to meet her maker": women, doctors, dying declarations, and the state's investigation of abortion, Chicago, 1867-1940. J Am Hist 1991;77:1240-64.
9. Joffe C. Physician provision of abortion before Roe v. Wade. Res Sociol Health Care 1991;9:21-32.
10. Santamarina BA, Smith SA. Septic abortion and septic shock. Clin Obstet Gynecol 1970;13:291-304.
11. Grimes DA, Gallo MF, Grigorieva V, Nanda K, Schulz KF. Fertility awareness-based methods for contraception: systematic review of randomized controlled trials. Contraception 2005;72:85-90.
12. Barno A. Criminal abortion deaths, illegitimate pregnancy deaths, and suicides in pregnancy. Minnesota, 1950-1965. Am J Obstet Gynecol 1967;98:356-67.
13. Barno A. Criminal abortion—deaths and suicides in pregnancy in Minnesota, 1950-1964. Minn Med 1967;50:11-6.
14. Stevenson LB. Maternal death and abortion, Michigan 1955-1964. Mich Med 1967;66:287-91.

15. Webster A. Maternal deaths at the Cook County Hospital. Fourteen-year survey (1952-1965). Am J Obstet Gynecol 1968;101:244-53.

16. Abernathy JR, Greenberg BG, Horvitz DG. Estimates of induced abortion in urban North Carolina. Demography 1970;7:19-29.

17. Calderone MS. Abortion in the United States. New York: Paul B. Hoeber; 1958.

18. Cates WJ, Rochat RW, Grimes DA, Tyler CWJ. Legalized abortion: effect on national trends of maternal and abortion-related mortality (1940 through 1976). Am J Obstet Gynecol 1978;132:211-4.

19. Studdiford W. A consideration of incomplete abortion complicated by fever. N Y State J Med 1939;39:1274.

20. Berman SM, MacKay HT, Grimes DA, Binkin NJ. Deaths from spontaneous abortion in the United States. JAMA 1985;253:3119-23.

21. Studdiford W, Douglas G. Placental bacteremia: a significant finding in septic abortion accompanied by vascular collapse. Am J Obstet Gynecol 1956;71:842-58.

22. Kopelman J, Douglas G. Abortions by resident physicians in a municipal hospital center. Am J Obstet Gynecol 1971;111:666-71.

23. Belsky JE. Medically indigent women seeking abortion prior to legalization: New York City, 1969-1970. Fam Plann Perspect 1992;24:129-34.

24. Polgar S, Fried ES. The bad old days: clandestine abortions among the poor in New York City before liberalization of the abortion law. Fam Plann Perspect 1976;8:125-7.

25. Grimes DA, Benson J, Singh S, et al. Unsafe abortion: the preventable pandemic. Lancet 2006;368:1908-19.

26. Salter C, Johnston H, Hengen N. Care for postabortion complications: saving women's lives. Popul Rep L 1997:1-31.

27. Sambhi JS. Abortion by massage—"bomoh". IPPF Med Bull 1977;11:3.

28. Liskin L. Complications of abortion in developing countries. Popul Rep 1980;Series F:F105-55.

29. Goyaux N, Yace-Soumah F, Welffens-Ekra C, Thonneau P. Abortion complications in Abidjan (Ivory Coast). Contraception 1999;60:107-9.

30. Thapa PJ, Thapa S, Shrestha N. A hospital-based study of abortion in Nepal. Stud Fam Plann 1992;23:311-8.

31. Ankomah A, Aloo-Obunga C, Chu M, Manlagnit A. Unsafe abortions: methods used and characteristics of patients attending hospitals in Nairobi, Lima, and Manila. Health Care Women Int 1997;18:43-53.

32. Okonofua F. Preventing unsafe abortion in Nigeria. Afr J Reprod Health 1997;1:25-36.

33. Burnhill MS. Treatment of women who have undergone chemically induced abortions. J Reprod Med 1985;30:610-4.

34. Carney BH. Vaginal burns from potassium permanganate. Am J Obstet Gynecol 1953;65:127-30.

35. Kobak AF, Wishnick S. Potassium permanganate burn of the vagina followed by bowel obstruction. Am J Obstet Gynecol 1955;70:409-11.

36. Lammert AC, Scott RB. Hazards in the intravaginal use of potassium permanganate. Am J Obstet Gynecol 1954;68:854-9.

37. O'Donnell RP. Vesicovaginal fistula produced by potassium permanganate. Obstet Gynecol 1954;4:122-3.

38. Radl S. Over our live bodies: preserving choice in America. Dallas, TX: Steve Davis; 1989.

39. Gold EM, Erhardt CL, Jacobziner H, Nelson FG. Therapeutic abortions in New York City: a 20-year review. Am J Public Health Nations Health 1965;55:964-72.

40. Cable News Network. FAA approves proposal for fixing Dreamliner battery woes. http://www.cnn.com/2013/03/12/travel/faa-dreamliner/index.html?iref=allsearch, accessed March 29, 2013.

41. Ahman E, Shah IH. New estimates and trends regarding unsafe abortion mortality. Int J Gynaecol Obstet 2011;115:121-6.

42. Grimes DA. Unsafe abortion: the silent scourge. Br Med Bull 2003;67:99-113.

43. Rogo KO. Induced abortion in sub-Saharan Africa. East Afr Med J 1993;70:386-95.

44. Koster-Oyekan W. Why resort to illegal abortion in Zambia? Findings of a community-based study in Western Province. Soc Sci Med 1998;46:1303-12.

45. Population Reference Bureau. Family planning worldwide. Washington, D.C.: Population Reference Bureau; 2008.

46. Oye-Adeniran BA, Umoh AV, Nnatu SN. Complications of unsafe abortion: a case study and the need for abortion law reform in Nigeria. Reprod Health Matters 2002;10:18-21.

47. Paxman JM, Rizo A, Brown L, Benson J. The clandestine epidemic: the practice of unsafe abortion in Latin America. Stud Fam Plann 1993;24:205-26.

48. Adetoro OO. A 15-year study of illegally induced abortion mortality at Ilorin, Nigeria. Int J Gynaecol Obstet 1989;29:65-72.

49. Ladipo OA. Preventing and managing complications of induced abortion in Third World countries. Int J Gynaecol Obstet 1989; 3:21-8.

50. Unuigbe JA, Oronsaye AU, Orhue AA. Abortion-related morbidity and mortality in Benin City, Nigeria: 1973-1985. Int J Gynaecol Obstet 1988;26:435-9.

51. Faundes A, Santos LC, Carvalho M, Gras C. Post-abortion complications after interruption of pregnancy with misoprostol. Adv Contracept 1996;12:1-9.

52. Saultes TA, Devita D, Heiner JD. The back alley revisited: sepsis after attempted self-induced abortion. West J Emerg Med 2009;10:278-80.

53. Ugboma HA, Akani CI. Abdominal massage: another cause of maternal mortality. Niger J Med 2004;13:259-62.

Chapter 2 - Mass medical tourism – and resulting health benefits

1. Levine PB, Staiger D, Kane TJ, Zimmerman DJ. Roe v Wade and American fertility. Am J Public Health 1999;89:199-203.

2. Institute of Medicine. Legalized abortion and the public health. Washington, D.C.: National Academy of Sciences; 1975.

3. Stewart GK, Goldstein PJ. Therapeutic abortion in California. Effects on septic abortion and maternal mortality. Obstet Gynecol 1971;37:510-4.

4. Russell K, Jackson E. Therapeutic abortion – the California experience. In: Osofsky H, Osofsky J, eds. The abortion experience Psychological and medical impact. Hagerstown, MD: Harper and Row; 1973:165-76.

5. Center for Disease Control. Abortion surveillance report – legal abortions, United States, annual summary, 1970. Atlanta, GA: Center for Disease Control; 1971.

6. Center for Disease Control. Abortion surveillance report – legal abortions, United States, annual summary, 1971. Atlanta, GA: Center for Disease Control; 1972.

7. Center for Disease Control. Abortion surveillance, annual summary, 1972. Atlanta, GA: Center for Disease Control; 1974.

8. Nathanson BN. Ambulatory abortion: experience with 26,000 cases (July 1, 1970, to August 1, 1971). N Engl J Med 1972;286:403-7.

9. Pakter J, O'Hare D, Nelson F, Svigir M. A review of two years' experience in New York City with the liberalized abortion law. In: Osofsky H, Osofsky J, eds. The abortion experience Psychological and medical impact. Hagerstown, MD: Harper and Row; 1973:47-93.

10. Rochat RW, Heath CW, Jr., Chu SY, Marchbanks PA. Maternal and child health epidemic-assistance investigations, 1946-2005. Am J Epidemiol 2011;174:S80-8.

11. Berger GS, Gibson JJ, Harvey RP, Tyler CW, Jr., Pakter J. One death and a cluster of febrile complications related to saline abortions. Obstet Gynecol 1973;42:121-6.

12. Nowicka W. The effects of the 1993 anti-abortion law in Poland. Entre Nous Cph Den 1996; Dec: 13-5.

13. Gilmartin M, White A. Interrogating medical tourism: Ireland, abortion, and mobility rights. Signs (Chic) 2011;36:275-79.

14. Payne D. Record numbers of Irish women visit Britain for abortion. BMJ 1999;319:593.

15. Grossman D, Garcia SG, Kingston J, Schweikert S. Mexican women seeking safe abortion services in San Diego, California. Health Care Women Int 2012;33:1060-9.

16. Angulo V, Guendelman S. Crossing the border for abortion services: the Tijuana-San Diego connection. Health Care Women Int 2002;23:642-53.

17. Palmer B. "Lonely, tragic, but legally necessary pilgrimages": transnational abortion travel in the 1970s. Can Hist Rev 2011;92:637-64.

18. Sethna C, Doull M. Far from home? A pilot study tracking women's journeys to a Canadian abortion clinic. J Obstet Gynaecol Can 2007;29:640-7.

19. Nickson C, Smith AM, Shelley JM. Travel undertaken by women accessing private Victorian pregnancy termination services. Aust N Z J Public Health 2006;30:329-33.

20. Nickson C, Shelley J, Smith A. Use of interstate services for the termination of pregnancy in Australia. Aust N Z J Public Health 2002;26:421-5.

21. Lanman JT, Kohl SG, Bedell JH. Changes in pregnancy outcome after liberalization of the New York State abortion law. Am J Obstet Gynecol 1974;118:485-92.

22. Kohl SG, Lanman JT. Further observations on changes in pregnancy outcome following liberalization of the New York State abortion law. Adv Plan Parent 1976;11:101-5.

23. Glass L, Evans HE, Swartz DP, Rajegowda BK, Leblanc W. Effects of legalized abortion on neonatal mortality and obstetrical morbidity at Harlem Hospital Center. Am J Public Health 1974;64:717-8.

24. Swartz D. The Harlem Hospital Center experience. In: Osofsky H, Osofsky J, eds. The abortion experience Psychological and medical impact. Hagerstown, MD: Harper and Row; 1973: 94-121.

25. McLean RA, Cochrane NE, Mattison ET. III. Abortion and mortality study. Twenty-year study in New York State exclusive of New York City. N Y State J Med 1979;79:49-52.

26. Kahan RS, Baker LD, Freeman MG. The effect of legalized abortion on morbidity resulting from criminal abortion. Am J Obstet Gynecol 1975;121:114-6.

27. Seward PN, Ballard CA, Ulene AL. The effect of legal abortion on the rate of septic abortion at a large county hospital. Am J Obstet Gynecol 1973;115:335-8.

28. Goldstein P, Stewart G. Trends in therapeutic abortion in San Francisco. Am J Public Health 1972;62:695-9.

29. Grimes DA, Schulz KF. Bias and causal associations in observational research. Lancet 2002;359:248-52.

30. Bracken MB, Freeman DHJ, Hellenbrand K. Hospitalization for medical-legal and other abortions in the United States 1970-1977. Am J Public Health 1982;72:30-7.

31. Tietze C. The effect of legalization of abortion on population growth and public health. Fam Plann Perspect 1975;7:123-7.

32. Tietze C, Bongaarts J. The demographic effect of induced abortion. Obstet Gynecol Surv 1976;31:699-709.

Chapter 3 - Legalization and medicalization

1. Thompson D. Health: A Setback for Pro-Life Forces. Time Magazine, March 27, 1989; http://content.time.com/time/magazine/article/0,9171,957302,00.html, accessed April 20,2014.

2. Grimes DA. Diagnostic dilation and curettage: a reappraisal. Am J Obstet Gynecol 1982;142:1-6.

3. Cates WJ, Schulz KF, Grimes DA, Tyler CWJ. 1. The effect of delay and method choice on the risk of abortion morbidity. Fam Plann Perspect 1977;9:266-8, 73.

4. Cates W, Grimes DA, Schulz KF. Abortion surveillance at CDC: creating public health light out of political heat. Am J Prev Med 2000;19:12-7.

5. Grimes DA, Cates WJ, Tyler CWJ. Comparative risk of death from legally induced abortion in hospitals and nonhospital facilities. Obstet Gynecol 1978;51:323-6.

6. Grimes DA, Cates WJ, Selik RM. Abortion facilities and the risk of death. Fam Plann Perspect 1981;13:30-2.

7. Garrett R. Philadelphia abortion clinic worker pleads guilty. http://www.cnn.com/2011/11/10/justice/pennsylvania-abortion-clinic/index.html?iref=allsearch, accessed April 1, 2013.
8. Kafrissen ME, Grimes DA, Hogue CJ, Sacks JJ. Cluster of abortion deaths at a single facility. Obstet Gynecol 1986;68:387-9.
9. Hakim-Elahi E, Tovell HM, Burnhill MS. Complications of first-trimester abortion: a report of 170,000 cases. Obstet Gynecol 1990;76:129-35.
10. Finks AA. Mid-trimester abortion. Lancet 1973;1:263-4.
11. Slome J. Termination of pregnancy. Lancet 1972;2:881-2.
12. Davis G, Liu DT. Mid-trimester abortion. Lancet 1972;2:1026.
13. Cates WJ, Grimes DA, Schulz KF, Ory HW, Tyler CWJ. World Health Organization studies of prostaglandins versus saline as abortifacients. A reappraisal. Obstet Gynecol 1978;52:493-8.
14. Comparison of intra-amniotic prostaglandin F2 alpha and hypertonic saline for induction of second-trimester abortion. Br Med J 1976;1:1373-6.
15. Grimes DA, Schulz KF, Cates WJ, Tyler CWJ. Mid-trimester abortion by dilatation and evacuation: a safe and practical alternative. N Engl J Med 1977;296:1141-5.
16. Grimes DA, Hulka JF, McCutchen ME. Midtrimester abortion by dilatation and evacuation versus intra-amniotic instillation of prostaglandin F2 alpha: a randomized clinical trial. Am J Obstet Gynecol 1980;137:785-90.
17. Kelly T, Suddes J, Howel D, Hewison J, Robson S. Comparing medical versus surgical termination of pregnancy at 13-20 weeks of gestation: a randomised controlled trial. BJOG 2010;117:1512-20.
18. Grimes DA, Schulz KF, Cates WJ, Tyler CWJ. Local versus general anesthesia: which is safer for performing suction curettage abortions? Am J Obstet Gynecol 1979;135:1030-5.
19. MacKay HT, Schulz KF, Grimes DA. Safety of local versus general anesthesia for second-trimester dilatation and evacuation abortion. Obstet Gynecol 1985;66:661-5.
20. Grimes DA, Cates WJ. Deaths from paracervical anesthesia used for first-trimester abortion, 1972-1975. N Engl J Med 1976;295:1397-9.
21. Atrash HK, Cheek TG, Hogue CJ. Legal abortion mortality and general anesthesia. Am J Obstet Gynecol 1988;158:420-4.
22. Jones RK, Finer LB, Singh S. Characteristics of U.S. abortion patients, 2008. New York: Guttmacher Institute; 2010.
23. Kumeh T. Conspiracy watch: is abortion black genocide? http://www.motherjones.com/media/2010/09/abortion-black-genocide, accessed April 1, 2013.

24. Cates WJ, Grimes DA, Schulz KF. The public health impact of legal abortion: 30 years later. Perspect Sex Reprod Health 2003;35:25-8.

25. Tietze C. The public health effects of legal abortion in the United States. Fam Plann Perspect 1984;16:26-8.

26. Cates WJ. Legal abortion: the public health record. Science 1982;215:1586-90.

27. Burkman RT, King TM, Burnett LS, Atienza MF. University abortion programs: one year later. Am J Obstet Gynecol 1974;119:131-6.

28. Hogberg U, Joelsson I. Maternal deaths related to abortions in Sweden, 1931-1980. Gynecol Obstet Invest 1985;20:169-78.

29. Idsoe O, Guthe T, Willcox RR, de Weck AL. Nature and extent of penicillin side-reactions, with particular reference to fatalities from anaphylactic shock. Bull World Health Organ 1968;38:159-88.

30. Rudolph AH, Price EV. Penicillin reactions among patients in venereal disease clinics. A national survey. JAMA 1973;223:499-501.

31. Dinman BD. The reality and acceptance of risk. JAMA 1980;244:1226-8.

32. Trussell J. The essentials of contraception: efficacy, safety, and personal considerations. In: Hatcher R, Trussell J, Stewart F, et al., eds. Contraceptive technology. 18th ed. New York: Ardent Media, Inc.; 2004:221-52.

33. Institute of Medicine. About the IOM. http://www.iom.edu/About-IOM.aspx, accessed April 1, 2013.

34. Institute of Medicine. Legalized abortion and the public health. Washington, D.C.: National Academy of Sciences; 1975.

35. National Center for Health Statistics. Vital Statistics of the United States, 1980, Volume II, Parts A and B, 1985. http://www.cdc.gov/nchs/products/vsus.htm, accessed September 28, 2005.

36. Strauss LT, Herndon J, Chang J, et al. Abortion surveillance—United States, 2001. MMWR Surveill Summ 2004;53:1-32.

37. Stam P. Woman's Right to Know Act: a legislative history. Issues Law Med 2012;28:3-67.

38. Kunins H, Rosenfield A. Abortion: a legal and public health perspective. Annu Rev Public Health 1991;12:361-82.

39. Council on Scientific Affairs. Induced termination of pregnancy before and after Roe v Wade. Trends in the mortality and morbidity of women. Council on Scientific Affairs, American Medical Association. JAMA 1992;268:3231-9.

Chapter 4 - Miscarriage: the healthy winnowing of pregnancy

1. Boklage CE. Survival probability of human conceptions from fertilization to term. Int J Fertil 1990;35:75,79-80, 81-94.
2. Roberts C, Lowe C. Letter: Where have all the conceptions gone? Lancet 1975;1:636-7.
3. Martin JA, Hamilton BE, Ventura SJ, et al. Births: final data for 2009. Natl Vital Stat Rep 2011;60:1-70.
4. Paulson R. Oocytes from development to fertilization. In: Lobo R, Mishell DJ, Paulson R, Shoupe D, eds. Mishell's textbook of infertility, contraception, and reproductive endocrinology. Fourth ed. Malden, MA: Blackwell Science; 1997.
5. Amann RP, Howards SS. Daily spermatozoal production and epididymal spermatozoal reserves of the human male. J Urol 1980;124:211-5.
6. Johnson L, Petty CS, Neaves WB. Further quantification of human spermatogenesis: germ cell loss during postprophase of meiosis and its relationship to daily sperm production. Biol Reprod 1983;29:207-15.
7. Saraiya M, Green CA, Berg CJ, Hopkins FW, Koonin LM, Atrash HK. Spontaneous abortion-related deaths among women in the United States—1981-1991. Obstet Gynecol 1999;94:172-6.
8. Conway K, Russell G. Couples' grief and experience of support in the aftermath of miscarriage. Br J Med Psychol 2000;73 Pt 4:531-45.
9. Geller PA, Klier CM, Neugebauer R. Anxiety disorders following miscarriage. J Clin Psychiatry 2001;62:432-8.
10. Short RV. When a conception fails to become a pregnancy. Ciba Found Symp 1978; 64:377-94.
11. American College of Obstetricians and Gynecologists. Preembryo research. Washington, D.C.: American College of Obstetricians and Gynecologists; 2004.
12. Hobcraft J. Fertility patterns and child survival: a comparative analysis. Popul Bull UN 1992; 33:1-31.
13. Forbes LS. The evolutionary biology of spontaneous abortion in humans. Trends Ecol Evol 1997;12:446-50.
14. Kozlowski J, Stearns S. Hypotheses for the production of excess zygotes: models of bet-hedging and selective abortion. Evolution 1989;43:1369-77.
15. Anonymous. Stedman's Medical Dictionary. 27th ed. Philadelphia: Lippincott Williams & Wilkins; 2000.

16. Bulletti C, Flamigni C, Giacomucci E. Reproductive failure due to spontaneous abortion and recurrent miscarriage. Hum Reprod Update 1996;2:118-36.

17. Stein Z, Susser M, Warburton D, Wittes J, Kline J. Spontaneous abortion as a screening device. The effect of fetal survival on the incidence of birth defects. Am J Epidemiol 1975;102:275-90.

18. Macintosh MC, Wald NJ, Chard T, et al. Selective miscarriage of Down's syndrome fetuses in women aged 35 years and older. Br J Obstet Gynaecol 1995;102:798-801.

19. Smith KE, Buyalos RP. The profound impact of patient age on pregnancy outcome after early detection of fetal cardiac activity. Fertil Steril 1996;65:35-40.

20. Wilcox AJ, Weinberg CR, O'Connor JF, et al. Incidence of early loss of pregnancy. N Engl J Med 1988;319:189-94.

21. Ellish NJ, Saboda K, O'Connor J, Nasca PC, Stanek EJ, Boyle C. A prospective study of early pregnancy loss. Hum Reprod 1996;11:406-12.

22. Zinaman MJ, Clegg ED, Brown CC, O'Connor J, Selevan SG. Estimates of human fertility and pregnancy loss. Fertil Steril 1996;65:503-9.

23. Wang X, Chen C, Wang L, Chen D, Guang W, French J. Conception, early pregnancy loss, and time to clinical pregnancy: a population-based prospective study. Fertil Steril 2003;79:577-84.

24. Benagiano G, Farris M, Grudzinskas G. Fate of fertilized human oocytes. Reprod Biomed Online 2010;21:732-41.

25. Hakim RB, Gray RH, Zacur H. Infertility and early pregnancy loss. Am J Obstet Gynecol 1995;172:1510-7.

26. Landy HJ, Keith LG. The vanishing twin: a review. Hum Reprod Update 1998;4:177-83.

27. Silver RK, Helfand BT, Russell TL, Ragin A, Sholl JS, MacGregor SN. Multifetal reduction increases the risk of preterm delivery and fetal growth restriction in twins: a case-control study. Fertil Steril 1997;67:30-3.

28. Wilcox AJ, Weinberg CR, Baird DD. Post-ovulatory ageing of the human oocyte and embryo failure. Hum Reprod 1998;13:394-7.

29. Wilcox AJ, Baird DD, Weinberg CR. Time of implantation of the conceptus and loss of pregnancy. N Engl J Med 1999;340:1796-9.

30. James PA, Rose K, Francis D, Norris F. High-level 46XX/46XY chimerism without clinical effect in a healthy multiparous female. Am J Med Genet A 2011;155A:2484-8.

31. Drexler C, Glock B, Vadon M, et al. Tetragametic chimerism detected in a healthy woman with mixed-field agglutination reactions in ABO blood grouping. Transfusion (Paris) 2005;45:698-703.

32. Sudik R, Jakubiczka S, Nawroth F, Gilberg E, Wieacker PF. Chimerism in a fertile woman with 46,XY karyotype and female phenotype. Hum Reprod 2001;16:56-8.

33. Bogdanova N, Siebers U, Kelsch R, et al. Blood chimerism in a girl with Down syndrome and possible freemartin effect leading to aplasia of the Mullerian derivatives. Hum Reprod 2010;25:1339-43.

34. Uehara S, Nata M, Nagae M, Sagisaka K, Okamura K, Yajima A. Molecular biologic analyses of tetragametic chimerism in a true hermaphrodite with 46,XX/46,XY. Fertil Steril 1995;63:189-92.

35. Grimes DA. Epidemiology of gestational trophoblastic disease. Am J Obstet Gynecol 1984;150:309-18.

36. Koonings PP, Grimes DA, Schlaerth JB. Acronym-associated illness [letter]. Crit Care Med 1990;18:1187.

37. Hughes E, Committee on Terminology of the American College of Obstetricians and Gynecologists. Obstetric-gynecologic terminology. Philadelphia: F. A. Davis; 1972.

38. World Health Organization. World Health Day 1998: Prevent unwanted pregnancy. http://www.who.int/mediacentre/factsheets/fs244/en/ accessed March 16, 2005.

39. U.S. Congress. Protection of human subjects. 45 CFR Sect 46 (1983).

40. Grimes DA. Emergency contraception—expanding opportunities for primary prevention. N Engl J Med 1997;337:1078-9.

Chapter 5 - Induced abortion: where and how

1. Anonymous. Stedman's Medical Dictionary. 27th ed. Philadelphia: Lippincott Williams & Wilkins; 2000.

2. Grimes DA, Creinin MD. Induced abortion: an overview for internists. Ann Intern Med 2004;140:620-6.

3. O'Connell K, Jones HE, Simon M, Saporta V, Paul M, Lichtenberg ES. First-trimester surgical abortion practices: a survey of National Abortion Federation members. Contraception 2009;79:385-92.

4. Jones RK, Jerman J. Abortion Incidence and Service Availability In the United States, 2011. Perspect Sex Reprod Health 2014; 46:3-14.

5. Pazol K, Creanga AA, Burley KD, Hayes B, Jamieson DJ. Abortion surveillance - United States, 2010. MMWR Surveill Summ 2013;62:1-44.

6. Vojta M. A critical view of vacuum aspiration: a new method for the termination of pregnancy. Obstet Gynecol 1967;30:28-34.

7. Kerslake D, Casey D. Abortion induced by means of the uterine aspirator. Obstet Gynecol 1967;30:35-45.

8. Finer LB, Wei J. Effect of mifepristone on abortion access in the United States. Obstet Gynecol 2009;114:623-30.

9. ACOG practice bulletin. Clinical management guidelines of obstetrician-gynecologists. Number 67, October 2005. Medical management of abortion. Obstet Gynecol 2005;106:871-82.

10. Kulier R, Gulmezoglu AM, Hofmeyr GJ, Cheng LN, Campana A. Medical methods for first trimester abortion. Cochrane Database Syst Rev 2004:CD002855.

11. Pazol K, Creanga AA, Zane SB. Trends in use of medical abortion in the United States: reanalysis of surveillance data from the Centers for Disease Control and Prevention, 2001-2008. Contraception 2012;86:746-51.

12. Warriner IK, Wang D, Huong NT, et al. Can midlevel health-care providers administer early medical abortion as safely and effectively as doctors? A randomised controlled equivalence trial in Nepal. Lancet 2011;377:1155-61.

13. Dehlendorf CE, Grumbach K. Medical liability insurance as a barrier to the provision of abortion services in family medicine. Am J Public Health 2008;98:1770-4.

14. Singh K, Fong YF, Prasad RN, Dong F. Evacuation interval after vaginal misoprostol for preabortion cervical priming: a randomized trial. Obstet Gynecol 1999;94:431-4.

15. Meirik O, My Huong NT, Piaggio G, Bergel E, von Hertzen H, W. H. O. Research Group on Postovulatory Methods of Fertility Regulation. Complications of first-trimester abortion by vacuum aspiration after cervical preparation with and without misoprostol: a multicentre randomised trial. Lancet 2012;379:1817-24.

16. Lichtenberg ES, Paul M, Jones H. First trimester surgical abortion practices: a survey of National Abortion Federation members. Contraception 2001;64:345-52.

17. Rubin GL, Cates WJ, Gold J, Rochat RW, Tyler CWJ. Fatal ectopic pregnancy after attempted legally induced abortion. JAMA 1980;244:1705-8.

18. Sawaya GF, Grady D, Kerlikowske K, Grimes DA. Antibiotics at the time of induced abortion: the case for universal prophylaxis based on a meta-analysis. Obstet Gynecol 1996;87:884-90.

19. Henry J. Kaiser Family Foundation. From the patient's perspective: the quality of abortion care. Menlo Park, California: Henry J. Kaiser Family Foundation; 1999.

20. Schulz KF, Cates WJ, Grimes DA, Selik RM, Tyler CWJ. Reducing classification errors in cohort studies: the approach and a practical application. Stat Med 1983;2:25-31.

21. Grimes DA, Schulz KF, Cates WJ, Tyler CWJ. Mid-trimester abortion by dilatation and evacuation: a safe and practical alternative. N Engl J Med 1977;296:1141-5.

22. Cates WJ, Grimes DA, Schulz KF, Ory HW, Tyler CWJ. World Health Organization studies of prostaglandins versus saline as abortifacients. A reappraisal. Obstet Gynecol 1978;52:493-8.

23. Cates WJ, Schulz KF, Grimes DA. The risks associated with teenage abortion. N Engl J Med 1983;309:621-4.

24. Dorfman SF, Grimes DA, Cates WJ, Binkin NJ, Kafrissen ME, O'Reilly KR. Ectopic pregnancy mortality, United States, 1979 to 1980: clinical aspects. Obstet Gynecol 1984;64:386-90.

25. Rubin GL, Peterson HB, Dorfman SF, et al. Ectopic pregnancy in the United States 1970 through 1978. JAMA 1983;249:1725-9.

26. Saraiya M, Berg CJ, Kendrick JS, Strauss LT, Atrash HK, Ahn YW. Cigarette smoking as a risk factor for ectopic pregnancy. Am J Obstet Gynecol 1998;178:493-8.

27. Atrash HK, Hogue CJ, Grimes DA. Epidemiology of hydatidiform mole during early gestation. Am J Obstet Gynecol 1986;154:906-9.

28. Grimes DA. Epidemiology of gestational trophoblastic disease. Am J Obstet Gynecol 1984;150:309-18.

29. Hopkins FW, MacKay AP, Koonin LM, Berg CJ, Irwin M, Atrash HK. Pregnancy-related mortality in Hispanic women in the United States. Obstet Gynecol 1999;94:747-52.

30. MacKay AP, Berg CJ, Atrash HK. Pregnancy-related mortality from preeclampsia and eclampsia. Obstet Gynecol 2001;97:533-8.

31. Chichakli LO, Atrash HK, MacKay AP, Musani AS, Berg CJ. Pregnancy-related mortality in the United States due to hemorrhage: 1979-1992. Obstet Gynecol 1999;94:721-5.

32. Cates WJ, Schulz KF, Grimes DA, Tyler CWJ. 1. The effect of delay and method choice on the risk of abortion morbidity. Fam Plann Perspect 1977;9:266-8, 73.

33. Hakim-Elahi E, Tovell HM, Burnhill MS. Complications of first-trimester abortion: a report of 170,000 cases. Obstet Gynecol 1990;76:129-35.

34. Lichtenberg E, Grimes D. Surgical complications: prevention and management. In: Paul M, Lichtenberg E, Borgatta L, Grimes D, Stubblefield P, Creinin MD, eds. Management of unintended and abnormal pregnancy:

Comprehensive abortion care. Chichester, England: Blackwell Publishing Ltd.; 2009:224-51.

35. Kaali SG, Szigetvari IA, Bartfai GS. The frequency and management of uterine perforations during first- trimester abortions. Am J Obstet Gynecol 1989;161:406-8.

36. Grimes DA, Cates WJ. Complications from legally-induced abortion: a review. Obstet Gynecol Surv 1979;34:177-91.

37. Schulz KF, Grimes DA, Cates WJ. Measures to prevent cervical injury during suction curettage abortion. Lancet 1983;1:1182-5.

38. Singh K, Fong YF, Prasad RN, Dong F. Vaginal misoprostol for pre-abortion cervical priming: is there an optimal evacuation time interval? BJOG 1999;106:266-9.

39. Elliot Institute. New explanation for Bobbitt mutilation points to abortion. http://www.abortionfacts.com/reardon/new-explanation-for-bobbitt-mutilation-points-to-abortion, accessed April 20 2014.

40. Barrett JM, Boehm FH, Killam AP. Induced abortion: a risk factor for placenta previa. Am J Obstet Gynecol 1981;141:769-72.

41. Grimes DA, Techman T. Legal abortion and placenta previa. Am J Obstet Gynecol 1984;149:501-4.

42. Gilliam M, Rosenberg D, Davis F. The likelihood of placenta previa with greater number of cesarean deliveries and higher parity. Obstet Gynecol 2002;99:976-80.

43. Hogue C, Boardman L, Stotland N. Answering questions about long-term outcomes. In: Paul M, Lichtenberg E, Borgatta L, Grimes D, Stubblefield P, Creinin M, eds. Management of unintended and abnormal pregnancy Comprehensive abortion care. Oxford: John Wiley & Sons; 2009:252-63.

Chapter 6 - Giving birth: still risky business

1. Hogan MC, Foreman KJ, Naghavi M, et al. Maternal mortality for 181 countries, 1980-2008: a systematic analysis of progress towards Millennium Development Goal 5. Lancet 2010;375:1609-23.

2. Hamilton BE, Martin JA, Ventura SJ. Births: preliminary data for 2012. http://www.cdc.gov/nchs/data/nvsr/nvsr62/nvsr62_03.pdf, accessed October 2, 2013 2013.

3. Nicholson WK, Frick KD, Powe NR. Economic burden of hospitalizations for preterm labor in the United States. Obstet Gynecol 2000;96:95-101.

4. Centers for Disease Control and Prevention. Pregnancy mortality surveillance system. http://www.cdc.gov/reproductivehealth/MaternalInfantHealth/PMSS.html, accessed October 2, 2013.

5. Berg CJ, Callaghan WM, Syverson C, Henderson Z. Pregnancy-related mortality in the United States, 1998 to 2005. Obstet Gynecol 2010;116:1302-9.
6. Atrash HK, Alexander S, Berg CJ. Maternal mortality in developed countries: not just a concern of the past. Obstet Gynecol 1995;86:700-5.
7. Horon IL. Underreporting of maternal deaths on death certificates and the magnitude of the problem of maternal mortality. Am J Public Health 2005;95:478-82.
8. Deneux-Tharaux C, Berg C, Bouvier-Colle MH, et al. Underreporting of pregnancy-related mortality in the United States and Europe. Obstet Gynecol 2005;106:684-92.
9. Berg CJ, Harper MA, Atkinson SM, et al. Preventability of pregnancy-related deaths: results of a state-wide review. Obstet Gynecol 2005;106:1228-34.
10. Villers MS, Jamison MG, De Castro LM, James AH. Morbidity associated with sickle cell disease in pregnancy. Am J Obstet Gynecol 2008;199:125 e1-5.
11. Bansil P, Kuklina EV, Meikle SF, et al. Maternal and fetal outcomes among women with depression. J Womens Health (Larchmt) 2010;19:329-34.
12. Zane SB, Kieke BAJ, Kendrick JS, Bruce C. Surveillance in a time of changing health care practices: estimating ectopic pregnancy incidence in the United States. Matern Child Health J 2002;6:227-36.
13. Saraiya M, Berg CJ, Shulman H, Green CA, Atrash HK. Estimates of the annual number of clinically recognized pregnancies in the United States, 1981-1991. Am J Epidemiol 1999;149:1025-9.
14. Grimes DA. Estimation of pregnancy-related mortality risk by pregnancy outcome, United States, 1991 to 1999. Am J Obstet Gynecol 2006;194:92-4.
15. LeBolt SA, Grimes DA, Cates WJ. Mortality from abortion and childbirth. Are the populations comparable? JAMA 1982;248:188-91.
16. Cates WJ, Smith JC, Rochat RW, Grimes DA. Mortality from abortion and childbirth. Are the statistics biased? JAMA 1982;248:192-6.
17. Raymond EG, Grimes DA. The comparative safety of legal induced abortion and childbirth in the United States. Obstet Gynecol 2012;119:215-9.
18. Bruce FC, Berg CJ, Hornbrook MC, et al. Maternal morbidity rates in a managed care population. Obstet Gynecol 2008;111:1089-95.
19. Bruce FC, Berg CJ, Joski PJ, et al. Extent of maternal morbidity in a managed care population in Georgia. Paediatr Perinat Epidemiol 2012;26:497-505.
20. Lyndon A, Lee HC, Gilbert WM, Gould JB, Lee KA. Maternal morbidity during childbirth hospitalization in California. J Matern Fetal Neonatal Med 2012;25:2529-35.

21. Berg CJ, Mackay AP, Qin C, Callaghan WM. Overview of maternal morbidity during hospitalization for labor and delivery in the United States: 1993-1997 and 2001-2005. Obstet Gynecol 2009;113:1075-81.

22. Danel I, Berg C, Johnson CH, Atrash H. Magnitude of maternal morbidity during labor and delivery: United States, 1993-1997. Am J Public Health 2003;93:631-4.

23. Say L, Pattinson RC, Gulmezoglu AM. WHO systematic review of maternal morbidity and mortality: the prevalence of severe acute maternal morbidity (near miss). Reprod Health 2004;1:3.

24. Gray KE, Wallace ER, Nelson KR, Reed SD, Schiff MA. Population-based study of risk factors for severe maternal morbidity. Paediatr Perinat Epidemiol 2012;26:506-14.

25. Mhyre JM, Bateman BT, Leffert LR. Influence of patient comorbidities on the risk of near-miss maternal morbidity or mortality. Anesthesiology 2011;115:963-72.

26. Callaghan WM, Mackay AP, Berg CJ. Identification of severe maternal morbidity during delivery hospitalizations, United States, 1991-2003. Am J Obstet Gynecol 2008;199:133 e1-8.

27. Bacak SJ, Callaghan WM, Dietz PM, Crouse C. Pregnancy-associated hospitalizations in the United States, 1999-2000. Am J Obstet Gynecol 2005;192:592-7.

28. Gazmararian JA, Petersen R, Jamieson DJ, et al. Hospitalizations during pregnancy among managed care enrollees. Obstet Gynecol 2002;100:94-100.

29. Hebert PR, Reed G, Entman SS, Mitchel EFJ, Berg C, Griffin MR. Serious maternal morbidity after childbirth: prolonged hospital stays and readmissions. Obstet Gynecol 1999;94:942-7.

30. Handa VL, Harris TA, Ostergard DR. Protecting the pelvic floor: obstetric management to prevent incontinence and pelvic organ prolapse. Obstet Gynecol 1996;88:470-8.

31. Oberwalder M, Connor J, Wexner SD. Meta-analysis to determine the incidence of obstetric anal sphincter damage. Br J Surg 2003;90:1333-7.

32. Callaghan WM, Kuklina EV, Berg CJ. Trends in postpartum hemorrhage: United States, 1994-2006. Am J Obstet Gynecol 2010;202:353 e1-6.

33. Eshkoli T, Weintraub AY, Sergienko R, Sheiner E. Placenta accreta: risk factors, perinatal outcomes, and consequences for subsequent births. Am J Obstet Gynecol 2013;208:219 e1-7.

34. Hou MY, Hurwitz S, Kavanagh E, Fortin J, Goldberg AB. Using daily text-message reminders to improve adherence with oral contraceptives: a randomized controlled trial. Obstet Gynecol 2010;116:633-40.

35. D'Anna LH, Korosteleva O, Warner L, et al. Factors associated with condom use problems during vaginal sex with main and non-main partners. Sex Transm Dis 2012;39:687-93.

Chapter 7 - Every third woman in America

1. Plotz D. Susan Carpenter-McMillan. The woman who ate Paula Jones. http://www.slate.com/articles/news_and_politics/assessment/1997/09/susan_carpentermcmillan.html, accessed August 19, 2005.
2. Shaw D. Abortion foes stereotyped, some in the media believe. Los Angeles Times, July 2, 1990.
3. Dean P. No more denial: voice of Right to Life League acknowledges abortion at 21. Los Angeles Times, April 2, 1990.
4. Rich F. Journal; Back to the future. http://www.nytimes.com/1998/01/21/opinion/journal-back-to-the-future.html, accessed October 21, 2013.
5. Rich F. Journal; Valentine's Day massacred. http://www.nytimes.com/1998/02/14/opinion/journal-valentine-s-day-massacred.html, accessed October 21, 2013.
6. MJM Entertainment Group. Designated conservative woman. http://www.mjmgroup.com/Transcripts/Susan_Carpenter.htm, accessed July 5, 2005.
7. Grove L. An antiabortion activist makes herself the unofficial mouthpiece for Paula Jones. The Washington Post, July 23, 1997, page C01.
8. Wiebe ER, Trouton KJ, Fielding SL, Grant H, Henderson A. Anxieties and attitudes towards abortion in women presenting for medical and surgical abortions. J Obstet Gynaecol Can 2004;26:881-5.
9. Wiebe ER, Trouton KJ, Fielding SL, Klippenstein J, Henderson A. Antichoice attitudes to abortion in women presenting for medical abortions. J Obstet Gynaecol Can 2005;27:247-50.
10. Jones RK, Jerman J. Abortion Incidence and Service Availability In the United States, 2011. Perspect Sex Reprod Health 2014;46:3-14.
11. Henshaw SK. Unintended pregnancy in the United States. Fam Plann Perspect 1998;30:24-9, 46.
12. Jones RK, Kooistra K. Abortion incidence and access to services in the United States, 2008. Perspect Sex Reprod Health 2011;43:41-50.
13. Pazol K, Creanga AA, Burley KD, Hayes B, Jamieson DJ. Abortion surveillance - United States, 2010. MMWR Surveill Summ 2013;62:1-44.
14. Finer LB, Zolna MR. Unintended pregnancy in the United States: incidence and disparities, 2006. Contraception 2011;84:478-85.

15. Jones RK, Kavanaugh ML. Changes in abortion rates between 2000 and 2008 and lifetime incidence of abortion. Obstet Gynecol 2011;117:1358-66.
16. Jones RK, Finer LB. Who has second-trimester abortions in the United States? Contraception 2012;85:544-51.
17. U.S. Department of Health and Human Services. The 2008 HHS Poverty Guidelines. http://aspe.hhs.gov/poverty/08poverty.shtml, accessed October 21, 2013.
18. FamiliesUSA. Federal poverty guidelines. http://familiesusa.org/product/federal-poverty-guidelines, accessed March 1, 2014.
19. Jones RK, Darroch JE, Henshaw SK. Patterns in the socioeconomic characteristics of women obtaining abortions in 2000-2001. Perspect Sex Reprod Health 2002;34:226-35.
20. Jones RK, Darroch JE, Henshaw SK. Contraceptive use among U.S. women having abortions in 2000-2001. Perspect Sex Reprod Health 2002;34:294-303.
21. Henshaw SK, Finer LB. The accessibility of abortion services in the United States, 2001. Perspect Sex Reprod Health 2003;35:16-24.
22. Finer LB, Frohwirth LF, Dauphinee LA, Singh S, Moore AM. Reasons U.S. women have abortions: quantitative and qualitative perspectives. Perspect Sex Reprod Health 2005;37:110-8.
23. Grimes DA. A 26-year-old woman seeking an abortion. JAMA 1999;282:1169-75.
24. Eligon J, Schwirtz M. Senate candidate provokes ire with 'legitimate rape' comment. http://www.nytimes.com/2012/08/20/us/politics/todd-akin-provokes-ire-with-legitimate-rape-comment.html?_r=1&, accessed March 1, 2014.
25. Holmes MM, Resnick HS, Kilpatrick DG, Best CL. Rape-related pregnancy: estimates and descriptive characteristics from a national sample of women. Am J Obstet Gynecol 1996;175:320-4.
26. Stewart FH, Trussell J. Prevention of pregnancy resulting from rape: a neglected preventive health measure. Am J Prev Med 2000;19:228-9.
27. McFarlane J. Pregnancy following partner rape: what we know and what we need to know. Trauma Violence Abuse 2007;8:127-34.
28. Munro ML, Foster Rietz M, Seng JS. Comprehensive care and pregnancy: the unmet care needs of pregnant women with a history of rape. Issues Ment Health Nurs 2012;33:882-96.
29. Evins G, Chescheir N. Prevalence of domestic violence among women seeking abortion services. Womens Health Issues 1996;6:204-10.
30. Saftlas AF, Wallis AB, Shochet T, Harland KK, Dickey P, Peek-Asa C. Prevalence of intimate partner violence among an abortion clinic population. Am J Public Health 2010;100:1412-5.

31. Glander SS, Moore ML, Michielutte R, Parsons LH. The prevalence of domestic violence among women seeking abortion. Obstet Gynecol 1998;91:1002-6.

32. Gessner BD, Perham-Hester KA. Experience of violence among teenage mothers in Alaska. J Adolesc Health 1998;22:383-8.

33. Woo J, Fine P, Goetzl L. Abortion disclosure and the association with domestic violence. Obstet Gynecol 2005;105:1329-34.

34. Wiebe ER, Janssen P. Universal screening for domestic violence in abortion. Womens Health Issues 2001;11:436-41.

35. Kaye D. Domestic violence among women seeking post-abortion care. Int J Gynaecol Obstet 2001;75:323-5.

36. Wu J, Guo S, Qu C. Domestic violence against women seeking induced abortion in China. Contraception 2005;72:117-21.

37. Moore AM, Frohwirth L, Miller E. Male reproductive control of women who have experienced intimate partner violence in the United States. Soc Sci Med 2010;70:1737-44.

38. Ganatra B, Hirve S. Induced abortions among adolescent women in rural Maharashtra, India. Reprod Health Matters 2002;10:76-85.

39. Bleil ME, Adler NE, Pasch LA, Sternfeld B, Reijo-Pera RA, Cedars MI. Adverse childhood experiences and repeat induced abortion. Am J Obstet Gynecol 2011;204:122 e1-6.

40. Jones RK, Frohwirth L, Moore AM. More than poverty: disruptive events among women having abortions in the USA. J Fam Plann Reprod Health Care 2013;39:36-43.

41. Hall EJ, Ferree MM. Race differences in abortion attitudes. Public Opin Q 1986;50:193-207.

42. Pazol K, Creanga AA, Zane SB, et al. Abortion surveillance—United States, 2009. MMWR Surveill Summ 2012;61:1-44.

43. Copen CE, Daniels K, Vespa J, Mosher WD. First marriages in the United States: data from the 2006-2010 National Survey of Family Growth. Natl Health Stat Report 2012:1-21.

44. Levitt SD, Dubner SJ. Freakonomics. New York: William Morrow; 2005.

Chapter 8 - Pro-natal, prenatal diagnosis

1. Henshaw SK, Forrest JD, Van Vort J. Abortion services in the United States, 1984 and 1985. Fam Plann Perspect 1987;19:63-70.

2. Finer LB, Frohwirth LF, Dauphinee LA, Singh S, Moore AM. Reasons U.S. women have abortions: quantitative and qualitative perspectives. Perspect Sex Reprod Health 2005;37:110-8.

3. Lyus R, Robson S, Parsons J, Fisher J, Cameron M. Second trimester abortion for fetal abnormality. BMJ 2013;347:f4165.

4. Thorp JMJ, Bowes WAJ. Prolife perinatologist—paradox or possibility? N Engl J Med 1992;326:1217-9.

5. Pryde PG, Drugan A, Johnson MP, Isada NB, Evans MI. Prenatal diagnosis: choices women make about pursuing testing and acting on abnormal results. Clin Obstet Gynecol 1993;36:469-509.

6. Farmakides G, Bracero L, Marion R, Fleischer A, Schulman H. Pregnancy termination after detection of fetal chromosomal or metabolic abnormalities. J Perinatol 1988;8:101-4.

7. Cook R. Legal abortion: limits and contributions to human life. In: Porter R, O'Connor M, eds. Abortion: medical progress and social implications. London: Pitman; 1985:211-27.

8. Hook EB. Rates of chromosome abnormalities at different maternal ages. Obstet Gynecol 1981;58:282-5.

9. Simpson JL. Cell-free fetal DNA and maternal serum analytes for monitoring embryonic and fetal status. Fertil Steril 2013;99:1124-34.

10. Wertz DC, Fletcher JC. Feminist criticism of prenatal diagnosis: a response. Clin Obstet Gynecol 1993;36:541-67.

11. Cuckle H, Wald N. The impact of screening for open neural tube defects in England and Wales. Prenat Diagn 1987;7:91-9.

12. Viljoen D, Oosthuizen C, van der Westhuizen S. Patient attitudes to prenatal screening and termination of pregnancy at Groote Schuur Hospital: a two year prospective study. East Afr Med J 1996;73:327-9.

13. Lowry DL, Campbell SA, Krivchenia EL, Dvorin E, Duquette D, Evans MI. Impact of abnormal second-trimester maternal serum single, double, and triple screening on patient choices about prenatal diagnosis. Fetal Diagn Ther 1995;10:286-9.

14. Kuppermann M, Learman LA, Gates E, et al. Beyond race or ethnicity and socioeconomic status: predictors of prenatal testing for Down syndrome. Obstet Gynecol 2006;107:1087-97.

15. Choi H, Van Riper M, Thoyre S. Decision making following a prenatal diagnosis of Down syndrome: an integrative review. J Midwifery Womens Health 2012;57:156-64.

16. Mennie ME, Gilfillan A, Compton ME, Liston WA, Brock DJ. Prenatal cystic fibrosis carrier screening: factors in a woman's decision to decline testing. Prenat Diagn 1993;13:807-14.

17. Furu T, Kaariainen H, Sankila EM, Norio R. Attitudes towards prenatal diagnosis and selective abortion among patients with retinitis pigmentosa or choroideremia as well as among their relatives. Clin Genet 1993;43:160-5.

18. Conway SP, Allenby K, Pond MN. Patient and parental attitudes toward genetic screening and its implications at an adult cystic fibrosis centre. Clin Genet 1994;45:308-12.

19. Ralston SJ, Wertz D, Chelmow D, Craigo SD, Bianchi DW. Pregnancy outcomes after prenatal diagnosis of aneuploidy. Obstet Gynecol 2001;97:729-33.

20. Skotko BG. Prenatally diagnosed Down syndrome: mothers who continued their pregnancies evaluate their health care providers. Am J Obstet Gynecol 2005;192:670-7.

21. Bauer PE. The abortion debate no one wants to have. The Washington Post, October 18, 2005, page A25; http://www.washingtonpost.com/wp-dyn/content/article/2005/10/17/AR2005101701311.html, accessed October 21, 2005.

22. Jeon KC, Chen LS, Goodson P. Decision to abort after a prenatal diagnosis of sex chromosome abnormality: a systematic review of the literature. Genet Med 2012;14:27-38.

23. Mansfield C, Hopfer S, Marteau TM. Termination rates after prenatal diagnosis of Down syndrome, spina bifida, anencephaly, and Turner and Klinefelter syndromes: a systematic literature review. European Concerted Action: DADA (Decision-making After the Diagnosis of a fetal Abnormality). Prenat Diagn 1999;19:808-12.

24. Drugan A, Greb A, Johnson MP, et al. Determinants of parental decisions to abort for chromosome abnormalities. Prenat Diagn 1990;10:483-90.

25. Evans MI, Pryde PG, Evans WJ, Johnson MP. The choices women make about prenatal diagnosis. Fetal Diagn Ther 1993;8:70-80.

26. Evans MI, Sobiecki MA, Krivchenia EL, et al. Parental decisions to terminate/continue following abnormal cytogenetic prenatal diagnosis: "what" is still more important than "when". Am J Med Genet 1996;61:353-5.

27. Verp MS, Bombard AT, Simpson JL, Elias S. Parental decision following prenatal diagnosis of fetal chromosome abnormality. Am J Med Genet 1988;29:613-22.

28. Kramer RL, Jarve RK, Yaron Y, et al. Determinants of parental decisions after the prenatal diagnosis of Down syndrome. Am J Med Genet 1998;79:172-4.

29. Hawkins A, Stenzel A, Taylor J, Chock VY, Hudgins L. Variables influencing pregnancy termination following prenatal diagnosis of fetal chromosome abnormalities. J Genet Couns 2013;22:238-48.

30. Faden RR, Chwalow AJ, Quaid K, et al. Prenatal screening and pregnant women's attitudes toward the abortion of defective fetuses. Am J Public Health 1987;77:288-90.

31. Drake H, Reid M, Marteau T. Attitudes towards termination for fetal abnormality: comparisons in three European countries. Clin Genet 1996;49:134-40.

32. Johnson CY, Honein MA, Dana Flanders W, Howards PP, Oakley GP, Jr., Rasmussen SA. Pregnancy termination following prenatal diagnosis of anencephaly or spina bifida: a systematic review of the literature. Birt Defects Res A Clin Mol Teratol 2012;94:857-63.

33. Smith RG, Gardner RW, Steinhoff P, Chung CS, Palmore JA. The effect of induced abortion on the incidence of Down's syndrome in Hawaii. Fam Plann Perspect 1980;12:201-5.

34. Cragan JD, Roberts HE, Edmonds LD, et al. Surveillance for anencephaly and spina bifida and the impact of prenatal diagnosis – United States, 1985-1994. Morb Mortal Wkly Rep 1995;44:1-13.

35. Cuckle HS, Wald NJ, Cuckle PM. Prenatal screening and diagnosis of neural tube defects in England and Wales in 1985. Prenat Diagn 1989;9:393-400.

36. Lorber J, Ward AM. Spina bifida—a vanishing nightmare? Arch Dis Child 1985;60:1086-91.

37. De Wals P, Trochet C, Pinsonneault L. Prevalence of neural tube defects in the province of Quebec, 1992. Can J Public Health 1999;90:237-9.

38. Gucciardi E, Pietrusiak MA, Reynolds DL, Rouleau J. Incidence of neural tube defects in Ontario, 1986-1999. CMAJ 2002;167:237-40.

39. Zimmer EZ, Avraham Z, Sujoy P, Goldstein I, Bronshtein M. The influence of prenatal ultrasound on the prevalence of congenital anomalies at birth. Prenat Diagn 1997;17:623-8.

40. Brett KM, Schoendorf KC, Kiely JL. Differences between black and white women in the use of prenatal care technologies. Am J Obstet Gynecol 1994;170:41-6.

41. Weiner J, Bernhardt BA. A survey of state Medicaid policies for coverage of abortion and prenatal diagnostic procedures. Am J Public Health 1990;80:717-20.

42. Khoshnood B, Pryde P, Blondel B, Lee KS. Socioeconomic and state-level differences in prenatal diagnosis and live birth prevalence of Down's syndrome in the United States. Rev Epidemiol Sante Publique 2003;51:617-27.

43. Cunningham GC, Tompkinison DG. Cost and effectiveness of the California triple marker prenatal screening program. Genet Med 1999;1:199-206.

44. Jones OW, Penn NE, Shuchter S, et al. Parental response to mid-trimester therapeutic abortion following amniocentesis. Prenat Diagn 1984;4:249-56.

45. White-Van Mourik MC, Connor JM, Ferguson-Smith MA. Patient care before and after termination of pregnancy for neural tube defects. Prenat Diagn 1990;10:497-505.

46. Dallaire L, Lortie G, Des Rochers M, Clermont R, Vachon C. Parental reaction and adaptability to the prenatal diagnosis of fetal defect or genetic disease leading to pregnancy interruption. Prenat Diagn 1995;15:249-59.

Chapter 9 - The economics of abortion: pay now or pay later

1. Boonstra HD. Insurance coverage of abortion: beyond the exceptions for life endangerment, rape and incest. http://www.guttmacher.org/pubs/gpr/16/3/gpr160302.html, accessed December 25, 2013.

2. Jones RK, Upadhyay UD, Weitz TA. At what cost? Payment for abortion care by U.S. women. Womens Health Issues 2013;23:e173-8.

3. Citizens for Responsible Government. Yes on #3. Denver, Colorado: Citizens for Responsible Government; 1984.

4. Urbina I. In the treatment of diabetes, success often does not pay. New York Times, http://www.nytimes.com/2006/01/11/nyregion/nyregionspecial5/11diabetes.html?pagewanted=all&module=Search&mabReward=relbias%3Ar%2C%7B%222%22%3A%22RI%3A18%22%7D&module=Search&mabReward=relbias%3Ar%2C%7B%222%22%3A%22RI%-3A18%22%7D accessed January 11, 2006.

5. Grimes DA. Prevention of cardiovascular disease in women: role of the obstetrician-gynecologist. Am J Obstet Gynecol 1988;158:1662-8.

6. Holmes KK, Levine R, Weaver M. Effectiveness of condoms in preventing sexually transmitted infections. Bull World Health Organ 2004;82:454-61.

7. Paz-Bailey G, Koumans EH, Sternberg M, et al. The effect of correct and consistent condom use on chlamydial and gonococcal infection among urban adolescents. Arch Pediatr Adolesc Med 2005;159:536-42.

8. Workowski KA, Berman S, Centers for Disease Control and Prevention. Sexually transmitted diseases treatment guidelines, 2010. MMWR Recomm Rep 2010;59:1-110.

9. DiCenso A, Guyatt G, Willan A, Griffith L. Interventions to reduce unintended pregnancies among adolescents: systematic review of randomised controlled trials. BMJ 2002;324:1426.

10. Chiou CF, Trussell J, Reyes E, et al. Economic analysis of contraceptives for women. Contraception 2003;68:3-10.

11. Winner B, Peipert JF, Zhao Q, et al. Effectiveness of long-acting reversible contraception. N Engl J Med 2012;366:1998-2007.

12. Peipert JF, Madden T, Allsworth JE, Secura GM. Preventing unintended pregnancies by providing no-cost contraception. Obstet Gynecol 2012;120:1291-7.

13. Levey LM, Nyman JA, Haugaard J. A benefit-cost analysis of family planning services in Iowa. Eval Health Prof 1988;11:403-24.

14. Forrest JD, Singh S. The impact of public-sector expenditures for contraceptive services in California. Fam Plann Perspect 1990;22:161-8.

15. Foster DG, Biggs MA, Amaral G, et al. Estimates of pregnancies averted through California's family planning waiver program in 2002. Perspect Sex Reprod Health 2006;38:126-31.

16. Foster DG, Klaisle CM, Blum M, Bradsberry ME, Brindis CD, Stewart FH. Expanded state-funded family planning services: estimating pregnancies averted by the Family PACT Program in California, 1997-1998. Am J Public Health 2004;94:1341-6.

17. Forrest JD, Singh S. Public-sector savings resulting from expenditures for contraceptive services. Fam Plann Perspect 1990;22:6-15.

18. Forrest JD, Samara R. Impact of publicly funded contraceptive services on unintended pregnancies and implications for Medicaid expenditures. Fam Plann Perspect 1996;28:188-95.

19. Cleland K, Peipert JF, Westhoff C, Spear S, Trussell J. Family planning as a cost-saving preventive health service. N Engl J Med 2011;364:e37.

20. Trussell J, Lalla AM, Doan QV, Reyes E, Pinto L, Gricar J. Cost effectiveness of contraceptives in the United States. Contraception 2009;79:5-14.

21. Sonnenberg FA, Burkman RT, Hagerty CG, Speroff L, Speroff T. Costs and net health effects of contraceptive methods. Contraception 2004;69:447-59.

22. Hughes D, McGuire A. The cost-effectiveness of family planning service provision. J Public Health Med 1996;18:189-96.

23. Cakir HV, Fabricant SJ, Kircalioglu FN. Comparative costs of family planning services and hospital-based maternity care in Turkey. Stud Fam Plann 1996;27:269-76.

24. Alan Guttmacher Institute. Issue Brief. Medicaid: a critical source of support for family planning in the United States. New York: Alan Guttmacher Institute; 2005.

25. Torres A, Donovan P, Dittes N, Forrest JD. Public benefits and costs of government funding for abortion. Fam Plann Perspect 1986;18:111-8.

26. Guttmacher Institute. State funding of abortion under Medicaid. http://www.guttmacher.org/statecenter/spibs/spib_SFAM.pdf, accessed December 25, 2013.

27. Robinson M, Pakter J, Svigir M. Medicaid coverage of abortions in New York City: costs and benefits. Fam Plann Perspect 1974;6:202-8.
28. Alan Guttmacher Institute. Safe and legal: experience with legal abortion in New York state. New York: Alan Guttmacher Institute; 1980.
29. Evans MI, Gleicher E, Feingold E, Johnson MP, Sokol RJ. The fiscal impact of the Medicaid abortion funding ban in Michigan. Obstet Gynecol 1993;82:555-60.
30. Trussell J, Menken J, Lindheim BL, Vaughan B. The impact of restricting Medicaid financing for abortion. Fam Plann Perspect 1980;12:120-3, 127-30.
31. Miller VL, Ransom SB, Ayoub MA, Krivchenia EL, Evans MI. Fiscal impact of a potential legislative ban on second trimester elective terminations for prenatally diagnosed abnormalities. Am J Med Genet 2000;91:359-62.
32. Cook PJ, Parnell AM, Moore MJ, Pagnini D. The effects of short-term variation in abortion funding on pregnancy outcomes. J Health Econ 1999;18:241-57.
33. Korenbrot CC, Brindis C, Priddy F. Trends in rates of live births and abortions following state restrictions on public funding of abortion. Public Health Rep 1990;105:555-62.
34. Hicken M. Average cost to raise a kid: $241,080. http://money.cnn.com/2013/08/14/pf/cost-children/, accessed December 25, 2013.
35. Meier KJ, McFarlane DR. State family planning and abortion expenditures: their effect on public health. Am J Public Health 1994;84:1468-72.
36. Dennis A, Blanchard K. A mystery caller evaluation of Medicaid staff responses about state coverage of abortion care. Womens Health Issues 2012;22:e143-8.
37. Dennis A, Blanchard K. Abortion providers' experiences with Medicaid abortion coverage policies: a qualitative multistate study. Health Serv Res 2013;48:236-52.
38. Bessett D, Gorski K, Jinadasa D, Ostrow M, Peterson MJ. Out of time and out of pocket: experiences of women seeking state-subsidized insurance for abortion care in Massachusetts. Womens Health Issues 2011;21:S21-5.
39. Dennis A, Blanchard K, Cordova D. Strategies for securing funding for abortion under the Hyde Amendment: a multistate study of abortion providers' experiences managing Medicaid. Am J Public Health 2011;101:2124-9.
40. Nichol KL. Cost-benefit analysis of a strategy to vaccinate healthy working adults against influenza. Arch Intern Med 2001;161:749-59.
41. Lieu TA, Cochi SL, Black SB, et al. Cost-effectiveness of a routine varicella vaccination program for US children. JAMA 1994;271:375-81.

42. Rosenthal P. Cost-effectiveness of hepatitis A vaccination in children, adolescents, and adults. Hepatology 2003;37:44-51.

43. Zhou F, Reef S, Massoudi M, et al. An economic analysis of the current universal 2-dose measles-mumps-rubella vaccination program in the United States. J Infect Dis 2004;189 Suppl 1:S131-45.

Chapter 10 - Stronger families, healthier babies

1. Abortion and patriarchy. Patriarch Magazine; http://en.wikipedia.org/wiki/Patriarch_%28magazine%29, accessed October 22, 2005.

2. Knight R. Girl Scouts national conclave to feature pro-abortion, pro-lesbian speakers. http://www.ahgonline.org/, accessed April 20, 2014.

3. Crary D. Conservatives turn on dollmaker that had been their darling / American Girl is helping group that backs legal abortion. SFGate, October 15, 2005; http://www.sfgate.com/news/article/Conservatives-turn-on-dollmaker-that-had-been-2602427.php accessed April 20,2014.

4. Sklar J, Berkov B. The effects of legal abortion on legitimate and illegitimate birth rates: the California experience. Stud Fam Plann 1973;4:281-92.

5. Sklar J, Berkov B. Abortion, illegitimacy, and the American birth rate. Science 1974;185:909-15.

6. Akerlof GA, Yellen JL, Katz ML. An analysis of out-of-wedlock childbearing in the United States. Q J Econ 1996;111:276-317.

7. Bauman KE, Koch GG, Udry JR, Freeman JL. The relationship between legal abortion and marriage. Soc Biol 1975;22:117-24.

8. Bauman KE, Anderson AE, Freeman JL, Koch GG. Legal abortions and trends in age-specific marriage rates. Am J Public Health 1977;67:52-3.

9. Bennett T. Marital status and infant health outcomes. Soc Sci Med 1992;35:1179-87.

10. Gordon RR, Sunderland R. Maternal age, illegitimacy, and postneonatal mortality. BMJ 1988;297:774.

11. Hollander D. Nonmarital childbearing in the United States: a government report. Fam Plann Perspect 1996;28:29-32, 41.

12. Filinson R. Illegitimate birth and deprivation: recent findings from an exploratory study. Soc Sci Med 1985;20:307-14.

13. Walsh A. Illegitimacy, child abuse and neglect, and cognitive development. J Genet Psychol 1990;151:279-85.

14. Robinson M, Pakter J, Svigir M. Medicaid coverage of abortions in New York City: costs and benefits. Fam Plann Perspect 1974;6:202-8.

15. Lundberg S, Plotnick RD. Effects of state welfare, abortion and family planning policies on premarital childbearing among white adolescents. Fam Plann Perspect 1990;22:246-51, 75.
16. Lichter DT, McLaughlin DK, Ribar DC. State abortion policy, geographic access to abortion providers and changing family formation. Fam Plann Perspect 1998;30:281-7.
17. Levine PB, Staiger D, Kane TJ, Zimmerman DJ. Roe v Wade and American fertility. Am J Public Health 1999;89:199-203.
18. Barfield WD, Committee on Fetus and Newborn. Standard terminology for fetal, infant, and perinatal deaths. Pediatrics 2011;128:177-81.
19. Grossman M, Jacobowitz S. Variations in infant mortality rates among counties of the United States: the roles of public policies and programs. Demography 1981;18:695-713.
20. Quick JD. Liberalized abortion in Oregon: effects on fertility, prematurity, fetal death, and infant death. Am J Public Health 1978;68:1003-8.
21. Bauman KE, Anderson AE. Legal abortions and trends in fetal and infant mortality rates in the United States. Am J Obstet Gynecol 1980;136:194-202.
22. Corman H, Grossman M. Determinants of neonatal mortality rates in the U.S. A reduced form model. J Health Econ 1985;4:213-36.
23. Joyce T. The impact of induced abortion on black and white birth outcomes in the United States. Demography 1987;24:229-44.
24. Joyce T, Grossman M. The dynamic relationship between low birthweight and induced abortion in New York City. An aggregate time-series analysis. J Health Econ 1990;9:273-88.
25. De Muylder X, Wesel S, Dramaix M, Candeur M. A woman's attitude toward pregnancy. Can it predispose her to preterm labor? J Reprod Med 1992;37:339-42.
26. Orr ST, Miller CA, James SA, Babones S. Unintended pregnancy and preterm birth. Paediatr Perinat Epidemiol 2000;14:309-13.
27. Kost K, Landry DJ, Darroch JE. The effects of pregnancy planning status on birth outcomes and infant care. Fam Plann Perspect 1998;30:223-30.
28. Bitler M, Zavodny M. Child abuse and abortion availability. Women, Children, and the Labor Market 2002;92:363-7.
29. Zuravin SJ. Unplanned childbearing and family size: their relationship to child neglect and abuse. Fam Plann Perspect 1991;23:155-61.
30. Bitler MP, Zavodny M. Child maltreatment, abortion availability, and economic conditions. Review of Economics of the Household 2004;2:119-41.

31. Gruber J, Levine P, Staiger D. Abortion legalization and child living circumstances: who is the "marginal child"? Q J Econ 1999;114:263-91.

32. Ananat EO, Gruber J, Levine PB, Staiger D. Abortion and selection. Review of Economics and Statistics 2009;91:124-36.

Chapter 11 - Missing criminals?

1. Donohue JJ, Levitt SD. The impact of legalized abortion on crime. Quarterly Journal of Economics 2001;CXVI:379-420.

2. Anonymous. Legalized abortion and crime effect. http://en.wikipedia.org/wiki/Legalized_abortion_and_crime_effect, accessed December 26, 2013.

3. CNN. Bennett under fire for remarks on blacks, crime. http://www.cnn.com/2005/POLITICS/09/30/bennett.comments/index.html?iref=allsearch, accessed September 30, 2005.

4. Goode E. Roe v. Wade resulted in unborn criminals, economists theorize. New York Times, August 20, 1999, http://www.wright.edu/~tdung/abortion.htm, accessed June 2, 2005.

5. Levitt SD, Dubner SJ. Freakonomics. New York: William Morrow; 2005.

6. U. S. Preventive Services Task Force. Guide to clinical preventive services. 2nd ed. Baltimore, MD: Williams & Wilkins; 1996.

7. Writing Group for the Women's Health Initiative Investigators. Risks and benefits of estrogen plus progestin in healthy postmenopausal women: principal results From the Women's Health Initiative randomized controlled trial. JAMA 2002;288:321-33.

8. Banta HD, Thacker SB. Electronic fetal monitoring. Lessons from a formative case of health technology assessment. Int J Technol Assess Health Care 2002;18:762-70.

9. Grimes DA, Schulz KF. Bias and causal associations in observational research. Lancet 2002;359:248-52.

10. Comanor W, Phillips L. The impact of income and family structure on delinquency. Department of Economics, UCSB, Paper 7'95R, Departmental Working Papers, http://library.softgenx.com/Children/crime/comanor%20father%20absent.pdf, accessed April 20,2014.

11. Rasanen P, Hakko H, Isohanni M, Hodgins S, Jarvelin MR, Tiihonen J. Maternal smoking during pregnancy and risk of criminal behavior among adult male offspring in the Northern Finland 1966 Birth Cohort. Am J Psychiatry 1999;156:857-62.

12. Joyce T. Did legalized abortion lower crime? J Hum Resour 2004;XXXIX:1-28.

13. Donohue JJ, Levitt SD. Further evidence that legalized abortion lowered crime: a reply to Joyce. J Hum Resour 2004;XXXIX:29-49.

14. Joyce T. The impact of induced abortion on black and white birth outcomes in the United States. Demography 1987;24:229-44.

15. Joyce T. Further tests of abortion and crime. NBER Working Paper No W10564, http://papers.ssrn.com/sol3/papers.cfm?abstract_id=557198, accessed June 2, 2005..

16. Charles KK, Stephens MJ. Abortion legalization and adolescent substance use. NBER Working Paper Series Number 9193, http://papers.ssrn.com/sol3/papers.cfm?abstract_id=330332, accessed June 2, 2005.

17. Sorenson SB, Wiebe DJ, Berk RA. Legalized abortion and the homicide of young children: an empirical investigation. Analyses of Social Issues and Public Policy 2002;2:239-56.

18. Berk RA, Sorenson SB, Wiebe DJ, Upchurch DM. The legalization of abortion and subsequent youth homicide: a time series analysis. Analyses of Social Issues and Public Policy 2003;3:45-64.

19. Lott JRJ, Whitley J. Abortion and crime: unwanted children and out-of-wedlock births. Yale Law School, Program for Studies in Law, Economics, and Public Policy Working Paper #254, http://papers.ssrn.com/sol3/papers.cfm?abstract_id=270126, accessed June 2, 2005.

20. Pop-Eleches C. The impact of an abortion ban on socio-economic outcomes of children: evidence from Romania. Journal of Political Economy 2006;114:744-73.

21. Leigh A, Wolfers J. Abortion and crime. AQ: Australian Quarterly 2000;August-September:28-31.

22. Sen A. Does increased abortion lead to lower crime? Evaluating the relationship between crime, abortion, and fertility. BE Journal of Economic Analysis and Policy 2007;7.

23. Levitt SD. Understanding why crime fell in the 1990s: four factors that explain the decline and six that do not. J Econ Perspect 2004;18:163-90.

24. Donohue JJ, Levitt SD. Measurement error, legalized abortion, and the decline in crime: a response to Foote and Goetz. Quarterly Journal of Economics 2008;123:425-40.

25. Joyce T. A simple test of abortion and crime. Review of Economics and Statistics 2009;91:112-23.

26. Joyce TJ. Abortion and crime: a review. http://papers.ssrn.com/sol3/papers.cfm?abstract_id=1422976, accessed December 26, 2013.

27. Hill AB. The environment and disease association or causation. Proc Royal Soc Med 1965;58:295-300.

28. Rothman KJ, Greenland S. Causation and causal inference in epidemiology. Am J Public Health 2005;95 Suppl 1:S144-50.

29. Shapiro S. Causation, bias and confounding: a hitchhiker's guide to the epidemiological galaxy Part 2. Principles of causality in epidemiological research: confounding, effect modification and strength of association. J Fam Plann Reprod Health Care 2008;34:185-90.

30. Susser M. What is a cause and how do we know one? A grammar for pragmatic epidemiology. Am J Epidemiol 1991;133:635-48.

Chapter 12 - Abortion on the Internet: Penile amputation and other dreaded complications

1. Scott W. Marmion: a tale of flodden field. Edinburgh: J. Ballantyne and Co.; 1808.

2. Spink A, Yang Y, Jansen J, et al. A study of medical and health queries to web search engines. Health Info Libr J 2004;21:44-51.

3. Ybarra ML, Suman M. Help seeking behavior and the Internet: A national survey. Int J Med Inform 2006; 75: 29-41.

4. Hanif F, Read JC, Goodacre JA, Chaudhry A, Gibbs P. The role of quality tools in assessing reliability of the internet for health information. Inform Health Soc Care 2009;34:231-43.

5. Kanuga M, Rosenfeld WD. Adolescent sexuality and the internet: the good, the bad, and the URL. J Pediatr Adolesc Gynecol 2004;17:117-24.

6. Guttmacher Institute. Facts on American teens' sources of information about sex. http://www.guttmacher.org/pubs/FB-Teen-Sex-Ed.pdf, accessed December 27, 2013.

7. Gupte CM, Hassan AN, McDermott ID, Thomas RD. The internet—friend or foe? A questionnaire study of orthopaedic out-patients. Ann R Coll Surg Engl 2002;84:187-92.

8. Crocco AG, Villasis-Keever M, Jadad AR. Analysis of cases of harm associated with use of health information on the internet. JAMA 2002;287:2869-71.

9. Sandvik H. Health information and interaction on the internet: a survey of female urinary incontinence. BMJ 1999;319:29-32.

10. Bock B, Graham A, Sciamanna C, et al. Smoking cessation treatment on the Internet: content, quality, and usability. Nicotine Tob Res 2004;6:207-19.

11. Moshirfar A, Campbell JT, Khasraghi FA, Wenz JFS. Evaluating the quality of Internet-derived information on plantar fasciitis. Clin Orthop Relat Res 2004:60-3.

12. Ernst E, Schmidt K. 'Alternative' cures for depression—how safe are web sites? Psychiatry Res 2004;129:297-301.
13. Impicciatore P, Pandolfini C, Casella N, Bonati M. Reliability of health information for the public on the World Wide Web: systematic survey of advice on managing fever in children at home. BMJ 1997;314:1875-9.
14. Haddow G, Watts R. Caring for a febrile child: the quality of Internet information. Collegian 2003;10:7-12.
15. Weiss E, Moore K. An assessment of the quality of information available on the internet about the IUD and the potential impact on contraceptive choices. Contraception 2003;68:359-64.
16. Sing A, Salzman JR, Sing H, Sing D. Evaluation of health information provided on the Internet by airlines with destinations in tropical and subtropical countries. Commun Dis Public Health 2000;3:195-7.
17. Klila H, Chatton A, Zermatten A, Khan R, Preisig M, Khazaal Y. Quality of Web-based information on obsessive compulsive disorder. Neuropsychiatr Dis Treat 2013;9:1717-23.
18. Ipser JC, Dewing S, Stein DJ. A systematic review of the quality of information on the treatment of anxiety disorders on the internet. Curr Psychiatry Rep 2007;9:303-9.
19. Dillon WA, Prorok JC, Seitz DP. Content and quality of information provided on canadian dementia websites. Can Geriatr J 2013;16:6-15.
20. Langille M, Bernard A, Rodgers C, Hughes S, Leddin D, van Zanten SV. Systematic review of the quality of patient information on the internet regarding inflammatory bowel disease treatments. Clin Gastroenterol Hepatol 2010;8:322-8.
21. Badarudeen S, Sabharwal S. Assessing readability of patient education materials: current role in orthopaedics. Clin Orthop Relat Res 2010;468:2572-80.
22. Nasser S, Mullan J, Bajorek B. Assessing the quality, suitability and readability of internet-based health information about warfarin for patients. Australas Med J 2012;5:194-203.
23. Reis BY, Brownstein JS. Measuring the impact of health policies using Internet search patterns: the case of abortion. BMC Public Health 2010;10:514.
24. Foster AM, Wynn L, Rouhana A, Diaz-Olavarrieta C, Schaffer K, Trussell J. Providing medication abortion information to diverse communities: use patterns of a multilingual web site. Contraception 2006;74:264-71.
25. Mashiach R, Seidman GI, Seidman DS. Use of mifepristone as an example of conflicting and misleading medical information on the internet. BJOG 2002;109:437-42.

26. Shanley C. Assessing credibility in online abortion information. https://cdr.lib.unc.edu/record/uuid:ac5c6931-834c-4a28-acbc-54e853af6934, accessed December 27, 2013.

27. Buhi ER, Daley EM, Oberne A, Smith SA, Schneider T, Fuhrmann HJ. Quality and accuracy of sexual health information web sites visited by young people. J Adolesc Health 2010;47:206-8.

28. Anonymous. Crisis pregnancy center. http://en.wikipedia.org/wiki/Crisis_pregnancy_center, accessed December 27, 2013.

29. Resnick S. Navigating anti-abortion online strategy. http://www.americanindependent.com/208240/navigating-anti-abortion-online-strategy, accessed December 27, 2013.

30. U.S. House of Representatives Committee on Government Reform. False and misleading health information provided by federally funded pregnancy resource centers. http://www.chsourcebook.com/articles/waxman2.pdf, accessed December 27, 2013.

31. Bryant AG, Levi EE. Abortion misinformation from crisis pregnancy centers in North Carolina. Contraception 2012;86:752-6.

32. Anonymous. David Reardon. http://en.wikipedia.org/wiki/David_Reardon, accessed December 28, 2013.

33. Anonymous. Pacific Western University (Hawaii). http://en.wikipedia.org/wiki/Pacific_Western_University_%28Hawaii%29, accessed December 28, 2013.

34. Reardon DC. Their deepest wound: an analysis. http://www.afterabortion.org/PAR/V4/n23/ANALYSIS.htm, accessed January 3, 2003.

35. Elliot Institute. New explanation for Bobbitt mutilation points to abortion. http://www.abortionfacts.com/reardon/new-explanation-for-bobbitt-mutilation-points-to-abortion, accessed April 20, 2014.

36. Elliot Institute. Tenth anniversary of Bobbitt mutilation brings new answers. http://www.afterabortion.org/bobbitt.html, accessed June 2, 2004.

37. Grimes DA, Cates WJ. Complications from legally-induced abortion: a review. Obstet Gynecol Surv 1979;34:177-91.

38. Paul M, Lichtenberg ES, Borgatta L, Grimes DA, Stubblefield PG, Creinin MD. Management of unintended and abnormal pregnancy. Comprehensive abortion care. Chichester, England: Blackwelll Publishing Ltd.; 2009.

39. Stam P. Woman's Right to Know Act: a legislative history. Issues Law Med 2012;28:3-67.

40. Elliot Institute. The aftereffects of abortion. http://www.afterabortion.info/complic.html, accessed October 22, 2005.

41. Diamond M, Palmore JA, Smith RG, Steinhoff PG. Abortion in Hawaii. Fam Plann Perspect 1973;5:54-60.
42. Council on Scientific Affairs. Induced termination of pregnancy before and after Roe v Wade. Trends in the mortality and morbidity of women. Council on Scientific Affairs, American Medical Association. JAMA 1992;268:3231-9.
43. Cates W, Grimes DA, Schulz KF. Abortion surveillance at CDC: creating public health light out of political heat. Am J Prev Med 2000;19:12-7.
44. Winker MA, Flanagin A, Chi-Lum B, et al. Guidelines for medical and health information sites on the internet: principles governing AMA web sites. American Medical Association. JAMA 2000;283:1600-6.
45. Bovi AM. Use of health-related online sites. Am J Bioeth 2003;3:W-IF3.
46. Papaevangelou G, Vrettos AS, Papadatos C, Alexiou D. The effect of spontaneous and induced abortion on prematurity and birthweight. J Obstet Gynaecol Br Commonw 1973;80:410-22.
47. Oliver-Williams C, Fleming M, Monteath K, Wood AM, Smith GC. Changes in association between previous therapeutic abortion and preterm birth in Scotland, 1980 to 2008: a historical cohort study. PLoS Med 2013;10:e1001481.
48. Grimes DA. Epidemiologic research using administrative databases: garbage in, garbage out. Obstet Gynecol 2010;116:1018-9.
49. Grimes DA, Schulz KF. False alarms and pseudo-epidemics: the limitations of observational epidemiology. Obstet Gynecol 2012;120:920-7.
50. Taubes G. Epidemiology faces its limits. Science 1995;269:164-9.
51. Hill AB. The environment and disease association or causation. Proc Royal Soc Med 1965;58:295-300.
52. Klemetti R, Gissler M, Niinimaki M, Hemminki E. Birth outcomes after induced abortion: a nationwide register-based study of first births in Finland. Hum Reprod 2012;27:3315-20.
53. North Carolina Family Policy Council. About us. http://www.ncfamily.org/about-north-carolina-family-policy-council/, accessed December 29, 2013.
54. Flat Earth Society. Flat Earth Society. http://www.alaska.net/~clund/e_djublonskopf/FlatHome.htm, accessed October 22, 2005.
55. Anonymous. The alchemy web site. http://www.levity.com/alchemy/home.html, accessed October 22, 2005.
56. Raymond EG, Grimes DA. The comparative safety of legal induced abortion and childbirth in the United States. Obstet Gynecol 2012;119:215-9.

57. Bruce FC, Berg CJ, Hornbrook MC, et al. Maternal morbidity rates in a managed care population. Obstet Gynecol 2008;111:1089-95.

58. Bruce FC, Berg CJ, Joski PJ, et al. Extent of maternal morbidity in a managed care population in Georgia. Paediatr Perinat Epidemiol 2012;26:497-505.

59. American College of Obstetricians and Gynecologists. Ethics in obstetrics and gynecology. Second ed. Washington, D.C.: American College of Obstetricians and Gynecologists; 2004.

Chapter 13 - Breast cancer: the jury is in

1. MSNBC.com. Govt. brochure wrongly links abortion, cancer. http://www.nbcnews.com/id/6446707/, accessed April 20, 2014.

2. National Cancer Institute. Summary report: early reproductive events and breast cancer. http://www.cancer.gov/cancertopics/causes/ere/workshop-report, accessed March 3, 2005.

3. Richardson CT, Nash E. Misinformed consent: the medical accuracy of state-developed abortion counseling methods. http://www.guttmacher.org/pubs/gpr/09/4/gpr090406.pdf, accessed January 10, 2014.

4. Guttmacher Institute. State mandated abortion counseling materials often medically inaccurate, biased.http://www.guttmacher.org/media/nr/2006/10/26/index.html, accessed January 10, 2014.

5. Texas Department of State Health Services. A woman's right to know. http://www.dshs.state.tx.us/layouts/contentpage.aspx?pageid=29816&id=29529&terms=Woman%27s+Right+to+Know, accessed January 10, 2014.

6. Thorp JMJ, Hartmann KE, Shadigian E. Long-term physical and psychological health consequences of induced abortion: review of the evidence. Obstet Gynecol Surv 2003;58:67-79.

7. American College of Obstetricians and Gynecologists. Ethics in obstetrics and gynecology. Second ed. Washington, D.C.: American College of Obstetricians and Gynecologists; 2004.

8. American Cancer Society. Estimated number of new cancer cases and deaths by sex, US, 2013. http://www.cancer.org/acs/groups/content/@epidemiologysurveilance/documents/document/acspc-037124.pdf, accessed January 10, 2014.

9. Grimes DA, Schulz KF. Cohort studies: marching towards outcomes. Lancet 2002;359:341-5.

10. U. S. Preventive Services Task Force. Guide to clinical preventive services. 2nd ed. Baltimore, MD: Williams & Wilkins; 1996.

11. Schulz KF, Grimes DA. Case-control studies: research in reverse. Lancet 2002;359:431-4.

12. Grimes DA, Schulz KF. Compared to what? Finding controls for case-control studies. Lancet 2005;365:1429-33.

13. Bartholomew LL, Grimes DA. The alleged association between induced abortion and risk of breast cancer: biology or bias? Obstet Gynecol Surv 1998;53:708-14.

14. Fu H, Darroch JE, Henshaw SK, Kolb E. Measuring the extent of abortion underreporting in the 1995 National Survey of Family Growth. Fam Plann Perspect 1998;30:128-33, 38.

15. Grimes DA, Schulz KF. Bias and causal associations in observational research. Lancet 2002;359:248-52.

16. Stuart GS, Grimes DA. Social desirability bias in family planning studies: a neglected problem. Contraception 2009;80:108-12.

17. Lindefors-Harris BM, Eklund G, Adami HO, Meirik O. Response bias in a case-control study: analysis utilizing comparative data concerning legal abortions from two independent Swedish studies. Am J Epidemiol 1991;134:1003-8.

18. Rookus MA, van Leeuwen FE. Induced abortion and risk for breast cancer: reporting (recall) bias in a Dutch case-control study. J Natl Cancer Inst 1996;88:1759-64.

19. Grimes DA, Schulz KF. An overview of clinical research: the lay of the land. Lancet 2002;359:57-61.

20. Beral V, Bull D, Doll R, Peto R, Reeves G. Breast cancer and abortion: collaborative reanalysis of data from 53 epidemiological studies, including 83?000 women with breast cancer from 16 countries. Lancet 2004;363:1007-16.

21. Melbye M, Wohlfahrt J, Olsen JH, et al. Induced abortion and the risk of breast cancer. N Engl J Med 1997;336:81-5.

22. Palmer JR, Wise LA, Adams-Campbell LL, Rosenberg L. A prospective study of induced abortion and breast cancer in African-American women. Cancer Causes Control 2004;15:105-11.

23. Lash TL, Fink AK. Null association between pregnancy termination and breast cancer in a registry-based study of parous women. Int J Cancer 2004;110:443-8.

24. Reeves GK, Kan SW, Key T, et al. Breast cancer risk in relation to abortion: Results from the EPIC study. Int J Cancer 2006;119:1741-5.

25. Rosenblatt KA, Gao DL, Ray RM, et al. Induced abortions and the risk of all cancers combined and site-specific cancers in Shanghai. Cancer Causes Control 2006;17:1275-80.

26. Michels KB, Xue F, Colditz GA, Willett WC. Induced and spontaneous abortion and incidence of breast cancer among young women: a prospective cohort study. Arch Intern Med 2007;167:814-20.

27. Henderson KD, Sullivan-Halley J, Reynolds P, et al. Incomplete pregnancy is not associated with breast cancer risk: the California Teachers Study. Contraception 2008;77:391-6.

28. Michels KB, Willett WC. Does induced or spontaneous abortion affect the risk of breast cancer? Epidemiology 1996;7:521-8.

29. Brind J, Chinchilli VM, Severs WB, Summy-Long J. Induced abortion as an independent risk factor for breast cancer: a comprehensive review and meta-analysis. J Epidemiol Community Health 1996;50:481-96.

30. Wingo PA, Newsome K, Marks JS, Calle EE, Parker SL. The risk of breast cancer following spontaneous or induced abortion. Cancer Causes Control 1997;8:93-108.

31. World Health Organization. Induced abortion does not increase breast cancer risk, Fact Sheet number 240. Geneva, Switzerland: World Health Organization; 2000.

32. American Cancer Society. Overview: breast cancer. What causes breast cancer?http://www.cancer.org/cancer/breastcancer/overviewguide/breast-cancer-overview-what-causes?sitearea=CRI&viewmode=print&, accessed March 3, 2005.

33. American College of Obstetricians and Gynecologists. ACOG committee opinion. Induced abortion and breast cancer risk. Number 285, August 2003. Int J Gynaecol Obstet 2003;83:233-5.

34. Royal College of Obstetricians and Gynaecologists. RCOG statement on abortion and breast cancer. London, England: Royal College of Obstetricians and Gynaecologists; 2004.

35. American Cancer Society. Is abortion linked to breast cancer? http://www.cancer.org/cancer/breastcancer/moreinformation/is-abortion-linked-to-breast-cancer, accessed January 10, 2014.

36. Canadian Cancer Society. Canadian Cancer Society's perspective on abortion and breast cancer. http://www.cancer.ca/en/about-us/news/national/2013/canadian-cancer-society-perspective-on-abortion-and-breast-cancer/?region=on, accessed January 10, 2014.

37. National Cancer Institute. Reproductive history and breast cancer risk. http://www.cancer.gov/cancertopics/factsheet/Risk/reproductive-history, accessed January 10, 2014.

38. American College of Obstetricians and Gynecologists. ACOG Committee Opinion No. 434: induced abortion and breast cancer risk. Obstet Gynecol 2009;113:1417-8.

39. Society of Obstetricians and Gynaecologists of Canada/Society of Gynecologic Oncologists of Canada. Breast cancer and abortion. http://sogc.org/guidelines/breast-cancer-and-abortion-committee-opinion/, accessed January 10, 2014.

40. Royal College of Obstetricians and Gynaecologists. Briefing note - scientific information on abortion.http://www.rcog.org.uk/what-we-do/campaigning-and-opinions/briefings-and-qas-/human-fertilisation-and-embryology-bill/brie-4, accessed January 10, 2014.

41. Union of Concerned Scientists. Scientific Integrity in Policymaking. An investigation in the Bush administration's misuse of science. Cambridge, Massachusetts: Union of Concerned Scientists; 2004.

42. Waxman HA. Breast cancer risks. http://oversight-archive.waxman.house.gov/documents/20080130103545.pdf, accessed March 3, 2005.

43. National Cancer Institute. Abortion, miscarriage, and breast cancer risk. http://www.cancer.gov/cancertopics/factsheet/Risk/abortion-miscarriage, accessed April 20, 2014.

44. New York Times. Abortion and breast cancer. http://www.nytimes.com/2003/01/06/opinion/06MON1Y.html, accessed March 3, 2005.

Chapter 14 - Abortion and mental health: apples and oranges, chickens and eggs

1. Grimes DA, Schulz KF. Compared to what? Finding controls for case-control studies. Lancet 2005;365:1429-33.

2. Stotland NL. Psychiatric aspects of induced abortion. J Nerv Ment Dis 2011;199:568-70.

3. American Psychological Association. The impact of abortion on women: what does the psychological research say? http://www.apa.org/ppo/issues/womenabortfacts.html, accessed July 13, 2005.

4. Major B. Psychological implications of abortion—highly charged and rife with misleading research. CMAJ 2003;168:1257-8.

5. Doane BK, Quigley BG. Psychiatric aspects of therapeutic abortion. Can Med Assoc J 1981;125:427-32.

6. Bradshaw Z, Slade P. The effects of induced abortion on emotional experiences and relationships: a critical review of the literature. Clin Psychol Rev 2003;23:929-58.

7. Zolese G, Blacker CV. The psychological complications of therapeutic abortion. Br J Psychiatry 1992;160:742-9.

8. Adler NE, David HP, Major BN, Roth SH, Russo NF, Wyatt GE. Psychological factors in abortion. A review. Am Psychol 1992;47:1194-204.

9. Romans-Clarkson SE. Psychological sequelae of induced abortion. Aust N Z J Psychiatry 1989;23:555-65.

10. Dagg PK. The psychological sequelae of therapeutic abortion—denied and completed. Am J Psychiatry 1991;148:578-85.

11. Adler NE, David HP, Major BN, Roth SH, Russo NF, Wyatt GE. Psychological responses after abortion. Science 1990;248:41-4.

12. Kendall T, Bird V, Cantwell R, Taylor C. To meta-analyse or not to meta-analyse: abortion, birth and mental health. Br J Psychiatry 2012;200:12-4.

13. Hill AB. The environment and disease association or causation. Proc Royal Soc Med 1965;58:295-300.

14. van Ditzhuijzen J, ten Have M, de Graaf R, van Nijnatten CH, Vollebergh WA. Psychiatric history of women who have had an abortion. J Psychiatr Res 2013;47:1737-43.

15. Munk-Olsen T, Laursen TM, Pedersen CB, Lidegaard O, Mortensen PB. Induced first-trimester abortion and risk of mental disorder. N Engl J Med 2011;364:332-9.

16. Grimes DA. Epidemiologic research using administrative databases: garbage in, garbage out. Obstet Gynecol 2010;116:1018-9.

17. Bland JM, Altman DG. Regression towards the mean. BMJ 1994;308:1499.

18. Freeman EW, Rickels K, Huggins GR, Garcia CR, Polin J. Emotional distress patterns among women having first or repeat abortions. Obstet Gynecol 1980;55:630-6.

19. Cohen L, Roth S. Coping with abortion. J Human Stress 1984;10:140-5.

20. Major B, Mueller P, Hildebrandt K. Attributions, expectations, and coping with abortion. J Pers Soc Psychol 1985;48:585-99.

21. Major B, Cozzarelli C, Cooper ML, et al. Psychological responses of women after first-trimester abortion. Arch Gen Psychiatry 2000;57:777-84.

22. Athanasiou R, Oppel W, Michelson L, Unger T, Yager M. Psychiatric sequelae to term birth and induced early and late abortion: a longitudinal study. Fam Plann Perspect 1973;5:227-31.

23. Zabin LS, Hirsch MB, Emerson MR. When urban adolescents choose abortion: effects on education, psychological status and subsequent pregnancy. Fam Plann Perspect 1989;21:248-55.

24. Eisen M, Zellman GL. Factors predicting pregnancy resolution decision satisfaction of unmarried adolescents. J Genet Psychol 1984;145:231-9.

25. Cohan CL, Dunkel-schetter C, Lydon J. Pregnancy decision making: predictors of early stress and adjustment. Psychol Women Q 1993;17:223-39.
26. Brewer C. Incidence of post-abortion psychosis: a prospective study. Br Med J 1977;1:476-7.
27. Russo NF, Zierk KL. Abortion, childbearing and women's well-being. Prof Psychol Res Pr 1992;23:269-80.
28. Pazol K, Creanga AA, Burley KD, Hayes B, Jamieson DJ. Abortion surveillance - United States, 2010. MMWR Surveill Summ 2013;62:1-44.
29. Gilchrist AC, Hannaford PC, Frank P, Kay CR. Termination of pregnancy and psychiatric morbidity. Br J Psychiatry 1995;167:243-8.
30. Anonymous. More on Koop's study of abortion. Fam Plann Perspect 1990;22:36-9.
31. American Psychological Association Task Force on Mental Health and Abortion. Abortion and mental health. Evaluating the evidence. http://www.apa.org/pi/women/resources/newsletter/2010/05/abortion-mental-health.aspx, accessed January 15, 2014.
32. Major B, Appelbaum M, Beckman L, Dutton MA, Russo NF, West C. Abortion and mental health: Evaluating the evidence. Am Psychol 2009;64:863-90.
33. Academy of Medical Royal Colleges. Induced abortion and mental health. http://www.nccmh.org.uk/publications_SR_abortion_in_MH.html, accessed January 13, 2014.
34. Charles VE, Polis CB, Sridhara SK, Blum RW. Abortion and long-term mental health outcomes: a systematic review of the evidence. Contraception 2008;78:436-50.
35. Robinson GE, Stotland NL, Russo NF, Lang JA, Occhiogrosso M. Is there an "abortion trauma syndrome"? Critiquing the evidence. Harv Rev Psychiatry 2009;17:268-90.
36. Steinberg JR. Later abortions and mental health: psychological experiences of women having later abortions—a critical review of research. Womens Health Issues 2011;21:S44-8.
37. Coleman PK. Abortion and mental health: quantitative synthesis and analysis of research published 1995-2009. Br J Psychiatry 2011;199:180-6.
38. Stroup DF, Berlin JA, Morton SC, et al. Meta-analysis of observational studies in epidemiology: a proposal for reporting. Meta-analysis Of Observational Studies in Epidemiology (MOOSE) group. JAMA 2000;283:2008-12.
39. Petitti DB. Meta-analysis, decision analysis, and cost-effectiveness analysis: methods for quantitative synthesis in medicine. Second ed. New York: Oxford University Press; 2000.

40. Reeves BC, Deeks JJ, Higgins JPT, Wells GA. Including non-randomised studies. In: Higgins JPT, Green S, eds. Cochrane handbook for systematic reviews of interventions. Chichester, West Sussex: John Wiley & Sons Ltd; 2008.

41. Shapiro S. Is meta-analysis a valid approach to the evaluation of small effects in observational studies? J Clin Epidemiol 1997;50:223-9.

42. Shapiro S. Meta-analysis/Shmeta-analysis. Am J Epidemiol 1994;140:771-8.

43. Egger M, Schneider M, Davey Smith G. Spurious precision? Meta-analysis of observational studies. BMJ 1998;316:140-4.

44. Petitti DB. Approaches to heterogeneity in meta-analysis. Stat Med 2001;20:3625-33.

45. Coleman PK, Coyle CT, Shuping M, Rue VM. Induced abortion and anxiety, mood, and substance abuse disorders: isolating the effects of abortion in the national comorbidity survey. J Psychiatr Res 2009;43:770-6.

46. Steinberg JR, Finer LB. Coleman, Coyle, Shuping, and Rue make false statements and draw erroneous conclusions in analyses of abortion and mental health using the National Comorbidity Survey. J Psychiatr Res 2012;46:407-8.

47. Kessler RC, Schatzberg AF. Commentary on abortion studies of Steinberg and Finer and Coleman. J Psych Res 2012;46:410-11.

48. Coleman PK, Reardon DC, Rue VM, Cougle J. State-funded abortions versus deliveries: a comparison of outpatient mental health claims over 4 years. Am J Orthopsychiatry 2002;72:141-52.

49. Coleman PK, Maxey CD, Spence M, al. e. Predictors and correlates of abortion in the Fragile Families and Well-Being Study: paternal behavior, substance use,and partner violence. International Journal of Mental Health Addiction 2009;7:405-22.

50. Coleman PK, Coyle CT, Rue VM. Late-term elective abortion and susceptibility to posttraumatic stress symptoms. Journal of Pregnancy 2010;2010:130519.

51. Reardon DC, Cougle JR. Depression and unintended pregnancy in the National Longitudinal Survey of Youth: a cohort study. BMJ 2002;324:151-2.

52. Reardon DC, Coleman PK, Cougle JR. Substance use associated with unintended pregnancy outcomes in the National Longitudinal Survey of Youth. Am J Drug Alcohol Abuse 2004;30:369-83.

53. Reardon DC, Ney PG, Scheuren F, Cougle J, Coleman PK, Strahan TW. Deaths associated with pregnancy outcome: a record linkage study of low income women. South Med J 2002;95:834-41.

54. Reardon DC, Cougle JR, Rue VM, Shuping MW, Coleman PK, Ney PG. Psychiatric admissions of low-income women following abortion and childbirth. CMAJ 2003;168:1253-6.
55. Seibel M, Graves WL. The psychological implications of spontaneous abortions. J Reprod Med 1980;25:161-5.
56. Turner MJ, Flannelly GM, Wingfield M, et al. The miscarriage clinic: an audit of the first year. Br J Obstet Gynaecol 1991;98:306-8.
57. Friedman T, Gath D. The psychiatric consequences of spontaneous abortion. Br J Psychiatry 1989;155:810-3.
58. Geller PA, Klier CM, Neugebauer R. Anxiety disorders following miscarriage. J Clin Psychiatry 2001;62:432-8.
59. Janssen HJ, Cuisinier MC, Hoogduin KA, de Graauw KP. Controlled prospective study on the mental health of women following pregnancy loss. Am J Psychiatry 1996;153:226-30.
60. Kaunitz AM, Grimes DA, Kaunitz KK. A physician's guide to adoption. JAMA 1987;258:3537-41.
61. Chandra A, Abma J, Maza P, Bachrach C. Adoption, adoption seeking, and relinquishment for adoption in the United States. Advance data from vital and health statistics; no. 306. Hyattsville, MD: National Center for Health Statistics; 1999.
62. Jones J. Adoption experiences of women and men and demand for children to adopt by women 18-44 years of age in the United States, 2002. Vital Health Stat 23 2008:1-36.
63. Condon JT. Psychological disability in women who relinquish a baby for adoption. Med J Aust 1986;144:117-9.
64. Thomas T, Tori CD. Sequelae of abortion and relinquishment of child custody among women with major psychiatric disorders. Psychol Rep 1999;84:773-90.
65. Sachdev P. Unlocking the adoption files. Lexington, MA: Lexington Books; 1989.
66. Elliot Institute. The aftereffects of abortion. http://www.afterabortion.info/complic.html, accessed October 22, 2005.
67. Stotland NL. The myth of the abortion trauma syndrome. JAMA 1992;268:2078-9.
68. Dadlez EM, Andrews WL. Post-abortion syndrome: creating an affliction. Bioethics 2010;24:445-52.
69. American Psychiatric Association. Diagnostic and statistical manual of mental disorders, 4th ed. Washington, D.C.: American Psychiatric Publishing, Inc.; 2000.

70. American Psychological Association. Mental health and abortion. http://www.apa.org/pi/women/programs/abortion/executive-summary.pdf, accessed December 30, 2013.

71. American Psychiatric Association. Abortion and women's reproductive health care rights. http://www.psychiatry.org/advocacy—newsroom/position-statements, accessed December 30, 2013.

72. Royal Australian and New Zealand College of Obstetricians and Gynaecologists. Termination of pregnancy: a resource for health professionals. http://www.ranzcog.edu.au/termination-of-pregnancy-booklet.html, accessed January 15, 2014.

73. Ioannidis JP. An epidemic of false claims. Competition and conflicts of interest distort too many medical findings. Sci Am 2011;304:16.

74. Ioannidis JP. Why most published research findings are false. PLoS Med 2005;2:e124.

Chapter 15 - Prematurity and abortion (and gum disease)

1. Goldenberg RL, Culhane JF, Iams JD, Romero R. Epidemiology and causes of preterm birth. Lancet 2008;371:75-84.

2. Hogue C, Boardman L, Stotland N. Answering questions about long-term outcomes. In: Paul M, Lichtenberg E, Borgatta L, Grimes D, Stubblefield P, Creinin M, eds. Management of unintended and abnormal pregnancy Comprehensive abortion care. Oxford: John Wiley & Sons; 2009:252-63.

3. Atrash HK, Hogue CJ. The effect of pregnancy termination on future reproduction. Baillieres Clin Obstet Gynaecol 1990;4:391-405.

4. Jones EF, Forrest JD. Underreporting of abortion in surveys of U.S. women: 1976 to 1988. Demography 1992;29:113-26.

5. Lindefors-Harris BM, Eklund G, Adami HO, Meirik O. Response bias in a case-control study: analysis utilizing comparative data concerning legal abortions from two independent Swedish studies. Am J Epidemiol 1991;134:1003-8.

6. Shapiro-Mendoza CK, Lackritz EM. Epidemiology of late and moderate preterm birth. Semin Fetal Neonatal Med 2012;17:120-5.

7. Simmons LE, Rubens CE, Darmstadt GL, Gravett MG. Preventing preterm birth and neonatal mortality: exploring the epidemiology, causes, and interventions. Semin Perinatol 2010;34:408-15.

8. Wen SW, Smith G, Yang Q, Walker M. Epidemiology of preterm birth and neonatal outcome. Semin Fetal Neonatal Med 2004;9:429-35.

9. Hogue CJ, Cates WJ, Tietze C. Impact of vacuum aspiration abortion on future childbearing: a review. Fam Plann Perspect 1983;15:119-26.
10. Zhou W, Sorensen HT, Olsen J. Induced abortion and subsequent pregnancy duration. Obstet Gynecol 1999;94:948-53.
11. Zhou W, Sorensen HT, Olsen J. Induced abortion and low birthweight in the following pregnancy. Int J Epidemiol 2000;29:100-6.
12. Raatikainen K, Heiskanen N, Heinonen S. Induced abortion: not an independent risk factor for pregnancy outcome, but a challenge for health counseling. Ann Epidemiol 2006;16:587-92.
13. Silver N. The signal and the noise. New York: The Penguin Press; 2012.
14. Swingle HM, Colaizy TT, Zimmerman MB, Morriss FH, Jr. Abortion and the risk of subsequent preterm birth: a systematic review with meta-analyses. J Reprod Med 2009;54:95-108.
15. Ioannidis JP, Lau J. Pooling research results: benefits and limitations of meta-analysis. Jt Comm J Qual Improv 1999;25:462-9.
16. Petitti DB. Meta-analysis, decision analysis, and cost-effectiveness analysis: methods for quantitative synthesis in medicine. Second ed. New York: Oxford University Press; 2000.
17. Grimes DA, Schulz KF. False alarms and pseudo-epidemics: the limitations of observational epidemiology. Obstet Gynecol 2012;120:920-7.
18. Taubes G. Epidemiology faces its limits. Science 1995;269:164-9.
19. Shah PS, Zao J, Knowledge Synthesis Group of Determinants of preterm/LBW births. Induced termination of pregnancy and low birthweight and preterm birth: a systematic review and meta-analyses. BJOG 2009;116:1425-42.
20. Klemetti R, Gissler M, Niinimaki M, Hemminki E. Birth outcomes after induced abortion: a nationwide register-based study of first births in Finland. Hum Reprod 2012;27:3315-20.
21. Grimes DA. Epidemiologic research using administrative databases: garbage in, garbage out. Obstet Gynecol 2010;116:1018-9.
22. Grimes DA, Raymond EG. Medical abortion for adolescents. BMJ 2011;342:d2185.
23. Behrman RE, Butler AS. Preterm birth: causes, consequences, and prevention. Washington, D.C.: National Academies Press; 2007.
24. Gestation, birth-weight, and spontaneous abortion in pregnancy after induced abortion. Report of Collaborative Study by W.H.O. Task Force on Sequelae of Abortion. Lancet 1979;1:142-5.

25. World Health Organization. Safe abortion: technical and policy guidance for health systems, second edition. http://www.who.int/reproductivehealth/publications/unsafe_abortion/9789241548434/en/, accessed December 28, 2013.

26. American College of Obstetricians and Gynecologists. Preterm labor and preterm birth FAQ087. http://www.acog.org/Search?Keyword=FAQ087, accessed January 8, 2014.

27. American College of Obstetricians and Gynecologists. Induced abortion. http://www.acog.org/Search?Sources=9fdcbde3-33d7-48a4-9a44-a4348f509988&Keyword=FAQ043, accessed January 1, 2014.

28. American College of Obstetricians and Gynecologists. Induced abortion AP 043. http://www.acog.org/Resources_And_Publications/Patient_Education_Pamphlets/Files/Induced_Abortion, accessed January 8, 2014.

29. American College of Obstetricians and Gynecologists Committee on Practice Bulletins-Obstetrics. Practice bulletin no. 130: prediction and prevention of preterm birth. Obstet Gynecol 2012;120:964-73.

30. Royal College of Obstetricians and Gynaecologists. The care of women requesting induced abortion. Evidence-based clinical guideline number 7. http://www.rcog.org.uk/womens-health/clinical-guidance/care-women-requesting-induced-abortion, accessed December 28, 2013.

31. American Academy of Pediatrics Committee on Adolescence. The adolescent's right to confidential care when considering abortion. Pediatrics 1996;97:746-51.

32. Centers for Disease Control and Prevention. Preterm birth. http://www.cdc.gov/reproductivehealth/MaternalInfantHealth/PretermBirth.htm, accessed December 28, 2013.

33. Centers for Disease Control and Prevention. Factors associated with preterm birth. http://www.cdc.gov/reproductivehealth/MaternalInfantHealth/PDF/PretermBirth-Infographic.pdf, accessed April 20, 2014.

34. American Public Health Association. Provision of abortion care by advanced practice nurses and physician assistants. http://www.apha.org/policies-and-advocacy/public-health-policy-statements/policy-database/2014/07/28/16/00/provision-of-abortion-care-by-advanced-practice-nurses-and-physician-assistants, accessed January 1, 2014.

35. March of Dimes. Finding the causes of prematurity. http://www.marchofdimes.com/research/finding-the-causes-of-prematurity.aspx#, accessed December 28, 2013.

36. McCaffrey MJ. Time to acknowledge the connection between abortion and later preterm birth. http://www.newsobserver.com/2013/05/22/2910986/time-to-acknowledge-the-connection.html, accessed January 8, 2014.

37. Aliaga SR, Smith PB, Price WA, et al. Regional variation in late preterm births in North Carolina. Matern Child Health J 2013;17:33-41.

38. Cunningham FG, Leveno KJ, Bloom SL, Hauth JC, Rouse DJ, Spong CY. Williams' Obstetrics, 23d ed. New York: McGraw-Hill Companies, Inc.; 2010.

39. Jefferson KK. The bacterial etiology of preterm birth. Adv Appl Microbiol 2012;80:1-22.

40. Michalowicz BS, Hodges JS, DiAngelis AJ, et al. Treatment of periodontal disease and the risk of preterm birth. N Engl J Med 2006;355:1885-94.

41. Lopez NJ, Smith PC, Gutierrez J. Periodontal therapy may reduce the risk of preterm low birth weight in women with periodontal disease: a randomized controlled trial. J Periodontol 2002;73:911-24.

42. Klebanoff M, Searle K. The role of inflammation in preterm birth—focus on periodontitis. BJOG 2006;113 Suppl 3:43-5.

43. Brody J. Gum disease in pregnancy linked to premature low-weight babies. http://www.nytimes.com/1996/10/09/us/gum-disease-in-pregnancy-linked-to-premature-low-weight-babies.html, accessed January 8, 2014.

44. Polyzos NP, Polyzos IP, Zavos A, et al. Obstetric outcomes after treatment of periodontal disease during pregnancy: systematic review and meta-analysis. BMJ 2010;341:c7017.

45. Uppal A, Uppal S, Pinto A, et al. The effectiveness of periodontal disease treatment during pregnancy in reducing the risk of experiencing preterm birth and low birth weight: a meta-analysis. J Am Dent Assoc 2010;141:1423-34.

46. George A, Shamim S, Johnson M, et al. Periodontal treatment during pregnancy and birth outcomes: a meta-analysis of randomised trials. Int J Evid Based Healthc 2011;9:122-47.

47. Chambrone L, Pannuti CM, Guglielmetti MR, Chambrone LA. Evidence grade associating periodontitis with preterm birth and/or low birth weight: II: a systematic review of randomized trials evaluating the effects of periodontal treatment. J Clin Periodontol 2011;38:902-14.

48. Rosa MI, Pires PD, Medeiros LR, Edelweiss MI, Martinez-Mesa J. Periodontal disease treatment and risk of preterm birth: a systematic review and meta-analysis. Cad Saude Publica 2012;28:1823-33.

49. Boutin A, Demers S, Roberge S, Roy-Morency A, Chandad F, Bujold E. Treatment of periodontal disease and prevention of preterm birth: systematic review and meta-analysis. Am J Perinatol 2013;30:537-44.

50. Jeffcoat MK, Hauth JC, Geurs NC, et al. Periodontal disease and preterm birth: results of a pilot intervention study. J Periodontol 2003;74:1214-8.

51. Offenbacher S, Beck JD, Jared HL, et al. Effects of periodontal therapy on rate of preterm delivery: a randomized controlled trial. Obstet Gynecol 2009;114:551-9.

52. Newnham JP, Newnham IA, Ball CM, et al. Treatment of periodontal disease during pregnancy: a randomized controlled trial. Obstet Gynecol 2009;114:1239-48.

53. Macones GA, Parry S, Nelson DB, et al. Treatment of localized periodontal disease in pregnancy does not reduce the occurrence of preterm birth: results from the Periodontal Infections and Prematurity Study (PIPS). Am J Obstet Gynecol 2010;202:147 e1-8.

54. Schulz KF, Chalmers I, Hayes RJ, Altman DG. Empirical evidence of bias. Dimensions of methodological quality associated with estimates of treatment effects in controlled trials. JAMA 1995;273:408-12.

55. Wood L, Egger M, Gluud LL, et al. Empirical evidence of bias in treatment effect estimates in controlled trials with different interventions and outcomes: meta-epidemiological study. BMJ 2008;336:601-5.

56. Schulz KF, Chalmers I, Grimes DA, Altman DG. Assessing the quality of randomization from reports of controlled trials published in obstetrics and gynecology journals. JAMA 1994;272:125-8.

57. Ahmed FU, Karim E. Children at risk of developing dehydration from diarrhoea: a case-control study. J Trop Pediatr 2002;48:259-63.

58. Abera B, Alem G, Yimer M, Herrador Z. Epidemiology of soil-transmitted helminths, Schistosoma mansoni, and haematocrit values among schoolchildren in Ethiopia. J Infect Dev Ctries 2013;7:253-60.

59. Mahmud MA, Spigt M, Mulugeta Bezabih A, Lopez Pavon I, Dinant GJ, Blanco Velasco R. Risk factors for intestinal parasitosis, anaemia, and malnutrition among school children in Ethiopia. Pathog Glob Health 2013;107:58-65.

60. Larsen CD, Stavisky E, Larsen MD, Rosenbaum MS. Children's fingernail hygiene and length as predictors of carious teeth. N Y State Dent J 2007;73:33-7.

61. Carroll ST, Riffenburgh RH, Roberts TA, Myhre EB. Tattoos and body piercings as indicators of adolescent risk-taking behaviors. Pediatrics 2002;109:1021-7.

62. Calhoun BC, Shadigian E, Rooney B. Cost consequences of induced abortion as an attributable risk for preterm birth and impact on informed consent. J Reprod Med 2007;52:929-37.

63. Thorp JMJ, Hartmann KE, Shadigian E. Long-term physical and psychological health consequences of induced abortion: review of the evidence. Obstet Gynecol Surv 2003;58:67-79.

Chapter 16 - Fetal feelings?

1. Goodman NW. Changing tactics in the abortion argument: does a fetus feel pain? Br J Hosp Med 1997;58:550.
2. Belluck P. Complex science at issue in politics of fetal pain.http://www.nytimes.com/2013/09/17/health/complex-science-at-issue-in-politics-of-fetal-pain.html, accessed January 6, 2014.
3. American College of Obstetricians and Gynecologists. Fetal pain and fetal pain legislation. http://www.acog.org/Search?Keyword=fetal+pain, accessed January 11, 2014.
4. Levy P. Can the myth of fetal pain topple Roe v. Wade? http://www.newsweek.com/can-myth-fetal-pain-topple-roe-v-wade-3077, accessed January 6, 2014.
5. International Association for the Study of Pain. IASP terminology. https://www.iasp-pain.org/Education/Content.aspx?ItemNumber=1698, accessed April 14, 2014.
6. Lloyd-Thomas AR, Fitzgerald M. Do fetuses feel pain? Reflex responses do not necessarily signify pain. BMJ 1996;313:797-8.
7. Derbyshire SW. Fetal pain: do we know enough to do the right thing? Reprod Health Matters 2008;16:117-26.
8. Royal College of Obstetricians and Gynaecologists. Fetal awareness. https://www.rcog.org.uk/globalassets/documents/guidelines/rcogfetalawarenesswpr0610.pdf accessed January 6, 2014.
9. Folger HT. The effects of mechanical shock on locomotion in Amoeba proteus. J Morphol 2005;42:359-70.
10. Brusseau R. Developmental perspectives: is the fetus conscious? Int Anesthesiol Clin 2008;46:11-23.
11. Vanhatalo S, van Nieuwenhuizen O. Fetal pain? Brain Dev 2000;22:145-50.
12. Lee SJ, Ralston HJ, Drey EA, Partridge JC, Rosen MA. Fetal pain: a systematic multidisciplinary review of the evidence. JAMA 2005;294:947-54.
13. Pazol K, Creanga AA, Burley KD, Hayes B, Jamieson DJ. Abortion surveillance - United States, 2010. MMWR Surveill Summ 2013;62:1-44.
14. Derbyshire SW. Fetal pain: an infantile debate. Bioethics 2001;15:77-84.
15. Klaus MH, Kennell JH. The doula: an essential ingredient of childbirth rediscovered. Acta Paediatr 1997;86:1034-6.
16. Medical Research Council. Report of the MRC expert group on fetal pain. London, England: Medical Research Council; 2001.

17. Szawarski Z. Do fetuses feel pain? Probably no pain in the absence of "self". BMJ 1996;313:796-7.

18. Derbyshire SW. Locating the beginnings of pain. Bioethics 1999;13:1-31.

19. Derbyshire SW, Furedi A. Do fetuses feel pain? "Fetal pain" is a misnomer. BMJ 1996;313:795.

20. American College of Obstetricians and Gynecologists. Facts are important: fetal pain. http://www.acog.org/Search?Keyword=fetal+pain, accessed January 11, 2014.

21. Brownback S. Unborn child pain awareness act of 2005, S. 51. Washington, D.C.: United States Senate; 2005.

22. Anonymous. Angry e-mail follows fetal pain article. Abortion opponents target JAMA editor. http://edition.cnn.com/TRANSCRIPTS/0508/26/acd.01.html, accessed August 25, 2005.

23. Anonymous. Stedman's Medical Dictionary. 27th ed. Philadelphia: Lippincott Williams & Wilkins; 2000.

24. Hughes E, Committee on Terminology of the American College of Obstetricians and Gynecologists. Obstetric-gynecologic terminology. Philadelphia: F. A. Davis; 1972.

25. Grimes DA, Stuart G. Abortion jabberwocky: the need for better terminology. Contraception 2010;81:93-6.

26. Spitz AM, Lee NC, Grimes DA, Schoenbucher AK, Lavoie M. Third-trimester induced abortion in Georgia, 1979 and 1980. Am J Public Health 1983;73:594-5.

27. Chervenak FA, McCullough LB, Campbell S. Is third trimester abortion justified? Br J Obstet Gynaecol 1995;102:434-5.

28. Chervenak FA, McCullough LB, Campbell S. Third trimester abortion: is compassion enough? Br J Obstet Gynaecol 1999;106:293-6.

29. Chervenak FA, McCullough LB. An ethically justified, clinically comprehensive management strategy for third-trimester pregnancies complicated by fetal anomalies. Obstet Gynecol 1990;75:311-6.

30. Dommergues M, Cahen F, Garel M, Mahieu-Caputo D, Dumez Y. Feticide during second- and third-trimester termination of pregnancy: opinions of health care professionals. Fetal Diagn Ther 2003;18:91-7.

31. Andersen PEJ. Prevalence of lethal osteochondrodysplasias in Denmark. Am J Med Genet 1989;32:484-9.

32. Barr MJ, Cohen MMJ. Holoprosencephaly survival and performance. Am J Med Genet 1999;89:116-20.

33. Glover V, Fisk NM. Fetal pain: implications for research and practice. Br J Obstet Gynaecol 1999;106:881-6.

34. Hook B, Kiwi R, Amini SB, Fanaroff A, Hack M. Neonatal morbidity after elective repeat cesarean section and trial of labor. Pediatrics 1997;100:348-53.

35. Levine EM, Ghai V, Barton JJ, Strom CM. Mode of delivery and risk of respiratory diseases in newborns. Obstet Gynecol 2001;97:439-42.

36. Taylor A, Fisk NM, Glover V. Mode of delivery and subsequent stress response. Lancet 2000;355:120.

37. Dixon S, Snyder J, Holve R, Bromberger P. Behavioral effects of circumcision with and without anesthesia. J Dev Behav Pediatr 1984;5:246-50.

38. Warnock F, Sandrin D. Comprehensive description of newborn distress behavior in response to acute pain (newborn male circumcision). Pain 2004;107:242-55.

39. Kmietowicz Z. Antiabortionists hijack fetal pain argument. BMJ 1996;313:188.

Chapter 17 - "Partial-birth abortion:" a distinct non-entity

1. Baxter JK, Weinstein L. HELLP syndrome: the state of the art. Obstet Gynecol Surv 2004;59:838-45.

2. Annas GJ. Partial-birth abortion, Congress, and the Constitution. N Engl J Med 1998;339:279-83.

3. Anonymous. Stedman's Medical Dictionary. 27th ed. Philadelphia: Lippincott Williams & Wilkins; 2000.

4. Esacove A. Dialogic framing: the framing/counterframing of "partial-birth" abortion. Sociological Inquiry 2004;74:70-101.

5. Johnson TR, Harris LH, Dalton VK, Howell JD. Language matters: legislation, medical practice, and the classification of abortion procedures. Obstet Gynecol 2005;105:201-4.

6. Pazol K, Creanga AA, Burley KD, Hayes B, Jamieson DJ. Abortion surveillance - United States, 2010. MMWR Surveill Summ 2013;62:1-44.

7. Annas GJ. "Partial-birth abortion" and the Supreme Court. N Engl J Med 2001;344:152-6.

8. President Bush signs Partial Birth Abortion Ban Act of 2003. http://www.whitehouse.gov/news/releases/2003/11/20031105-1.html, accessed July 20, 1995.

9. Dailard C. Courts strike 'partial-birth' abortion ban; decisions presage future debates. The Guttmacher Report on Public Policy 2004;7:1-6.

10. Anonymous. Partial-birth abortion ban act. http://en.wikipedia.org/wiki/Partial-Birth_Abortion_Ban_Act, accessed January 24, 2014.
11. Chasen ST, Kalish RB, Gupta M, Kaufman JE, Rashbaum WK, Chervenak FA. Dilation and evacuation at >or=20 weeks: comparison of operative techniques. Am J Obstet Gynecol 2004;190:1180-3.
12. Chasen ST, Kalish RB, Gupta M, Kaufman J, Chervenak FA. Obstetric outcomes after surgical abortion at > or = 20 weeks' gestation. Am J Obstet Gynecol 2005;193:1161-4.
13. Sprang ML, Neerhof MG. Rationale for banning abortions late in pregnancy. JAMA 1998;280:744-7.
14. S.3 Partial Birth Abortion Ban Act of 2003. http://www.gpo.gov/fdsys/pkg/BILLS-108s3enr/pdf/BILLS-108s3enr.pdf, accessed July 19, 2005.
15. Porto M. A call for an evidence-based evaluation of late midtrimester abortion. Am J Obstet Gynecol 2004;190:1175-6.
16. Oransky I. US Congress passes "partial-birth abortion" ban. Lancet 2003;362:1464.
17. American College of Obstetricians and Gynecologists. Abortion policy. http://www.acog.org/~/media/Statements%20of%20Policy/sop069.pdf?dmc=1&ts=20140106T0936096483, accessed January 6, 2014.
18. American College of Obstetricians and Gynecologists. Statement on so-called "partial birth abortion" law. http://www.acog.org/About-ACOG/News-Room/News-Releases/2013/Ob-Gyns-Speak-Out-for-Womens-Health, accessed July 19, 2005.
19. Greene MF, Ecker JL. Abortion, health, and the law. N Engl J Med 2004;350:184-6.
20. Kassirer JP. Practicing medicine without a license—the new intrusions by Congress. N Engl J Med 1997;336:1747.
21. Miles SH, August A. Courts, gender and "the right to die". Law Med Health Care 1990;18:85-95.
22. Meier DE, Emmons CA, Wallenstein S, Quill T, Morrison RS, Cassel CK. A national survey of physician-assisted suicide and euthanasia in the United States. N Engl J Med 1998;338:1193-201.
23. Parks JA. Why gender matters to the euthanasia debate. On decisional capacity and the rejection of women's death requests. Hastings Cent Rep 2000;30:30-6.
24. Botelho G. Hospital, family concur that brain-dead Texas woman's fetus 'not viable'. http://www.cnn.com/2014/01/24/health/pregnant-brain-dead-woman-texas/index.html?hpt=hp_t1, accessed January 24, 2014.

25. Blendon RJ, Benson JM, Herrmann MJ. The American public and the Terri Schiavo case. Arch Intern Med 2005;165:2580-4.
26. Cunningham FG, Leveno KJ, Bloom SL, Hauth JC, Rouse DJ, Spong CY. Williams' Obstetrics, 23d ed. New York: McGraw-Hill Companies, Inc.; 2010.
27. Haddad L, Yanow S, Delli-Bovi L, Cosby K, Weitz TA. Changes in abortion provider practices in response to the Partial-Birth Abortion Ban Act of 2003. Contraception 2009;79:379-84.
28. Weitz TA, Yanow S. Implications of the Federal Abortion Ban for Women's Health in the United States. Reprod Health Matters 2008;16:99-107.
29. Goldberg C. Shots assist in aborting fetuses. Lethal injections offer legal shield. http://www.boston.com/yourlife/health/women/articles/2007/08/10/shots_assist_in_aborting_fetuses/?page=1, accessed January 24, 2014.
30. Dean G, Colarossi L, Lunde B, Jacobs AR, Porsch LM, Paul ME. The safety of digoxin for fetal demise before second-trimester abortion by dilation and evacuation. Contraception 2012;85:144-9.
31. Jackson RA, Teplin VL, Drey EA, Thomas LJ, Darney PD. Digoxin to facilitate late second-trimester abortion: a randomized, masked, placebo-controlled trial. Obstet Gynecol 2001;97:471-6.
32. Grimes DA, Stuart GS, Raymond EG. Feticidal digoxin injection before dilation and evacuation abortion: evidence and ethics. Contraception 2012;85:140-3.
33. American College of Obstetricians and Gynecologists. Ethics in obstetrics and gynecology. Second ed. Washington, D.C.: American College of Obstetricians and Gynecologists; 2004.

Chapter 18 - State legislatures: practicing medicine without a license

1. Johnson B. Pro-life laws and clinic closures lowered abortion rate: CDC. http://www.lifesitenews.com/news/abortion-laws-and-clinic-closures-lowered-abortion-rate-cdc, accessed February 23, 2014.
2. Deprez E. Abortion clinics close at record pace after states tighten rules. http://www.bloomberg.com/news/2013-09-03/abortion-clinics-close-at-record-pace-after-states-tighten-rules.html, accessed February 23, 2014.
3. Frammolino R. 2 get prison for trying to bomb abortion clinic. http://articles.latimes.com/print/1988-05-06/local/me-2616_1_bomb-abortion-clinic, accessed February 23, 2014.

4. Grimes DA, Forrest JD, Kirkman AL, Radford B. An epidemic of antiabortion violence in the United States Am J Obstet Gynecol 1991;165:1263-8.
5. Russo JA, Schumacher KL, Creinin MD. Antiabortion violence in the United States. Contraception 2012;86:562-6.
6. Kafrissen ME, Grimes DA, Hogue CJ, Sacks JJ. Cluster of abortion deaths at a single facility. Obstet Gynecol 1986;68:387-9.
7. Grimes DA. Estimation of pregnancy-related mortality risk by pregnancy outcome, United States, 1991 to 1999. Am J Obstet Gynecol 2006;194:92-4.
8. Bruce FC, Berg CJ, Hornbrook MC, et al. Maternal morbidity rates in a managed care population. Obstet Gynecol 2008;111:1089-95.
9. Council on Scientific Affairs. Induced termination of pregnancy before and after Roe v Wade. Trends in the mortality and morbidity of women. Council on Scientific Affairs, American Medical Association. JAMA 1992;268:3231-9.
10. Charo RA. Physicians and the (woman's) body politic. N Engl J Med 2014;370:193-5.
11. Bruce FC, Berg CJ, Joski PJ, et al. Extent of maternal morbidity in a managed care population in Georgia. Paediatr Perinat Epidemiol 2012;26:497-505.
12. Leslie L. Proposed abortion rules could close NC clinics. http://www.wral.com/proposed-abortion-rules-could-close-nc-clinics/12227713/, accessed February 25, 2014.
13. Medoff MH, Dennis C. TRAP abortion laws and partisan political party control of state government. Am J Econ Sociol 2011;70:951-73.
14. Blue M. Pastor who claims he's behind new South Carolina abortion bill admits goal is to 'get the costs to go up,' close clinics. http://www.rightwingwatch.org/content/pastor-who-claims-hes-behind-new-south-carolina-abortion-bill-admits-goal-get-costs-go-close, accessed January 6, 2014.
15. Marcotte A. The Texas abortion law is a trial run for the right wing's strategy across America. http://www.theguardian.com/commentisfree/2013/oct/29/texas-abortion-ruling-shot-down-strategy/print, accessed February 23, 2014.
16. Reitman J. The stealth war on abortion. http://www.rollingstone.com/politics/news/the-stealth-war-on-abortion-20140115?print=true, accessed January 25, 2014.
17. Bassett L. Anti-abortion laws take dramatic toll on clinics nationwide. http://www.huffingtonpost.com/2013/08/26/abortion-clinic-closures_n_3804529.html, accessed February 23, 2014.

18. Coles MS, Makino KK, Stanwood NL, Dozier A, Klein JD. How are restrictive abortion statutes associated with unintended teen birth? J Adolesc Health 2010;47:160-7.
19. Grimes DA, Cates WJ, Tyler CWJ. Comparative risk of death from legally induced abortion in hospitals and nonhospital facilities. Obstet Gynecol 1978;51:323-6.
20. Cates WJ, Grimes DA, Smith JC, Tyler CWJ. Legal abortion mortality in the United States. Epidemiologic surveillance, 1972-1974. JAMA 1977;237:452-5.
21. Cates WJ, Smith JC, Rochat RW, Grimes DA. Mortality from abortion and childbirth. Are the statistics biased? JAMA 1982;248:192-6.
22. Guttmacher Institute. State policies in brief, as of February 1, 2014. Targeted regulation of abortion providers. http://www.guttmacher.org/statecenter/spibs/index.html, accessed February 23, 2014.
23. Institute of Medicine. Legalized abortion and the public health. Washington, D.C.: National Academy of Sciences; 1975.
24. Guttmacher Institute. State policies in brief, as of February 1, 2014. An overview of abortion laws. http://www.guttmacher.org/statecenter/spibs/index.html, accessed February 23, 2014.
25. Warriner IK, Wang D, Huong NT, et al. Can midlevel health-care providers administer early medical abortion as safely and effectively as doctors? A randomised controlled equivalence trial in Nepal. Lancet 2011;377:1155-61.
26. Picardo C. Pharmacist-administered depot medroxyprogesterone acetate. Contraception 2006;73:559-61.
27. American Academy of Pediatrics Committee on Adolescence. The adolescent's right to confidential care when considering abortion. Pediatrics 1996;97:746-51.
28. Kritz F. Doctors not fans of Tom Cruise's baby gift. http://www.nbcnews.com/id/10309963/ns/health-womens_health/t/doctors-not-fans-tom-cruises-baby-gift/#Uw1jyoWmWMM, accessed February 25, 2014.
29. Kristoff N. When states abuse women. http://www.nytimes.com/2012/03/04/opinion/sunday/kristof-when-states-abuse-women.html?_r=0, accessed February, 25, 2014.
30. Mungin L, Sutto J. Judge strikes down North Carolina ultrasound abortion law. http://www.cnn.com/2014/01/18/justice/north-carolina-abortion-law, accessed March 1, 2014.
31. CBS News. Judge strikes down Oklahoma ultrasound law. http://www.cbsnews.com/news/judge-strikes-down-oklahoma-ultrasound-law/, accessed March 1, 2014.
32. Kenney A. Second suspected sedative death a 'huge concern' for NC dental board. http://www.newsobserver.com/2013/12/12/3455728/second-suspected-sedative-death.html, accessed February 23, 2014.

33. Medlin D. Eight more flu-related deaths reported in NC; total climbs to 74. http://www.wral.com/eight-more-flu-related-deaths-reported-in-nc-total-climbs-to-74/13412997/, accessed February 23, 2014.

34. Browder C. Five more flu deaths reported in NC. http://www.wral.com/five-more-flu-deaths-reported-in-nc/13284576/, accessed February 23, 2014.

35. United States Coast Guard. 2012 Recreational boating statistics. http://www.uscgboating.org/statistics/default.aspx, accessed February 23, 2014.

36. N.C. General Assembly. Senate Bill 636. http://www.ncga.state.nc.us/gascripts/BillLookUp/BillLookUp.pl?Session=2013&BillID=S636&submitButton=Go, accessed March 1, 2014.

37. N.C. General Assembly. Session Law 2013-366 Senate Bill 353. http://www.ncleg.net/Sessions/2013/Bills/Senate/PDF/S353v5.pdf, accessed February 23, 2014.

38. New York Times. The decline of North Carolina. http://www.nytimes.com/2013/07/10/opinion/the-decline-of-north-carolina.html, accessed February 23, 2014.

39. Robertson GD. Multitude at "Moral March" protest NC GOP policies. http://www.washingtontimes.com/news/2014/feb/8/annual-march-rally-builds-upon-nc-moral-mondays/print/, accessed February 23, 2014.

40. Hulka JF. Editorial: Eugenic sterilization in North Carolina. N C Med J 1973;34:950.

41. Reilly PR. Involuntary sterilization in the United States: a surgical solution. Q Rev Biol 1987;62:153-70.

42. Anonymous. Eugenics board of North Carolina.http://en.wikipedia.org/wiki/Eugenics_Board_of_North_Carolina, accessed February 25, 2014.

43. Weinberger SE, Lawrence HC, 3rd, Henley DE, Alden ER, Hoyt DB. Legislative interference with the patient-physician relationship. N Engl J Med 2012;367:1557-9.

44. American College of Obstetricians and Gynecologists. Ethics in obstetrics and gynecology. Second ed. Washington, D.C.: American College of Obstetricians and Gynecologists; 2004.

45. Reagan R. http://www.goodreads.com/quotes/tag/government, accessed March 1, 2014.

46. Cates WJ, Grimes DA. And the poor get buried. Fam Plann Perspect 1977;9:2.

Chapter 19 - Turning back the clock in Romania

1. Aspinal D. Romania: queues and personality cults, May 16, 1984. Soviet Analyst:4.
2. Anonymous. Vlad the Impaler. http://en.wikipedia.org/wiki/Vlad_the_Impaler, accessed January 12, 2014.
3. Anonymous. Impalement. http://en.wikipedia.org/wiki/Impalement, accessed January 12, 2014.
4. Stewart GK, Goldstein PJ. Therapeutic abortion in California. Effects on septic abortion and maternal mortality. Obstet Gynecol 1971;37:510-4.
5. Seward PN, Ballard CA, Ulene AL. The effect of legal abortion on the rate of septic abortion at a large county hospital. Am J Obstet Gynecol 1973;115:335-8.
6. Kahan RS, Baker LD, Freeman MG. The effect of legalized abortion on morbidity resulting from criminal abortion. Am J Obstet Gynecol 1975;121:114-6.
7. Nunes FE, Delph YM. Making abortion law reform work: steps and slips in Guyana. Reprod Health Matters 1997;9:66-76.
8. United Nations Secretariat. Romania. http://www.un.org/esa/population/publications/abortion/doc/romania.doc, accessed May 10, 2005.
9. Hord C, David HP, Donnay F, Wolf M. Reproductive health in Romania: reversing the Ceausescu legacy. Stud Fam Plann 1991;22:231-40.
10. Johnson BR, Horga M, Andronache L. Women's perspectives on abortion in Romania. Soc Sci Med 1996;42:521-30.
11. Jacobson JL. Out from behind the contraceptive Iron Curtain. World Watch 1990;3:29-34.
12. David HP. Eastern Europe: pronatalist policies and private behavior. Popul Bull 1982;36:1-49.
13. David HP. Abortion in Europe, 1920-91: a public health perspective. Stud Fam Plann 1992;23:1-22.
14. Stephenson P, Wagner M, Badea M, Serbanescu F. Commentary: the public health consequences of restricted induced abortion—lessons from Romania. Am J Public Health 1992;82:1328-31.
15. Benson J, Andersen K, Samandari G. Reductions in abortion-related mortality following policy reform: evidence from Romania, South Africa and Bangladesh. Reprod Health 2011;8:39.
16. Serbanescu F, Morris L, Stupp P, Stanescu A. The impact of recent policy changes on fertility, abortion, and contraceptive use in Romania. Stud Fam Plann 1995;26:76-87.

17. Vladutiu DS, Spanu C, Patiu IM, Neamtu C, Gherman M, Manasia M. Abortion prohibition and acute renal failure: the tragic Romanian experience. Ren Fail 1995;17:605-9.

18. Horga M, Gerdts C, Potts M. The remarkable story of Romanian women's struggle to manage their fertility. J Fam Plann Reprod Health Care 2013;39:2-4.

19. Wright NH. Restricting legal abortion: Some maternal and child health effects in Romania. Am J Obstet Gynecol 1975;121:246-56.

20. David HP, Wright NH. Abortion legislation: the Romanian experience. Stud Fam Plann 1971;2:205-10.

21. Keil TJ, Andreescu V. Fertility policy in Ceausescu's Romania. J Fam Hist 1999;24:478-92.

22. Kligman G. The politics of reproduction in Ceausescu's Romania: a case study in political culture. East Eur Polit Soc 1992;6:364-418.

23. Morrison L. Ceausescu's legacy: family struggles and institutionalization of children in Romania. J Fam Hist 2004;29:168-82.

24. David HP, Titkow A. Abortion and women's rights in Poland, 1994. Stud Fam Plann 1994;25:239-42.

25. Guttmacher Institute. Facts on induced abortion worldwide. http://www.guttmacher.org/pubs/fb_IAW.html, accessed January 12, 2014.

26. Becker D, Diaz Olavarrieta C. Decriminalization of abortion in Mexico City: the effects on women's reproductive rights. Am J Public Health 2013;103:590-3.

27. Anonymous. Abortion: one Romania is enough. Lancet 1995;345:137-8.

Chapter 20 - On the road again?

1. Berger GS, Gibson JJ, Harvey RP, Tyler CW, Jr., Pakter J. One death and a cluster of febrile complications related to saline abortions. Obstet Gynecol 1973;42:121-6.

2. Joyce T, Tan R, Zhang Y. Abortion before & after Roe. J Health Econ 2013;32:804-15.

3. Jones RK, Jerman J. How far did US women travel for abortion services in 2008? J Womens Health (Larchmt) 2013;22:706-13.

4. Joyce TJ, Tan R, Zhang Y. Back to the future? Abortion before and after Roe. NBER working paper #18338. http://www.nber.org/papers/w18338 accessed January 22, 2014.

5. Gold RB. Lessons from before Roe: will past be prologue? http://www.guttmacher.org/pubs/tgr/06/1/gr060108.html, accessed January 20, 2014.
6. Center for Reproductive Rights. What if Roe fell? The state-by-state consequences of overturning Roe v. Wade. New York: Center for Reproductive Rights; 2004.
7. Nash E, Benson RB, Rowan A, Rathbun G, Vierboom Y. Laws affecting reproductive health and rights: 2013 state policy review. http://www.guttmacher.org/statecenter/updates/2013/statetrends42013.html, accessed January 20, 2014.
8. Guttmacher Institute. Abortion policy in the absence of Roe. http://www.guttmacher.org/statecenter/spibs/spib_APAR.pdf, accessed January 20, 2014.
9. Joyce T. The supply-side economics of abortion. N Engl J Med 2011;365:1466-9.
10. National Women's Law Center. Health care report card. http://hrc.nwlc.org/states, accessed January 22, 2014.
11. Holland J. Misogyny: the world's oldest prejudice. New York: Carroll and Graf; 2006.
12. Goldberg M. The means of reproduction: sex, power, and the future of the world. New York: Penguin Press HC; 2009.
13. Grimes DA, Benson J, Singh S, et al. Unsafe abortion: the preventable pandemic. Lancet 2006;368:1908-19.
14. Shelton JD, Brann EA, Schulz KF. Abortion utilization: does travel distance matter? Fam Plann Perspect 1976;8:260-2.
15. Gober P. The role of access in explaining state abortion rates. Soc Sci Med 1997;44:1003-16.
16. Matthews S, Ribar D, Wilhelm M. The effects of economic conditions and access to reproductive health services on state abortion rates and birthrates. Fam Plann Perspect 1997;29:52-60.
17. Jewell RT, Brown RW. An economic analysis of abortion: the effect of travel cost on teenagers. Soc Sci J 2000;37:113-24.
18. Henshaw SK. Factors hindering access to abortion services. Fam Plann Perspect 1995;27:54-9, 87.
19. Joyce T, Henshaw SK, Skatrud JD. The impact of Mississippi's mandatory delay law on abortions and births. JAMA 1997;278:653-8.
20. Althaus FA, Henshaw SK. The effects of mandatory delay laws on abortion patients and providers. Fam Plann Perspect 1994;26:228-33.

21. Cates WJ, Schulz KF, Grimes DA, Tyler CWJ. 1. The effect of delay and method choice on the risk of abortion morbidity. Fam Plann Perspect 1977;9:266-8, 73.
22. Joyce T, Kaestner R. The impact of Mississippi's mandatory delay law on the timing of abortion. Fam Plann Perspect 2000;32:4-13.
23. Cates WJ, Kimball AM, Gold J, et al. The health impact of restricting public funds for abortion. October 10, 1977—June 10, 1978. Am J Public Health 1979;69:945-7.
24. Binkin N, Gold J, Cates WJ. Illegal-abortion deaths in the United States: why are they still occurring? Fam Plann Perspect 1982;14:163-7.
25. Cates WJ, Tietze C. Standardized mortality rates associated with legal abortion: United States, 1972-1975. Fam Plann Perspect 1978;10:109-12.
26. Korenbrot CC, Brindis C, Priddy F. Trends in rates of live births and abortions following state restrictions on public funding of abortion. Public Health Rep 1990;105:555-62.
27. Haas-Wilson D. Women's reproductive choices: the impact of Medicaid funding restrictions. Fam Plann Perspect 1997;29:228-33.
28. Levine PB, Trainor AB, Zimmerman DJ. The effect of Medicaid abortion funding restrictions on abortions, pregnancies and births. J Health Econ 1996;15:555-78.
29. Blank RM, George CC, London RA. State abortion rates. The impact of policies, providers, politics, demographics, and economic environment. J Health Econ 1996;15:513-53.
30. Cook PJ, Parnell AM, Moore MJ, Pagnini D. The effects of short-term variation in abortion funding on pregnancy outcomes. J Health Econ 1999;18:241-57.
31. Cartoof VG, Klerman LV. Parental consent for abortion: impact of the Massachusetts law. Am J Public Health 1986;76:397-400.
32. Rogers JL, Boruch RF, Stoms GB, DeMoya D. Impact of the Minnesota Parental Notification Law on abortion and birth. Am J Public Health 1991;81:294-8.
33. Henshaw SK. The impact of requirements for parental consent on minors' abortions in Mississippi. Fam Plann Perspect 1995;27:120-2.
34. Ellertson C. Mandatory parental involvement in minors' abortions: effects of the laws in Minnesota, Missouri, and Indiana. Am J Public Health 1997;87:1367-74.
35. Joyce T, Kaestner R. State reproductive policies and adolescent pregnancy resolution: the case of parental involvement laws. J Health Econ 1996;15:579-607.

36. Joyce T, Kaestner R, Colman S. Changes in abortions and births and the Texas parental notification law. N Engl J Med 2006;354:1031-8.

37. Ventura SJ, Hamilton BE, Mathews TJ, Centers for Disease C, Prevention. Pregnancy and childbirth among females aged 10-19 years - United States, 2007-2010. MMWR Surveill Summ 2013;62 Suppl 3:71-6.

38. Ohsfeldt RL, Gohmann SF. Do parental involvement laws reduce adolescent abortion rates? Contemp Econ Policy 1994;12:65-76.

39. Haas-Wilson D. The economic impact of state restrictions on abortion: parental consent and notification laws and Medicaid funding restrictions. J Policy Anal Manage 1993;12:498-511.

40. Haas-Wilson D. The impact of state abortion restrictions on minors' demand for abortions. J Hum Resour 1996;31:140-58.

41. Bitler M, Zavodny M. The effect of abortion restrictions on the timing of abortions. J Health Econ 2001;20:1011-32.

42. Levine PB. Parental involvement laws and fertility behavior. J Health Econ 2003;22:861-78.

43. Kane TJ, Staiger D. Teen motherhood and abortion access. Q J Econ 1996;111:467-506.

44. American College of Obstetricians and Gynecologists. Ethics in obstetrics and gynecology. Second ed. Washington, D.C.: American College of Obstetricians and Gynecologists; 2004.

45. Council on Scientific Affairs. Induced termination of pregnancy before and after Roe v Wade. Trends in the mortality and morbidity of women. Council on Scientific Affairs, American Medical Association. JAMA 1992;268:3231-9.

Chapter 21 - Abortion denied

1. David HP, Dytrych Z, Matejcek Z, Schuller V. Born unwanted. New York: Springer Publishing Company; 1988.

2. Conseur A, Rivara FP, Barnoski R, Emanuel I. Maternal and perinatal risk factors for later delinquency. Pediatrics 1997;99:785-90.

3. Rantakallio P, Laara E, Isohanni M, Moilanen I. Maternal smoking during pregnancy and delinquency of the offspring: an association without causation? Int J Epidemiol 1992;21:1106-13.

4. Kost K, Forrest JD. Intention status of U.S. births in 1988: differences by mothers' socioeconomic and demographic characteristics. Fam Plann Perspect 1995;27:11-7.

5. Raine A, Brennan P, Mednick SA. Birth complications combined with early maternal rejection at age 1 year predispose to violent crime at age 18 years. Arch Gen Psychiatry 1994;51:984-8.

6. Forssman H, Thuwe I. One hundred and twenty children born after application for therapeutic abortion refused. Their mental health, social adjustment and educational level up to the age of 21. Acta Psychiatr Scand 1966;42:71-88.
7. Forssman H, Thuwe I. Continued follow-up study of 120 persons born after refusal of application for therapeutic abortion. Acta Psychiatr Scand 1981;64:142-9.
8. Blomberg S. Influence of maternal distress during pregnancy on postnatal development. Acta Psychiatr Scand 1980;62:405-17.
9. Dytrych Z, Matejcek Z, Schuller V, David HP, Friedman HL. Children born to women denied abortion. Fam Plann Perspect 1975;7:165-71.
10. David HP, Dytrych Z, Matejcek Z. Born unwanted. Observations from the Prague Study. Am Psychol 2003;58:224-9.
11. Kubicka L, Roth Z, Dytrych Z, Matejcek Z, David HP. The mental health of adults born of unwanted pregnancies, their siblings, and matched controls: a 35-year follow-up study from Prague, Czech Republic. J Nerv Ment Dis 2002;190:653-62.
12. Matejcek Z, Dytrych Z, Schuller V. Follow-up study of children born to women denied abortion. Ciba Found Symp 1985;115:136-49.
13. Cameron P, Tichenor JC. The Swedish "Children born to women denied abortion" study: a radical criticism. Psychol Rep 1976;39:391-4.
14. Forssman H, Thuwe I. Comments on Cameron and Tichenor's remarks on our 1966 paper. Psychol Rep 1976;39:400.
15. Hook K. Refused abortion. A follow-up study of 249 women whose applications were refused by the National Board of Health in Sweden. Acta Psychiatr Scand 1963;39(Suppl 168):1-156.
16. David HP, Matejcek Z. Children born to women denied abortion: an update. Fam Plann Perspect 1981;13:32-4.
17. Binkin N, Mhango C, Cates WJ, Slovis B, Freeman M. Women refused second-trimester abortion: correlates of pregnancy outcome. Am J Obstet Gynecol 1983;145:279-84.
18. Stampar D. Croatia: outcome of pregnancy in women whose requests for legal abortion have been denied. Stud Fam Plann 1973;4:267-9.
19. Hunton RB. Patients denied abortion at a private early pregnancy termination service in Auckland. N Z Med J 1977;85:424-5.
20. Drower SJ, Nash ES. Therapeutic abortion on psychiatric grounds. Part I. A local study. S Afr Med J 1978;54:604-8.
21. Drower SJ, Nash ES. Therapeutic abortion on psychiatric grounds. Part II. The continuing debate. S Afr Med J 1978;54:643-7.

22. Pare CM, Raven H. Follow-up of patients referred for termination of pregnancy. Lancet 1970;1:635-8.

23. Foster DG, Roberts SCM, Mauldon J. Socioeconomic consequences of abortion compared to unwanted birth. https://apha.confex.com/apha/140am/webprogram/Paper263858.html, accessed December 24, 2013 2012.

24. Lang J. What happens to women who are denied abortions? http://www.nytimes.com/2013/06/16/magazine/study-women-denied-abortions.html?pagewanted=all&src=ISMR_AP_LO_MST_FB&_r=1, accessed January 18, 2014.

25. Dagg PK. The psychological sequelae of therapeutic abortion—denied and completed. Am J Psychiatry 1991;148:578-85.

Chapter 22 - Bad old days redux: self-induced abortion

1. Grossman D, Holt K, Pena M, et al. Self-induction of abortion among women in the United States. Reprod Health Matters 2010;18:136-46.

2. Rosing MA, Archbald CD. The knowledge, acceptability, and use of misoprostol for self-induced medical abortion in an urban US population. J Am Med Womens Assoc 2000;55:183-5.

3. Jones RK. How commonly do US abortion patients report attempts to self-induce? Am J Obstet Gynecol 2011;204:23 e1-4.

4. Briozzo L, Vidiella G, Rodriguez F, Gorgoroso M, Faundes A, Pons JE. A risk reduction strategy to prevent maternal deaths associated with unsafe abortion. Int J Gynaecol Obstet 2006;95:221-6.

5. Costa SH, Vessey MP. Misoprostol and illegal abortion in Rio de Janeiro, Brazil. Lancet 1993;341:1258-61.

6. Netland KE, Martinez J. Abortifacients: toxidromes, ancient to modern—a case series and review of the literature. Acad Emerg Med 2000;7:824-9.

7. Coles MS, Koenigs LP. Self-induced medical abortion in an adolescent. J Pediatr Adolesc Gynecol 2007;20:93-5.

8. Ensign J. Reproductive health of homeless adolescent women in Seattle, Washington, USA. Women Health 2000;31:133-51.

9. Friedman A. Mail-order abortions. http://www.motherjones.com/politics/2006/11/mail-order-abortions, accessed January 18, 2014.

10. Buchsbaum HJ, Staples PP, Jr. Self-inflicted gunshot wound to the pregnant uterus: report of two cases. Obstet Gynecol 1985;65:32S-5S.

11. Anderson IB, Mullen WH, Meeker JE, et al. Pennyroyal toxicity: measurement of toxic metabolite levels in two cases and review of the literature. Ann Intern Med 1996;124:726-34.

12. Anonymous. Woman who shot herself is charged in fetus' death. http://articles.latimes.com/1994-09-10/news/mn-36919_1_charge-death-fetus, accessed January 19, 2014.

13. Gold J, Cates W, Jr. Herbal abortifacients. JAMA 1980;243:1365-6.

14. Von Allmen SD, Cates WJ, Schulz KF, Grimes DA, Tyler CWJ. Abortion methods: morbidity, costs and emotional impact. 2. Costs of treating abortion-related complications. Fam Plann Perspect 1977;9:273-6.

15. Sullivan JB, Jr., Rumack BH, Thomas H, Jr., Peterson RG, Bryson P. Pennyroyal oil poisoning and hepatotoxicity. JAMA 1979;242:2873-4.

16. Honigman B, Davila G, Petersen J. Reemergence of self-induced abortions. J Emerg Med 1993;11:105-12.

17. Saultes TA, Devita D, Heiner JD. The back alley revisited: sepsis after attempted self-induced abortion. West J Emerg Med 2009;10:278-80.

18. Smith JP. Risky choices: the dangers of teens using self-induced abortion attempts. J Pediatr Health Care 1998;12:147-51.

19. Ballou BR, Ryan A. DA: young mother botched abortion with ulcer medication. http://forums.azbilliards.com/showthread.php?t=52543, accessed January 18, 2014.

20. Lee J, Buckley C. For privacy's sake, taking risks to end pregnancy. http://www.nytimes.com/2009/01/05/nyregion/05abortion.html?pagewanted=all&_r=0, accessed January 18, 2014.

21. Ramirez P. Self abortion: woman took Tylenol, Motrin. http://blog.syracuse.com/news/2007/04/west_monroe_woman_charged_with.html, accessed January 18, 2014.

22. Goldberg M. The return of back-alley abortions. http://www.thedailybeast.com/articles/2011/06/03/abortions-return-to-back-alleys-amid-restrictive-new-state-laws.html, accessed January 18, 2014.

23. Stormcoming. Repro-hell: ending pregnancy via gunshot (self-inflicted). http://www.dailykos.com/story/2006/03/03/191252/-Repro-hell-ending-pregnancy-via-gunshot-self-inflicted#, accessed January 19, 2014.

24. Anonymous. Self-induced abortion. http://en.wikipedia.org/wiki/Self-induced_abortion, accessed January 18, 2014.

25. Yanow S. Positive legal precedent set in case against a woman who self-induced abortion: a rare victory for women in the United States. Reprod Health Matters 2012;20:175-6.

26. Grimes DA. Estimation of pregnancy-related mortality risk by pregnancy outcome, United States, 1991 to 1999. Am J Obstet Gynecol 2006;194:92-4.

Chapter 23 - Children and barbarians

1. Santayana G. The life of reason. Amherst, NY: Prometheus Books; 1998.
2. Cates WJ, Grimes DA. And the poor get buried. Fam Plann Perspect 1977;9:2.
3. Abernathy JR, Greenberg BG, Horvitz DG. Estimates of induced abortion in urban North Carolina. Demography 1970;7:19-29.
4. Calderone MS. Abortion in the United States. New York: Paul B. Hoeber; 1958.
5. Cates WJ, Rochat RW, Grimes DA, Tyler CWJ. Legalized abortion: effect on national trends of maternal and abortion-related mortality (1940 through 1976). Am J Obstet Gynecol 1978;132:211-4.
6. Grimes DA. Unsafe abortion: the silent scourge. Br Med Bull 2003;67:99-113.
7. Center for Disease Control. Abortion surveillance report – legal abortions, United States, annual summary, 1970. Atlanta, GA: Center for Disease Control; 1971.
8. Seward PN, Ballard CA, Ulene AL. The effect of legal abortion on the rate of septic abortion at a large county hospital. Am J Obstet Gynecol 1973;115:335-8.
9. Stewart GK, Goldstein PJ. Therapeutic abortion in California. Effects on septic abortion and maternal mortality. Obstet Gynecol 1971;37:510-4.
10. Cates WJ, Grimes DA, Schulz KF. The public health impact of legal abortion: 30 years later. Perspect Sex Reprod Health 2003;35:25-8.
11. Cates WJ, Grimes DA, Hogue LL. Justice Blackmun and legal abortion—a besieged legacy to women's reproductive health. Am J Public Health 1995;85:1204-6.
12. Grimes DA. Estimation of pregnancy-related mortality risk by pregnancy outcome, United States, 1991 to 1999. Am J Obstet Gynecol 2006;194:92-4.
13. Raymond EG, Grimes DA. The comparative safety of legal induced abortion and childbirth in the United States. Obstet Gynecol 2012;119:215-9.
14. Edwards RG. Causes of early embryonic loss in human pregnancy. Hum Reprod 1986;1:185-98.
15. Stein Z, Susser M, Warburton D, Wittes J, Kline J. Spontaneous abortion as a screening device. The effect of fetal survival on the incidence of birth defects. Am J Epidemiol 1975;102:275-90.

16. Boklage CE. Survival probability of human conceptions from fertilization to term. Int J Fertil 1990;35:75, 79-80, 81-94.
17. O'Connell K, Jones HE, Simon M, Saporta V, Paul M, Lichtenberg ES. First-trimester surgical abortion practices: a survey of National Abortion Federation members. Contraception 2009;79:385-92.
18. Institute of Medicine. Legalized abortion and the public health. Washington, D.C.: National Academy of Sciences; 1975.
19. Council on Scientific Affairs. Induced termination of pregnancy before and after Roe v Wade. Trends in the mortality and morbidity of women. Council on Scientific Affairs, American Medical Association. JAMA 1992;268:3231-9.
20. Bruce FC, Berg CJ, Hornbrook MC, et al. Maternal morbidity rates in a managed care population. Obstet Gynecol 2008;111:1089-95.
21. Bruce FC, Berg CJ, Joski PJ, et al. Extent of maternal morbidity in a managed care population in Georgia. Paediatr Perinat Epidemiol 2012;26:497-505.
22. Hebert PR, Reed G, Entman SS, Mitchel EFJ, Berg C, Griffin MR. Serious maternal morbidity after childbirth: prolonged hospital stays and readmissions. Obstet Gynecol 1999;94:942-7.
23. Jones RK, Kooistra K. Abortion incidence and access to services in the United States, 2008. Perspect Sex Reprod Health 2011;43:41-50.
24. Jones RK, Finer LB. Who has second-trimester abortions in the United States? Contraception 2012;85:544-51.
25. Henshaw SK, Wallisch LS. The Medicaid cutoff and abortion services for the poor. Fam Plann Perspect 1984;16:171-2, 177-80.
26. Robinson M, Pakter J, Svigir M. Medicaid coverage of abortions in New York City: costs and benefits. Fam Plann Perspect 1974;6:202-8.
27. Alan Guttmacher Institute. Safe and legal: experience with legal abortion in New York state. New York: Alan Guttmacher Institute; 1980.
28. Bauman KE, Anderson AE, Freeman JL, Koch GG. Legal abortions and trends in age-specific marriage rates. Am J Public Health 1977;67:52-3.
29. Sklar J, Berkov B. The effects of legal abortion on legitimate and illegitimate birth rates: the California experience. Stud Fam Plann 1973;4:281-92.
30. Corman H, Grossman M. Determinants of neonatal mortality rates in the U.S. A reduced form model. J Health Econ 1985;4:213-36.
31. Wertz DC, Fletcher JC. Feminist criticism of prenatal diagnosis: a response. Clin Obstet Gynecol 1993;36:541-67.
32. Cragan JD, Roberts HE, Edmonds LD, et al. Surveillance for anencephaly and spina bifida and the impact of prenatal diagnosis — United States, 1985-1994. Morb Mortal Wkly Rep 1995;44:1-13.

33. Anonymous. David Reardon. http://en.wikipedia.org/wiki/David_Reardon, accessed December 28, 2013.
34. Steinberg JR, Finer LB. Coleman, Coyle, Shuping, and Rue make false statements and draw erroneous conclusions in analyses of abortion and mental health using the National Comorbidity Survey. J Psychiatr Res 2012;46:407-8.
35. Elliot Institute. The aftereffects of abortion. http://www.afterabortion.info/complic.html, accessed October 22, 2005.
36. Winker MA, Flanagin A, Chi-Lum B, et al. Guidelines for medical and health information sites on the internet: principles governing AMA web sites. American Medical Association. JAMA 2000;283:1600-6.
37. Melbye M, Wohlfahrt J, Olsen JH, et al. Induced abortion and the risk of breast cancer. N Engl J Med 1997;336:81-5.
38. Lindefors-Harris BM, Eklund G, Adami HO, Meirik O. Response bias in a case-control study: analysis utilizing comparative data concerning legal abortions from two independent Swedish studies. Am J Epidemiol 1991;134:1003-8.
39. Chasen ST, Kalish RB, Gupta M, Kaufman JE, Rashbaum WK, Chervenak FA. Dilation and evacuation at >or=20 weeks: comparison of operative techniques. Am J Obstet Gynecol 2004;190:1180-3.
40. Kelly T, Suddes J, Howel D, Hewison J, Robson S. Comparing medical versus surgical termination of pregnancy at 13-20 weeks of gestation: a randomised controlled trial. BJOG 2010;117:1512-20.
41. Lee SJ, Ralston HJ, Drey EA, Partridge JC, Rosen MA. Fetal pain: a systematic multidisciplinary review of the evidence. JAMA 2005;294:947-54.
42. Major B, Appelbaum M, Beckman L, Dutton MA, Russo NF, West C. Abortion and mental health: Evaluating the evidence. Am Psychol 2009;64:863-90.
43. Anonymous. More on Koop's study of abortion. Fam Plann Perspect 1990;22:36-9.
44. Horga M, Gerdts C, Potts M. The remarkable story of Romanian women's struggle to manage their fertility. J Fam Plann Reprod Health Care 2013;39:2-4.
45. Thornton M. Alcohol prohibition was a failure. Cato Policy Analysis No 157, http://www.cato.org/pubs/pas/pa-157.html, accessed October 31, 2005.
46. Anonymous. The jazz age: the American 1920's. Prohibition. http://www.digitalhistory.uh.edu/era.cfm?eraID=13&smtID=2, accessed October 31, 2005.

47. Ahman E, Shah IH. New estimates and trends regarding unsafe abortion mortality. Int J Gynaecol Obstet 2011;115:121-6.

48. Center for Reproductive Rights. What if Roe fell? The state-by-state consequences of overturning Roe v. Wade. New York: Center for Reproductive Rights; 2004.

49. Joyce T. The supply-side economics of abortion. N Engl J Med 2011;365:1466-9.

50. Jones RK. How commonly do US abortion patients report attempts to self-induce? Am J Obstet Gynecol 2011;204:23 e1-4.

51. Rosing MA, Archbald CD. The knowledge, acceptability, and use of misoprostol for self-induced medical abortion in an urban US population. J Am Womens Assoc 2000;55:183-5.

52. Coles MS, Koenigs LP. Self-induced medical abortion in an adolescent. J Pediatr Adolesc Gynecol 2007;20:93-5.

53. Cook PJ, Parnell AM, Moore MJ, Pagnini D. The effects of short-term variation in abortion funding on pregnancy outcomes. J Health Econ 1999;18:241-57.

54. David HP, Dytrych Z, Matejcek Z, Schuller V. Born unwanted. New York: Springer Publishing Company; 1988.

55. Forssman H, Thuwe I. Continued follow-up study of 120 persons born after refusal of application for therapeutic abortion. Acta Psychiatr Scand 1981;64:142-9.

56. Holland J. Misogyny: the world's oldest prejudice. New York: Carroll and Graf; 2006.

57. Goldberg M. The means of reproduction: sex, power, and the future of the world. New York: Penguin Press HC; 2009.

58. Kristoff ND, WuDunn S. Half the sky: turning oppression into opportunity for women worldwide. New York: Vintage; 2010.

59. Russo JA, Schumacher KL, Creinin MD. Antiabortion violence in the United States. Contraception 2012;86:562-6.

60. Stumpe J, Davey M. Abortion doctor shot to death in Kansas church. http://www.nytimes.com/2009/06/01/us/01tiller.html?pagewanted=all, accessed December 30, 2013.

61. Grimes DA, Forrest JD, Kirkman AL, Radford B. An epidemic of antiabortion violence in the United States Am J Obstet Gynecol 1991;165:1263-8.

62. Forrest JD, Henshaw SK. The harassment of U.S. abortion providers. Fam Plann Perspect 1987;19:9-13.

63. Weinberger SE, Lawrence HC, 3rd, Henley DE, Alden ER, Hoyt DB. Legislative interference with the patient-physician relationship. N Engl J Med 2012;367:1557-9.

64. Medoff MH, Dennis C. TRAP abortion laws and partisan political party control of state government. Am J Econ Sociol 2011;70:951-73.

65. Resnick S. Anti-abortion scholar: restrictions should be designed to raise costs for women. http://www.motherjones.com/politics/2012/09/anti-abortion-restrictions-raise-costs-women, accessed December 27, 2013.

66. Resnick S. Navigating anti-abortion online strategy. http://www.americanindependent.com/208240/navigating-anti-abortion-online-strategy, accessed December 27, 2013.

67. Cates W, Grimes DA, Schulz KF. Abortion surveillance at CDC: creating public health light out of political heat. Am J Prev Med 2000;19:12-7.

Index

A

B

C

D

E

F

G

H

I

J

K

L

M

N

O

P

Q

R

S

T

U

V

W

Z

Made in the USA
San Bernardino, CA
16 April 2016